The Schoolhouse Door

E. Culpepper Clark

The Schoolhouse Door

*Segregation's
Last Stand
at the
University
of Alabama*

*New York
Oxford
Oxford University Press
1993*

Oxford University Press

Oxford   New York   Toronto
Delhi   Bombay   Calcutta   Madras   Karachi
Kuala Lumpur   Singapore   Hong Kong   Tokyo
Nairobi   Dar es Salaam   Cape Town
Melbourne   Auckland   Madrid

and associated companies in
Berlin   Ibadan

Published by Oxford University Press, Inc.,
200 Madison Avenue, New York, New York 10016

Oxford is a registered trademark of Oxford University Press

Library of Congress Cataloging-in-Publication Data
Clark, E. Culpepper.
The schoolhouse door : segregation's last stand
at the University of Alabama / E. Culpepper Clark.
p. cm.   Includes bibliographical references (p.) and index.
ISBN 0-19-507417-3
1. University of Alabama—History.
2. College integration—Alabama—History.
3. Wallace, George C. (George Corley),
1919—Views on college integration.
4. Kennedy, John F. (John Fitzgerald),
1917–1963—Views on college integration.
5. Civil rights movements—Alabama—History.
I. Title.
LD73.C57   1993   378.761'84—dc20
91-48106

1 2 3 4 5 6 7 8 9

Printed in the United States of America
on acid-free paper

. . . dedicated
to Autherine Lucy Foster

and for Mary

Contents

Introduction

On May 5, 1843, one hundred years before this story begins, the University of Alabama buried Jack, one of its slaves, in a cemetery set aside for students. A year later, Boysey, a seven-year-old boy owned by President Basil Manly, died of whooping cough and also found his resting place in the student plot. Thus did two slaves come to occupy melancholy ground on the university campus, buried alongside two students whose deaths far from home abandoned them to lonely graves, a common resting place for master and slave that spoke worlds about their separation in life.

This is the story of how the University of Alabama experienced the end of segregation and how events on that campus influenced both the civil rights movement and massive resistance to the changes it promised. It is the story of two confrontations, the Autherine Lucy episode and the stand in the school-house door, confrontations that transformed Tuscaloosa into an international dateline and gave the nation symbols for an age of moral struggle. It is the story of courageous black applicants and reactionary trustees, lawyers and judges, of cautious university officials, fist-shaking demonstrators and fiery crosses; of brave, bewildered students and their worried parents; of powerful men and their low cunning, also of high-minded men and women struggling almost without hope; and, in the end, of George Wallace, whose confrontation with the Kennedys changed America's political landscape. For all its drama, no one dies in this story. Behind these scenes violence simmers, but no one dies.

A perspective on violence

It is plausible to attribute America's success, at least in part, to a relative absence of violence, or certainly the absence of killing on a scale experienced in other parts of the world. Thus, the success of our revolution and the comparative failure of others, the French and the Russian, for example, owes importantly to our avoidance of convulsive violence. Even our worst bloodbath, the American Civil War, except for its disproportional effect upon white southerners, pales in comparison with the millions who died in the pogroms of eastern Europe or the Stalinist purges of the thirties. But violence is a matter of perception. We often think of ourselves as a violent society. Perhaps because we can count our dead, as at the Vietnam Memorial, we comprehend it in ways that societies which have undergone paroxysms of killing cannot.

Between 1955 and 1968 only forty people were murdered in the American South for reasons related to the civil rights movement. Hundreds had been slain in a comparable span of time, 1865–77, as former Confederates fought to restore white supremacy. In the 1890s, their sons and grandsons hanged, burned, or shot, sometimes all three plus mutilation for sport, more than 150 black southerners each year to enforce the new segregation codes. In 1919 racial violence spread over the whole country as more people were killed in race riots that year than died in the urban riots that began in the Watts suburb of Los Angeles in 1965 and ended two summers later in Detroit—the sixties' rioting being a televised season of burning and shooting that occurred north of strict segregation and took more than a hundred African-American lives, forty-one in Detroit alone.

Why did so few die in America's Second Reconstruction? In years to come teachers will no doubt ask students to list reasons, as they now list causes for the First World War. Thus a student might answer: (1) The media made it more difficult for a tribal South to carry out lynchings behind a wall of silence; (2) Civil rights was a church-based movement and preached Christian brotherhood; (3) Non-violence struck a responsive chord among African-Americans who likened their own travail to that of the Children of Israel; (4) Southern businessmen read the *Wall Street Journal* and became uneasy with a system that retarded growth and cut profits; (5) Rapid economic growth made whites generally less prone to risk losing the comfort of their new ranch-style houses in order to maintain the less tangible value of white legal supremacy; (6) Three hundred years of loud protestations about white charity for the inferior race had become self-hypnotic, thereby neutralizing hatred for black neighbors who exposed, for all to see, the visible scars of slavery and segregation; (7) Southerners were actually Americans who at a higher ideological level subscribed to the American Creed of equality, especially after waging war against Nazi Germany; (8) Intimacy of contact between the races created, for example, natural sympathies in a rigidly patriarchal society between white, middle-class women and their colored help; or perhaps (9) Southern whites were used to losing battles.

Whatever the combination of reasons, the image that lingers is not one of acquiescence but of massive, often brutal resistance. The image is more than illusion. To say that only forty lives were cut down ignores the hideous morbidity and mortality rates of a people relegated to society's hindmost. Violence was instrumental. If fewer and fewer blacks were lynched in the twenties, thirties, and forties, it owed in no small part to the success of terror in the three decades from 1890 to 1920. Erected on a foundation of fear, segregation exacted an enormous psychological toll, requiring patterns of servile behavior from blacks incompatible with life in a free society, much less a capitalistic economy. Bent and twisted coping behavior, at one level proof of a remarkable adaptability, robbed blacks at a more fundamental level of meaningful equipment for living. Whites insisted that blacks adopt manners and modes of communication that whites could not respect; and if blacks adopted any other forms of social address, they risked death.

The image of Emmett Till's mutilated body, of police dogs and firehoses, of mounted deputies charging peaceful demonstrators, of burning buses and bloodied Freedom Riders, of black children marching through lines of screaming white parents, of bodies under earthen dams—all testify to the stunning fact that only forty died; that when evil systems collapse, they fall, sometimes simply because the oppressed refuse to be oppressed any longer. Black southerners stood up by going to their knees. They resembled every parable of Christian suffering ever told in a white church. They prayed while their tormentors swirled about them in frenzied rage. With cameras clicking and whirring, white rage looked like impotent fury as blacks took the blows and kept their "eyes on the prize." Their martyrdom won converts throughout a nation not yet aware of its own racial passions.

Why Alabama?

Nowhere did the images of freedom's struggle or the faces of evil show themselves in sharper relief than in Alabama. Alabama was to the civil rights movement what Virginia was to the Civil War—its significance lending itself to enlargement in the public mind because the most memorable engagements occurred on its soil. Of course, a general history of the South's Second Reconstruction must include all of Dixie and nowhere more intensely experienced than in Mississippi; but still, Alabama occupies a special place. The Montgomery bus boycott inaugurated the movement; Birmingham catalyzed the Southern Christian Leadership Conference (SCLC) and led in a direct line to the 1964 Civil Rights Act; the road from Selma led across the Edmund Pettus Bridge to the Voting Rights Act of 1965; and Tuscaloosa dramatized the demise of states' rights as George Wallace stood jaw-to-jaw with the Kennedy administration at the schoolhouse door and stepped aside. Geographically these four cities form a rectangle, roughly sixty by ninety miles, in west-central Alabama. In the political economy of the 1950s, they also formed a St. Andrew's cross inside this rectangle; education and politics creating one axis, cotton and steel the other.

Demographically Alabama also represented the quintessential South in all its diversity. The state had hillbillies and industrial workers, small farmers and big planters, a large black population concentrated in its Black Belt (named for its rich alluvial soil) and identifiable ethnic and religious groupings in cities like Birmingham and Mobile, rocket scientists and rednecks, aristocrats and populists, a politically influential KKK and a politically active NAACP, lintheads and longshoremen, bankers and shopkeepers. Such diversity created political alliances that approximated liberalism and conservatism as conventionally defined in the nation's two major parties; though for historic and racial reasons the state remained solidly Democratic. Unlike Mississippi, Alabama had real cities; but unlike Atlanta or Charlotte, they were southern, not cosmopolitan, in outlook or ambition. Thus Alabama was a microcosm of the larger South, as ardently committed to white supremacy as Mississippi, but more vulnerable to change by virtue of its social and economic composition. Recognizing this, an increasingly media-sophisticated SCLC would win smashing victories in the cities of Alabama while Voter Education Project workers suffered in Mississippi's county jails, out of sight and, until the murders of Goodman, Schwerner, and Cheney, out of mind.

Tuscaloosa and Alabama's university

Like Alabama, Tuscaloosa also lacked a center, other than its southernness. Not really a university town, it was more an industrial and manufacturing town. Goodrich employed 1,000 workers in its rubber plant. Gulf States Paper Corporation provided jobs and fouled the air with noxious fumes from its paper mill, while Reichold Chemical and Central Foundry added to Tuscaloosa's industrial make-up. Many of the coal miners who worked the strip-pits and deep mines that dotted Tuscaloosa County to the north and east also lived in the city or in adjoining communities such as Northport, Holt, and Cottondale. The county also had a farming population, but only a few of the farms to the west and south could be called large, and these, along the Warrior River basin, quickly gave way to hard-scrabble farms in the hill country. The state's mental hospital and home for the retarded were also in Tuscaloosa, but the staffs of those institutions, along with a nearby V.A. hospital, did not mingle with the university community. Blacks occupied much of the city's west side, but of course they remained in their place. Only in the two country clubs did Tuscaloosa's elite mix with the university's top administrators and a select few from the faculty.

Though isolated within the larger Tuscaloosa community, the university lent an air of charm to the city and provided most of its students with happy memories. Like most state universities its faculty was good but not distinguished. A handful cultivated national reputations but drifted somewhere outside the major currents of scholarship. Hudson Strode taught creative writing and wrote a biography of Jefferson Davis. Clarence Cason, a talented professor of journalism, came close to a thorough-going critique of his native Southland in *90° in the Shade* but fell short of fellow journalist Wilbur Cash's landmark, *The Mind of the South*. Only in his suicide did Cason completely prefigure the brilliant Cash. The historian Frank Owsley was there for a time but made most of his contributions at Vanderbilt, and the university's School of Public Administration had a hand in launching V.O. Key's pathbreaking *Southern Politics in State and Nation*. Teaching was the primary mission.

The university earned its most distinguished reputation in football, playing in six Rose Bowls between 1927 and 1946, winning four, tying another, and losing only one. When in 1947 the Rose Bowl decided to invite only the champions of the Big 10 Conference to play the winners of the Pacific 8, everybody in Alabama knew it was because folks on the West Coast got tired of seeing their boys whipped by a bunch of southern boys.

The students who strolled beneath the oaks and hackberry trees in the 1950s belonged more to the generation of their parents than to a future of accelerating material change and spiritual alienation. The roads they took to school were still two-lane, though more were paved; the farms they passed were still dotted with tenant shacks and pocked with gullies from misuse and disuse, though even here agribusiness was beginning to improve the land's appearance. Compared with the present, there were fewer cars by a half, fewer people by a third, no glass and steel skyscrapers, no fast-food franchises. Motels had not yet replaced motor courts, home phones shared party lines, and in many small towns all calls began with a pleasant "number please" from an operator everyone knew and who in turn knew everyone's business—privacy, an affordable urban luxury, had not yet replaced community in the hierarchy of social values.

In the fifties only a small fraction of high-school students went to college in the first place, and it was still respectable to return to Luverne, or Arab, or Monroeville and take over Daddy's business—television was yet to explode the primacy of place. If students did not major in commerce, they might pick up some culture in Dr. Perry's popular classics course or in Dr. Ramsey's equally popular Western Civilization; but if they missed those courses, it mattered little. The main streets and farms they left behind provided all the cultural assurance they would ever need. College was at best a happy time for extending the adolescence of a lucky few, at worst, a miserable time for living up to parental expectations. A handful did come to campus with more in mind. Ambitious sons of the prosperous middle class often arrived intending to advance their political fortunes. Student politics proved a ready route to success in state politics. With few exceptions the state's more notable congressmen, senators, and governors bore University of Alabama credentials. In 1939 George Wallace had no trouble in determining which school best suited his political career.

Of course, profound changes swept the postwar campus. Enrollments fluctuated wildly as draft-depleted class rosters filled with returning veterans. From a low of 1,800 in 1944, the figures jumped to 10,000 by 1950 before settling around 7,000 in the mid-fifties. Five of every seven students were male, and a large majority of these either attended on the G.I. bill or wore ROTC uniforms. The annual Governor's Day was still an occasion for parading and saluting. Increasingly students brought automobiles to campus as the age of chrome heralded a pervasive consumerism. Fashion-conscious students wore pleated skirts, bobby socks, and saddle oxfords. The Hit Parade went from smiling, dance-along tunes like "Shrimp Boats Are a'Coming" to the hip-gyrating "You Ain't Nothin' but a Hounddog." Parents shook their heads, but everybody seemed to be doing so well that enduring a few of prosperity's perils seemed a small price.

In thinking about the University of Alabama, two things cannot be overlooked. First, it is a place of unusual beauty. Located on a plateau that rises above the Warrior River, the campus centers on a large, picturesque quadrangle where oaks vault skyward to filter sunlight, creating colors reminiscent of Watteau's ethereal French landscapes. Several generations of columned, classical-revival architecture surround this broad expanse. Along its walks and inside its buildings, young men and women made friendships and acquired experiences to shape a lifetime, and that, too, should not be forgotten about the state's first university.

The other Alabama

On the eve of World War II, the state Department of Education commissioned a survey of higher education for Negroes in Alabama. The conclusions were revealing, though not surprising. Negroes suffered disproportionately from the ills of poverty. Their death rate in Alabama was 13 out of every 1,000, as compared with eight for whites. Blacks also suffered the ravages of diseases in greater numbers. Like an ancient plague, tuberculosis, syphilis, malaria, pellagra, pneumonia, nephritis, and the infections and ailments associated with childbirth visited tenant shacks and rows of shotgun houses nestled against blackened steel mills. Additionally, crime rates were twice those of whites, rates which the study attributed to poor health, low income, the absence of medical facilities, and "the almost total lack of constructive recreational facilities."

Impossible conditions forced a move from farm to city and from south to north. Every decade showed a decline in the number of blacks living in Alabama. Still, 37 percent of the state's population in 1940 could trace an African ancestry. Of those who found gainful employment, half were in agriculture, 20 percent in domestic or personal service, 14 percent in manufacturing, 5 percent in transportation, and 4 percent in mining. Over 80 percent held jobs classified as unskilled or semi-skilled. Cash income from these jobs amounted to a pittance. The average annual wage for Negro workers in seventeen southern states was $329 as compared with $745 for white workers.

Interestingly the survey never attributed the Negroes' condition to segregation, nor did it explore the psychological effects of subordination. Still, despite a debilitating system, the Negro made remarkable progress after slavery—from 12,000 homes in 1866 to 750,000 in 1936, from 20,000 farms in 1866 to 880,000 in 1936, from 700 churches to 45,000, from 10 to 90 percent literacy, from 100,000 enrolled in public schools to 2,500,000. The survey's persuasive intent was to demonstrate the Negroes' capacity for citizenship, and this demonstration of hope was never more compelling than in Alabama, where the percentage of average daily attendance in public schools for Negroes doubled between 1920 and 1939, exhibiting a thirst for education's promise of a brighter tomorrow.

When these children arrived at school, however, they found dilapidated buildings and too few teachers. Of the 2,407 structures used for Negro schools in 1939, some 515 were church buildings and 825 were privately owned. Most would have been condemned under any conscionable set of safety standards. Estimates showed that a minimum of 1,300 new teachers was required to relieve overcrowding. Turnover rates were high. Four hundred teachers in the 1939–40 school year had no previous experience, and 200 new teachers were needed annually to replace those who gave up and quit. Still, they labored with the little they had. The old buildings have long since tumbled in; a few, filled with hay or covered with kudzu, echo a past when barefoot children laughed and played, slept and learned while their parents worked nearby fields or cleaned and cooked for whites.

If a child cleared the final hurdle into higher education, only the buildings looked better. For every $100 spent by the state in educating white youth, Negroes received $6.24, a differential compounded by tuition and endowment resources available to white institutions. Nor did the state care what was taught. Though 70 percent of all blacks worked either in agriculture or in domestic and personal services, Alabama neglected these pursuits in its only land-grant college for blacks, the Agricultural and Mechanical Institute in Huntsville. In 1939 the Institute in Huntsville graduated only 14 students in agriculture and another 21 in home economics. Making matters worse, A&M offered only two-year degrees while the state required four-year diplomas to qualify for teaching certificates in agriculture and home economics. Moreover, after 1941 the state discontinued the issuance of two-year certificates for academic teachers generally, meaning that A&M's "graduates could not find employment in their own state." In 1940 the Agricultural and Mechanical Institute was the only land-grant college in America not offering four-year education. Not a single Ph.D. sat on its faculty.

Conditions at the State Teachers College in Montgomery were not much better, though by concentrating on teacher education and having a four-year curriculum, the 102 students who graduated in 1939 were better prepared for the jobs they sought. The Teachers College hired no Ph.D.s either, and its library holdings were below minimum standards. The bottom line was money. Total state appropriations for higher education in the 1939–40 academic year directed $1,975,962 toward white colleges and universities, with only $131,500 allocated for blacks, including a $10,000 allocation for the Tuskegee Institute.

The failure of the state to meet its responsibilities forced a greater reliance on non-public agencies. Private and denominational organizations maintained 6 colleges, 6 high schools, 26 "elementary high schools," and 25 elementary schools for blacks in Alabama. Of the private colleges only Talledega and the famous Tuskegee Institute operated with anything like an adequate endowment, and these, because of their eminence, served a large out-of-state student population. Of Tuskegee's 1,282 students in 1939–40, only 351 were from Alabama; of Talladega's 281, only 26 percent were in-state. The other colleges limped along miserably. The Stillman Institute in Tuscaloosa offered a junior college program on a $100,000 endowment and an operating budget of $65,000. Miles Memorial College in Birmingham was in worse shape. Maintained by the Colored (now Christian) Methodist Episcopal Church, the college had no endowment, no accreditation, and an annual budget of $29,000. In 1939 Miles awarded the bachelor of arts degree to 14 students and diplomas in teacher training to another 25 persons.

The real problem was not the age-old one of resources, but a settled conviction among whites that higher education was wasted on Negroes. The 1940 study devoted most of its rhetorical energy to combating this notion. Observing that Negroes in Alabama are "the custodians of two basic resources: land and children," the report warned that "an acre of land wasted by a Negro is no more nor less a loss to society than is one wasted by a white man." "Likewise," the report continued, "the care of children, the preparation of food, and the care of the household require preparation which cannot be left to chance." Disproving "the idea that a trained Negro is a handicap to whites with training," the study pointed to Massachusetts as proof "that the higher the percentage of skilled workers in the total group the greater the opportunities for all." Moreover, self-interest alone dictated that the Negro leadership class be trained in the South rather than be sent elsewhere. Finally, if regional self-interest were not enough, national security made higher education for Negroes imperative. "National defense cannot be guaranteed by mere physical equipment—guns, planes, tanks; it requires a citizenship trained in body and mind and, equally important, determined and willing. In modern warfare every corner of the nation is a front line. . . . A nation cannot defend itself to the maximum of its ability unless every individual, Negro and white, is trained to do his utmost."

Despite the urgency of the report, neglect remained the general policy. As segregation staggered toward its legal demise, the state did pump more money into its two Negro colleges in an effort to show a commitment to "parallel" educational opportunities for blacks. It also sought to strengthen its contractual relationship with Tuskegee. But these were desperate attempts to prop up a system under attack from without. Majority white sentiment in Alabama still held that the term "educated Negro" was an oxymoron.

A blot on the escutcheon

The two Alabamas, white and black, confronted each other not once, but twice on the campus of the University of Alabama; a campus that refers to itself as the capstone of higher education in the state. Unfortunately, the confrontations put a lasting stain on the image of the university. No blood was shed in Tuscaloosa as occurred during the Ole Miss crisis. But the image holds. To this day the recruitment of new faculty often begins with the disclaimer: "It's not what you think," and usually ends in agreement: "It's better than I thought." The permanence of the image problem owes to the highly visible, if symbolic, nature of the two confrontations and to their timing, both serving to punctuate dramatic campaigns in the movement for civil rights.

The first confrontation, often referred to as the Autherine Lucy episode, occurred in early February 1956. Only two months before, in December 1955, the Montgomery bus boycott had started. A new television generation watched as Montgomery's blacks adopted a new approach to protest and a new leader. However, the name Martin Luther King, Jr., remained relatively obscure while the nation fastened onto the name of a young black woman who came to represent martyred innocence. Driven from campus after three days of tumultuous demonstrations, Autherine Lucy's name enlisted world-wide sympathy for a civil rights cause that was yet to be called a movement. At the same time, her suspension and later expulsion gave heart to the massive resistance movement and led in a direct line to the crisis at Little Rock a year and a half later. The University of Alabama now had the dubious distinction of being the first educational institution ordered to desegregate under the *Brown v. Board of Education* implementation decree and the first where a court order was effectively flouted by a determined show of massive resistance.

The second confrontation also occurred two months into a major campaign of the civil rights movement. In April 1963, Martin Luther King brought his direct action tactics to Birmingham. After continuous demonstrations through April and May, unforgettably marked by Bull Connor's firehoses and police dogs, the camera's eye again switched to Tuscaloosa. On June 11, 1963, George Wallace used the entrance to Foster Auditorium to cement his image as the South's most determined champion of segregation. Ever after known as "the stand in the schoolhouse door," Wallace's confrontation with the Kennedy administration captured white racial passions and resentment against the welfare state and combined them into a winning formula that permanently disrupted the old New Deal coalition in the national Democratic party. On that same day John F. Kennedy committed his presidency to civil rights as a "moral issue" and guaranteed his place among America's great emancipators with his death by assassination six months later. For its part, the University of Alabama, despite its late-but-willing cooperation with the Kennedy administration, shared the national opprobrium cast on Alabama as the last state in the South to accomplish desegregation in higher education.

Narrative approach

The story of the university's struggle with desegregation divides naturally into two parts. In the first period, 1943–57, it is important to observe that the university made choices that steered her onto the shoals of massive resistance. The ugly scenes of 1956 and 1963 were not necessary outcomes. The state that gave the nation Montgomery, Birmingham, Selma, and Tuscaloosa had a different reputation in the late forties and early fifties. In these years Alabama twice elected a populist governor who held the region's most progressive views on racial accommodation. One of the state's senators, Lister Hill, was a leading New Deal Democrat, and the other, John Sparkman, became Adlai Stevenson's running mate in 1952. Alabama had a large industrial labor population, and urban boosters continued to debate whether Birmingham or Atlanta would win the race to become the deep South's biggest city. Moreover, during this time the state's reputable legal community told the university more than once that it had no legal remedies for maintaining segregation in higher education.

If the laws provided the university no comfort, the customs of Alabama created an inertia that forced the issue into the federal courts. Unwilling to act until compelled, attorneys for the university's board of trustees and the NAACP Legal Defense Fund slogged their way through legal stratagems and the courts for three and a half long years, from September 1952 until Autherine Lucy was finally ordered admitted at the worst possible time, the winter of 1956. By then the forces of massive resistance had gathered for the storm. Put another way, if the university had allowed Autherine Lucy to enroll during any term before February 1956, the chances for peaceful desegregation would have improved measurably. When the storm broke, everyone, and no one, was surprised. The wreckage it left forced the university further into the arms of massive resistance; and in the interest of survival, university officials took steps that undermined the institution's moral authority.

The second period, spanning the years 1957 through 1963, is in many ways the story of George Wallace's rise to power. It is also the story of a remnant band of liberals who held out hope for an alternative Alabama and of university officials who continued to straddle the fence in fear of a great fall. It charts the Kennedy administration's move toward accepting civil rights as a moral as well as a legal issue. It looks at white students who foreshadowed the sixties' generation and young blacks who brought an end to the old ways. Above all it is the story of how Tuscaloosa became the Appomattox of segregation.

Point of view

This book is filled with the names of many people whose lives and actions occupied my thoughts over a considerable period of time. As is usually the case, however, a story begins with one name. In 1971 I drove a U-Haul truck along Highway 78 into Alabama to take my first university teaching job. I yet had a dissertation to write and spent much of the rest of the decade shaping it into a book about South Carolina during Reconstruction. Somewhere along the Appalachian contours leading toward Anniston, I thought about the name Autherine Lucy and with no settled determination thought vaguely about doing my next book on her.

I knew a little about Lucy from my doctoral studies at the University of North Carolina, but very little and that through the dark veil of historiography. I had one keen recollection of her from childhood. In the winter of 1956, I was a twelve-year-old seventh grader at Southside Elementary School in Cairo, Georgia. I remember running around barefoot, hollering with my buddies "two, four, six, eight, we ain't gonna integrate" (pronounced "intergrate"), and "hey, hey, ho, ho, Autherine's gotta go!" Neither the concept of integration nor the name Autherine Lucy meant much beyond something Yankees tried to force on us. Fortunately, I had good Methodist parents and two older brothers, one of whom was at Duke University, who made it easy for me to accept and soon to promote the changing racial order. Thus did my years in high school and college come to be filled with purpose. Forever after, the name Autherine Lucy stayed with me, a little reminder of childhood shame and a constant source of gratitude—thus, the dedication of this book.

In the middle and late seventies I began to collect, in desultory fashion, information about Lucy. I had not met her but was aided by a remarkably fine thesis about her ordeal, written by Ann Mitchell and completed in 1970 at Georgia Southern College (now University). I learned that she had married the Reverend Hugh Lawrence Foster, and in 1975, discovered that she had recently moved back to Birmingham from Texas. Fearful of having overtures for an interview rejected, I procrastinated, hoping to find a third party to make the connection for me. In the meantime, a young black student named Sam Hughes, who was enrolled in a history class at the university, learned of her whereabouts, picked up the phone, and asked her if she would talk to his class. She agreed.

On a dark and stormy morning in March 1975, Howard Jones, a history professor, and I sat in a car across from the Student Center waiting for young Mr. Hughes. A little before six o'clock, he dashed toward the car through heavy rain. We arrived at Mrs. Foster's house in Birmingham shortly after seven and made the journey back in tornadic weather, typical for Alabama at that time of year. She was even more than I expected. I attempted not to be too effusive in expressing gratitude for what she had done and, I hope, restrained myself sufficiently. She was unassuming and pleasant. She recalled her experiences with credible ease; not too clearly but with the kind of narrative fidelity that comes from putting the events in perspective and telling the story many times.

I continued to interview people off and on over the next few years, as did a number of my students working on oral history projects, but I did not turn seriously to the task until the spring of 1985. During these years, I came to know many other people, some whose social and political points of view contrasted with my own. Shortly after interviewing Mrs. Foster, I interviewed Leonard Ray Wilson, leader of the student opposition to Autherine Lucy's enrollment and the only person other than Lucy to be expelled as a result of the demonstrations. Though I did not stress my own convictions, he must have suspected that I operated from a different set of beliefs. Still, he was unfailingly courteous, agreeing on a later occasion to talk with my students on campus.

Without exception, I left all the interviews listed in the bibliography liking the people with whom I had spoken and grateful for their contribution of time. I also felt guilty, knowing that the story I would tell was not always the story as they remembered it, or at least not as they would tell it. I knew that the fragments I would select from their narratives robbed them of the ability to control the context in which they spoke their memories. Yet, not only because I like them and owe each a personal debt of gratitude but equally because it is my professional and moral obligation, I have endeavored to be faithful to their intentions in relating attitudes, opinions, and beliefs. In the end, all their narrative voices came together to give me my own voice.

Of course, one cannot write history, and certainly not that of the civil rights movement, without a point of view. Martin Luther King, Jr., articulated my view best, and my sympathies and value judgments lie completely with his dream. Though I reject the social and moral assumptions that lie behind arguments for racial separation and especially separation that leads to subordination, I do not say that people holding these views operate from no moral compass or possess no charitable impulses. They are guilty, I believe, of taking what they see in nature and converting it into an "ought" condition of social organization. King's vision strains against the natural tendency to separation and from separation to dominion. Built from the same cultural materials as arguments for segregation, his morality is the more difficult, for people do not incline toward it naturally. It requires for its maintenance constant persuasion. To fulfill King's dream, people are urged to forgo powerful advantages (wealth, strength, intelligence), whether endowed by nature or acquired, in order to make common cause with all people. At the very least, they are urged to forgo distinctions of color. For King, the ideal of equality is not a necessary outcome of liberalizing tendencies over the last three hundred years, but a problematic aspiration to be enacted and re-enacted in succeeding generations. Without extraordinary moral commitment and persuasion, it yields to baser, anthropological imperatives. Equality is pure ideology, not inductively derived from nature, but a product of humankind's loftier impulses to free itself from brute, natural conditions.

Because I see the story from a moral point of view, albeit one commonly held, the drama has good characters and bad, people who do good and people who do evil. There are a few instances where people sacrifice almost without recompense (Dan Whitsett comes to mind), and a few instances where Klansmen or their fellow travelers wreak violence. The vast majority, of course, are people who try to do good but fall short, or people whose desire to do good is compromised by cultural blinders or pragmatic considerations of the moment. Certain university trustees, for example, show themselves in a bad light, but their turn-of-the-century education blinded them to the human- ness of Negro Americans almost as completely as a color-blind person is denied normal spectral vision. Of course, their constricted vision does not excuse them, but it does explain them.

If I sense correctly the way I have told this story, my least charitable depictions are not of segregationists and massive resisters nor of civil rights advocates and lawyers whose feet were made of clay, but are reserved instead for people like myself—university people, faculty and administrators, who ought to have known better and should have done better. This edge to the story possesses a sharpness that is not completely fair and serves as a reminder of how autobiographical all writing is.

A different light

My initial impressions of this story, born of interviews and first readings, came from the generation that lived the events. Their accounts would have made the story read something like this: young blacks with the aid of NAACP lawyers pushed, not always with appropriate discretion, to gain admission; the university, victimized by intolerant elements in the state, struggled to do what was right and in the end prevailed; George Wallace was a scheming demagogue with few scruples; while the Kennedys represented the nation's better side and cut a deal with Wallace to stage the schoolhouse-door drama. Of course this is the dominant view. Other perspectives from a variety of participating factions, white supremacists for instance, would change key elements.

What I subsequently learned differs only in degree from the dominant view. The young black women who first applied were not carefully selected by the NAACP. As will be seen, the NAACP supported their applications with reservations as to their qualifications and fitness for the task, though it would be difficult to imagine two more courageous or emotionally stable candidates for the subsequent ordeal. Local committees formed to support them, but the rigidly hierarchical national NAACP sapped local energy by insisting on its share of the proceeds from state and branch fund-raising efforts. The lawyers for the NAACP Legal Defense Fund, including local counsel Arthur Shores, were certainly able, though at times less attentive to detail than they should have been. The fact that the Legal Defense Fund was going to win the University of Alabama case under almost any conceivable circumstances made their occasional carelessness disconcerting—and in one instance catastrophic for Autherine Lucy's chances.

Anyone who has read or heard tapes of George Wallace's speeches from that era knows him to be appropriately labeled a "racist demagogue." Though demagogic, he believed what he said. He did not pander to white resentment, knowing it to be wrong in principle or policy. For Wallace, political advantage and personal philosophy functioned in perfect alignment. When Wallace pledged in March 1962 to stand in the schoolhouse door, he made no idle threat. After the Ole Miss debacle in September 1962, Wallace charted a consistent, well-thought-out course to the door at Foster Auditorium. He said early and believed completely that if he did not show up at the schoolhouse door, the mob would wreck the university. He said early and believed that only he could keep the peace and that peace would come only if he stood good on his pledge to resist and renounce force-induced desegregation. Wallace showed admirable qualities in a reprehensible cause.

The Kennedys have undergone a thorough debunking in recent times, chided in civil rights for being late and cautious. My personal view is that they took action more quickly than any other politicians in America could or would have; that within less than two and a half years, John Kennedy and his brother Robert moved the presidency and the Democratic party behind a movement that destroyed the party's presidential base for the next thirty years, the elections of '64 and '76 notwithstanding. Wallace warned that it would happen, that the Democratic party would lose the solid South, and the Kennedy administration accepted the risk. What the Kennedys did in a thousand days doomed their ambitions and those of like-minded politicians for the next ten thousand. The ultimate political fall-out from the stand in the schoolhouse door and all it symbolized was the presidency of Ronald Reagan.

Perhaps the major interpretive shift resulting from my studies comes in the university's own self-assessment. In this accounting Oliver Cromwell Carmichael, alumnus and distinguished educator, returned to his alma mater as its president only to become a victim of the growing racial crisis. In 1957, the university hired Frank Anthony Rose and gave him a secret charge to bring about peaceful desegregation, which he finally accomplished in June 1963. My own findings reveal that between 1943 and 1955, the university had numerous opportunities to set a different course on desegregation (three in Carmichael's presidency alone) and each time forsook alternatives that might have allowed it to avoid the blot that was ultimately stamped on its escutcheon. Carmichael's segregationist sympathies and his style condemned the university to rule by its board of trustees, a condition that aligned the administration with white majority sentiment in Alabama. Moreover, despite Rose's fine talk in

1963, his administration also walked in lock-step with political sentiment in the state. University administrators wanted to do the right thing but never were a meaningful part of the solution. Only after black students, the NAACP Legal Defense Fund, and the federal government forced the issue did the university move to cooperate in its desegregation. Actions taken by university officials in 1963 were courageous, but they only contributed to the inevitable. To be sure, a case can be made that a state university is powerless to move beyond public sentiment, but that is not the view perferred by those who believe the university played a major role in accomplishing its desegregation. They cannot have it both ways.

ACKNOWLEDGMENTS

Few historical events attract writers with talent equal to their importance. As H. L. Mencken observed, "It is the misfortune of humanity that its history is chiefly written by third-rate men." If I have brought a particular talent to the present subject, it is my ability to leverage friendships. Indeed, because of friends, this book is far better than it would have been if left to my own powers. They will not acknowledge their contributions, such is my knack for attracting only the best of friends, but I know their influence.

Three merit special mention. Jim Rachels's penchant for simplicity and clarity of expression inspired the early going. My brother Charles read every chapter, made me feel good about myself, and put me in active voice. Harold Davis had the rare gift of flattering my work, then, with the surgical precision of his own talent, showing me how to make it better by pruning a third of the original. Together these friends saw the *David* in my block of granite and, in so doing, placed both me and the reader in their debt.

I am also indebted to two historians who give the lie to Mencken's observation. Their prize-winning accomplishments are earned dividends of the scholarly virtues. Robert J. Norrell and Dan T. Carter gave me the confidence that I had placed my text in reasonable context. In this same vein, I should like to thank William D. Barnard, David Garrow, J. Mills Thornton III, and Harry J. Knopke for useful comments and helpful insights.

For the documentary trail that produced the text, I am grateful for the assistance of archivists, among them: Joyce Lamont, the University of Alabama Library; Marvin Whiting, Birmingham Public Library; Henry Gwiazda, John F. Kennedy Presidential Library; Paul Chestnut, Library of Congress; Miriam Jones, Alabama Department of Archives and History; and Leon Spencer, Talladega College Historical Collections.

The manuscript itself has been tended carefully by Lisa Taylor, Gloria Keller, and Roxanne Gregg. Their thoughtfulness saved me worry. My copyeditor, Stephanie Sakson-Ford, spared me details of punctuation, abbreviation, and other requirements of publication I should have known.

Having held four different jobs in three universities during the development of this book, I have acquired other debts too numerous to mention, especially to friends and colleagues. However, I can and must note the complete freedom given me by the University of Alabama to explore a troubled part of its past. I also wish to thank Joe Loveland whose hospitality during the year I was at Georgia State University facilitated the completion of Chapters 3, 4, and 5.

So, what the reader finds of merit is the product of friends, named and unnamed, whose conversations, laughter, and shared experiences have brought me thus far. In this last connection, I must mention Lynn and Bill Carter and Tom and Ina Leonard. Our children grew up together in the years that the manuscript grew up.

To my sons, Stephen and Culpepper, thanks for all the basketball games, for keeping me in stitches with complete renderings of *Raising Arizona* and the Monty Python corpus, and for your general irreverence. Who knows? You and your Uncle Ken may have spurred the project to completion by refusing to hear anymore about my ever-receding deadlines—but, then, maybe not.

Finally, this book is for Mary. Each page reflects her knowledge of books, her patient readings, her willing suspension of other priorities, and her judgment.

Tuscaloosa, Ala.　　　　　　　　　　　　　　　　　　　　　　　　E. C. C.
June 1992

The Schoolhouse Door

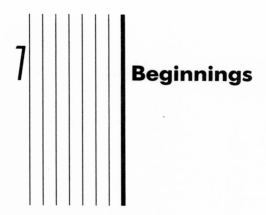

1 Beginnings

Understanding how "the system" works comes earlier for children of the disadvantaged. For African-Americans growing up in the segregated South, it is remembered from particular moments. It comes sharply, recalling what it meant to be ugly, to look in the mirror and wish to be different. Pollie Anne Myers knew what it meant to be ugly. Even in her all-black high school it always seemed that "the lighter-skinned girl got all the breaks." She heard them call her "a black so and so" in the same way whites called her "nigger." She saw it in the skin of her mother. Alice Myers did things Pollie's darker-skinned father could not. Pollie remembered the bill collectors who called at their home and said, "Alice, where's Henry?" It always began cheerfully enough. Alice would say, "He's not here," even though he was hiding in the house. "Alice," the collector frowned, "you tell that boy now that I'm looking for him and he owes money on his account." And always Alice would go and pay.[1]

Pollie never forgot the day her father tried to pay his own bill. "I don't know why," Pollie reflected, "but he went and he came back, and he said, 'I will never go again.' " The next day Alice went to see the collector. She took the same money her husband had taken, which was not enough. But this time it was different. "Alice, this all you got? You tell Henry to save some more when he can." Pollie surmised that her father had been humiliated—"chewed out or beaten up"—but whatever happened, her mother had succeeded where her father failed, and the young girl, who looked more like her father, ever after believed that her mother had succeeded because of her complexion. Of course, it also had to do with gender, but color bound Pollie to her father in ways that only colored Americans could understand.

Pollie was born to Alice Mae Lamb and Henry Myers on July 14, 1932, on the Naman Lamb plantation near Robinson Springs, Alabama, north and east of Montgomery in Elmore County. She was named Pollie for her grandmother and Anner (later changed to Anne) for an aunt. She spent her first five years on a farm in the pleasant surroundings of a large family of uncles, aunts,

3

cousins, and grandparents. The grandfather on her mother's side was a Methodist minister, and through him she acquired a sense of what it meant to have status in the community. She experienced separation and loss when she and her brother, Henry, Jr., moved with their parents to Birmingham. It would be years before Birmingham was anything more than the "staying place" and Robinson Springs "the living quarters." "We were farmers," she recalled, "and we would go back to farm."

Like many blacks coming to Birmingham in the thirties, Pollie's father took a job in one of the foundries as a cotton tie bundler. He was dependable, and the income was steady. The family moved into a house on an alley on the city's west side. With her "nickel allowance per week," Pollie never thought of herself as poor—at least, not until she was grown. She attended Cameron Elementary School, started high school at Ullman, and graduated from Parker High School in January 1949. The good memories included the principal's bragging about her dependability at selling lunchroom tickets in elementary school, and later of being asked to deliver one of the commencement speeches. The bad memories centered on walking the two blocks from her house to the handsome West End School only to wait in the cold for one of two buses that took Pollie and her friends to all-black Parker High School; then, returning in the afternoon on a city bus where they stood in the back despite the availability of seats up front in the white section. Life was interminable waiting in a white-first world; going to stores and "having to wait until your time came." Each and every affront, she remembered, "were things that said to me that I was someone different."

Pollie's first personal setback came in her choice of colleges. The friends she admired all talked of going away to college, so she wrote to the same schools. But her mother would have none of that. A practical woman who had spent three years in the eighth grade "waiting for the crops to get better," Alice Myers had little patience with fanciful ambitions. She told Pollie to "get to work now—you've finished." Pollie did get a job, at a restaurant in Pratt City. Her mother also worked in cafes, but for Pollie it was a means to another end. "As foolish as I was, I put my pennies in a jar; a regular canning jar, a fruit jar. I did that through 'til August." Then, unable to consider a school away from home, Pollie set her sights on Miles Memorial College in Birmingham. The night before she was to go for testing and orientation she asked her mother to wash her hair and "press" it: "It had to be straight." Alice Myers reluctantly worked on her daughter's hair until late that night, and Pollie remembers saying to her, "If I don't wake up in the morning, wake me up." But Pollie believed her mother had no intention of awakening her, and so that night she prayed not to oversleep. The next thing she knew, "I was up bright and early . . . and gone. My mother didn't even know I had left the house."

Pollie arrived at Miles with a brown paper bag containing her jar full of money. She took the tests and spent the next few hours in the school library waiting for the results. Finally, the head of the English Department called her in to say that she had done well, but she might have to take the spelling part again. When the department head brought up the subject of fees, Pollie presented her paper bag. The woman peeked inside and said, "You don't expect me to count all this?" Pollie did expect them to count it, but when they had finished, she was still short. Arrangements were made. She would enroll and make up the difference by working on Saturdays as a housekeeper.

Pollie was never without jobs. Two became important. The first involved maintaining a clipping file for Ruby Hurley, regional secretary for the NAACP. The NAACP offices were in the Masonic Temple Building, a hub of activity in the black community, located a block south of Kelly Ingram Park in downtown Birmingham. The second job bore more directly on Pollie's ambitions. She became a writer for the *Birmingham World*. The paper's editor, Emory Overton Jackson, an expansive and energetic gadfly in the black community, liked the young woman who, despite her problems with spelling, wanted to be a reporter. She was soon contributing to the society page and developed "quite a little list" of important people. Jackson himself gave her a boost when in one of his editorials he commented on "a signed story by Miss Pollie Myers, herself a Miles College student and aspiring creative writer."[2]

Pollie also caught some of Jackson's spirit, especially in lambasting the evils of segregation. In her senior year, 1951–52, as president of the state NAACP Youth Council, she helped map out an educational program for youth groups throughout Alabama. With the advice of senior officials, she concentrated on segregation laws, voting procedures, and court rulings on education. "It is through this program," she declared in one speech, "that the youths of Alabama intend to jar the smug security of those persons prospering in the wake of segregation which stifles and exploits the masses." She warned black leaders who placed their "people on the market to the highest bidder in exchange for a few years of indulgence" and called them "Judases of our own people, who are now wallowing in the reflected glory of segregation and its rewards to them."[3]

Though not the best student at Miles, Pollie did make the honor role occasionally and possessed extraordinary determination. Nor was she the ugly child she remembered being. She was in fact good-looking, even striking. A sleek five feet, eight inches and 125 pounds, her complexion was dark, a blend of deep earth tones. Often looking aside, seldom fixing a person directly, her large and open eyes betrayed an underlying tenacity. Appearing chic when she pulled her hair straight back from her oval face, she could have stepped from the pages of *Ebony;* the magazine for which she dreamed of writing, not modeling. Pollie was always more than a college student. She moved among adults.

Autherine Juanita Lucy

Pollie made one friend in a public speaking class and the intersection of their lives gave the nation a name to remember. Autherine Juanita Lucy and Pollie

Anne Myers proved that opposites attract. Lucy was shy and not a little uncomfortable, having grown up on a farm and now living in a large city. Pollie was everything Juanita, as she was then called, would like to have been: popular, a good speaker, savvy. And given Pollie's routine, she needed a friend like Juanita: unassuming, understanding, alert to people. The two women could be found in the front room of the Myers house, acting like the girls they had recently been. Pollie played the piano, Juanita sang, and they mixed it all with laughter. They shared their lives as only good friends of that age can.[4]

Autherine Lucy was not always called Juanita. She grew up in the Shiloh community of Marengo County in southwest Alabama being called "Reney," a diminutive that along with the original Autherine she grew to hate. She was the last of ten children born to Minnie Maud Hosea and Milton Cornelius Lucy. Autherine arrived October 5, 1929, just in time for the Great Depression, but living on the 110 acres her father owned and farmed, she never felt deprived. To supplement his income, Milton Lucy split white-ash to make baskets, whittled chairs and axe handles, and did blacksmithing on farm implements. For a time, he also ground sugar cane for syrup.

Autherine attended Shiloh Elementary School, about a mile and a half from her home. Reading, grammar, and spelling were her favorite subjects. During spelling bees, she always asked to start at the end of the line. "I liked the idea that I could . . . end up at the head of the line." But that was the only competition with which she felt comfortable. Her gangliness on the playground instilled a natural reserve. Even on the farm, she would fall behind the others. When her oldest brother and his wife were killed in an automobile accident in 1938, their children came to live with the Lucys. For a time she resented their intrusion on her status as the "baby." When she finally came to terms with their presence, even to the point of liking them, there remained the inevitable competition with the new arrivals. One day her father challenged Autherine to compete with her niece to see who could pick the most cotton. At the end of a back-breaking day, Autherine weighed her haul in at 107 pounds, only to have her niece steal the prize with 109. Even at that, Autherine "didn't do too much fussing," because her "mother could pick at least 200 pounds when she'd go out."

Autherine took her last two years of high school at Linden Academy in Linden, Alabama, about twelve miles from her home. The now defunct academy can boast alumni such as Ralph David Abernathy, but prospects of future fame meant little to the girl from Shiloh who boarded there during the week. Her memories would not all be fond. On one occasion she was even dismissed, the incident starting because Autherine was a "picky" eater. One evening she asked for collard greens instead of pork and beans. Before she could get them, the bell rang, emptying the dining hall. Autherine went upstairs where a cousin had just received some food from home, and ate there. A moment later, there came a knock. A young girl had been sent to tell Autherine that the greens were ready in the kitchen. Something in the tone of Autherine's voice, as reported by the girl, caused the cook to take Autherine's refusal as sass. Soon everything got out of hand. The principal sent for Autherine's father, who had never heard a word of sass from his "baby," a fact he made clear. The principal eventually backed down but not before Autherine had been sent home to talk it over with her parents.

The "great collard green incident" was just the kind of thing to convince a shy girl that she was a troublemaker. Fortunately, the rest of her days at Linden were not so eventful, and upon graduation the principal recommended that she go to Selma University for two years of study in preparation for being a teacher. An older sister had completed two years of college and, as was common then, started teaching with the promise of finishing later. Autherine liked Selma, particularly being around all the young men studying to be Baptist preachers.

The decision to go to Miles came only after Autherine tried and failed to find a teaching job for the fall of 1949. (The state no longer accepted two-year degrees for full-time positions, though temporary appointments remained possible.) She wrote a brother and a sister living in Birmingham to see if they would "sponsor" her at Miles. She soon moved in with her sister at 714 Attalla Place in Wylam, a suburb west of Birmingham. To help pay for her education, Autherine worked the first summer as a maid on a hospital ward. "We'd have to make the beds and pass out the bed pans and serve the food and this kind of thing," she recalled. The next summer she moved up to lab assistant.

Miles was not a hospitable place for Milton and Minnie Lucy's baby. Though she was elected president of the YWCA and passed an audition for the choir, she still suffered rejections that hurt. One club, Rho Nu, led Autherine to believe that she would become a pledge, but when the roster appeared, her name was not listed. That rejection stung but did not compare with being turned down for a position on the school paper. The head of the English Department had said that all English majors "should get the experience of working with this paper," but when Autherine went to sign up, she was told that membership was selective.

There were happier memories. One day, the speech instructor told the students it was time to check their standings. The grades were generally poor. When Autherine at last summoned sufficient courage to go by, she was told, "You most of all don't have anything to worry about." That simple line, passed off casually by the instructor, brought a thrill of excitement and self-esteem that still radiates. Autherine went on to give her final speech on the "Four Essentials of Success": courage, enthusiasm, self-mastery, and determination. When she finished, "the audience was quite receptive." The only thing that could have added more, she thought, would have been the presence of Dr. W.A. Bell, president of Miles College, who was out of town. There was always a little cloud.

The public speaking class contributed one other thing to Autherine's education: Pollie Anne Myers. The two shared a seat and became fast friends. Autherine completed her work in the winter quarter of 1952 and came back for spring commencement exercises. In the meantime, she took a job at the

Britling Cafeteria, a favorite spot for Birmingham's whites, and began looking for work as a teacher, trying the schools in Birmingham first but with no success. She then put in an application with the Dixie Teachers Agency and was awaiting word when she got a call from Pollie. It was a beautiful, early summer day, and Pollie asked if she would like to go to the University of Alabama. "I thought she was joking at first, I really did. But she said, 'I'm not joking.' " With that simple request and with mutual pledges of support, Pollie and Autherine set in motion a train of events that would not end until June 11, 1963, almost eleven years to the day after the two young women, bubbling with nervous enthusiasm, bargained their future against the policy, laws, and customs of the state of Alabama.

The changing legal environment

Today, neither Myers nor Lucy remembers clearly how the decision evolved. Of the two, Myers was the natural leader. Her involvement with the NAACP Youth Council, her employment on the newspaper, and her general pluck and energy put her in better position to know the risks and the possibilities. Yet Myers was and is a woman of the present. She remembers only vaguely the people she knew then and not at all her involvement with the Youth Council. More important to her are her present accomplishments: her two master's degrees from Wayne State University, her teaching experience in Nigeria, her children. The crisis in Tuscaloosa would bear Lucy's name, but Myers initiated the decision to go.

The decision emerged from a combination of ambition and circumstances. The year 1952 was not a "target" for the NAACP in Alabama nor were Lucy and Myers hand-picked candidates. Things just happened. The actions making it possible began to take shape long before, about the time Lucy and Myers were born. In 1930 the NAACP hired Nathan R. Margold, a New York attorney, to attack the broad spectrum of "handicaps facing the Negro." Charles H. Houston of Howard Law School took on the specific fight against segregation in education. When the Depression dried up resources, the broad-spectrum attack on general handicaps was narrowed to Houston's strategy against segregated education, and in 1934 Houston became the first full-time head of the NAACP legal department. In these labors Houston was assisted by a former student, Thurgood Marshall, who succeeded him as head of what became known as the NAACP Legal Defense Fund.[5]

The general plan called for forcing an end to inequalities in education, to make more equal those opportunities that had been lawfully separated by *Plessy v. Ferguson*. Decided in 1896, *Plessy* tested the constitutionality of a Louisiana law requiring separate railway accommodations for blacks and whites. The Supreme Court ruled that separate cars could be provided as long as the accommodations were "substantially equal," thereby giving the Court's imprimatur to the notion that segregation and the equal protection clause of the Fourteenth Amendment were compatible. Three years later the Court ruled that a Georgia county, strapped for finances, did not have to provide a high school for sixty black children. The alternative of providing *no* high school for either race was ruled an inappropriate remedy, because Negro children would not benefit by closing the white school.

By forcing states to make separate facilities for blacks "substantially equal," the NAACP hoped to destroy Jim Crow education by making it too expensive. When Donald Murray, a graduate of Amherst, applied to study law at the University of Maryland in 1936, Thurgood Marshall scored the first major victory. To keep Murray out of the all-white university, Maryland offered to pay for his education outside the state. However, the Maryland Court of Appeals was persuaded that forcing a student to go out of state failed to meet the requirement of equal protection. Then, in 1938, the Supreme Court in an opinion written by Chief Justice Charles Evans Hughes said that Missouri had to provide Lloyd L. Gaines either a "separate but equal" law school within Missouri or allow him to enter the university's law school. The law was clear. Out-of-state tuition grants, a dodge which southern states would adopt to avoid desegregation but which only the border states had implemented in the thirties, would work only if blacks agreed voluntarily to take a state's money. If demanded, a state would have to establish a separate law school equal to that provided for whites or admit blacks on an equal footing. "Separate but equal" had been challenged but not overthrown.

Ten years later the Supreme Court reaffirmed the *Gaines* decision in *Sipuel v. Board of Regents* (1948), a similar case involving the University of Oklahoma Law School. Although neither *Gaines* nor *Sipuel* represented a direct attack on segregation, the stage had been set. In late 1949 the Court agreed to hear cases arising in Texas and Oklahoma testing whether separate schools could be equal. When Herman Sweatt applied for law school at the University of Texas, the state proceeded to establish a Negro law school. Sweatt's lawyers argued that such a school would be inferior and thus a violation of the equal protection clause. Because virtually all black schools in the South were inferior, the case threatened to undermine the whole system of segregation. Consequently, all southern states joined Texas in friends-of-the-court briefs, while Sweatt was joined by the United States Attorney General and a committee of law professors.

Without overturning *Plessy,* the Court in *Sweatt v. Painter* (1950) ruled that "the University of Texas Law School possesses to a far greater degree those qualities which are incapable of objective measurement but which make for greatness in a law school." On the same day, June 5, the Court extended its view of equality in the *McLaurin v. Board of Regents* case. The University of Oklahoma had sought to avoid its responsibility under *Sipuel* by admitting blacks but segregating them on campus. Upholding the complaint of G. W. McLaurin, that such discrimination denied him equality before the law, the Court again pointed to intangibles that worked to the disadvantage of students forced into separate facilities. A third decision the same day drove the point home. Like the original *Plessy, Henderson v. United States* treated separate accommodations in rail transportation, but this time ruled them unconstitutional. Thus, after June 5, 1950, no respectable jurist believed segregation had any remaining legal life. The Court had not overturned the doctrine per se, but there was no doubt that "separate but equal" was fiction in fact as well as law.

The Supreme Court was reacting to an underlying current of change. Just as *Plessy* reflected sociological shifts in race relations since Reconstruction, so too did decisions leading to *Brown* mirror changing modes of thought since the turn of the century, south and north. In the South, journalists and scholars began to break from the Victorian equation of civilization with elite culture; an equation that necessitated the expenditure of considerable energy on the part of elites to protect themselves from the lesser classes, when by virtue of their higher endowments elites should have needed no protection. Still, this equation supported a hierarchical arrangement of society with whites at the top and blacks at the bottom. Other gradations of caste and class lay between that basic racial divide. In place of Victorian thought appeared a doctrine of cultural relativism that nudged the nation toward modernist thought. The new view saw cultures as different, but not in terms of more or less civilized, just different. Anthropology, which formerly stood in service of segregation by suggesting genetic hypotheses to explain cultural differences, was used now to discredit notions of racial superiority and inferiority. The new science favored environmental explanations, thereby remaining neutral on race.[6]

The new day also promised to be more than an intellectual awakening. Social forces seemed to work in the same direction as modernist thought. The need for unskilled agricultural labor dropped sharply in the thirties and forties. Farm tenantry, at its peak in Alabama around 1935, began to decline as New Deal policies and improved farm machinery forced sharecroppers and small farmers off the land. The emigration of blacks to northern industrial centers created an economic demand to upgrade their skills and a political demand to respond to their votes. Moreover, churches which ministered to the needs of Depression America came to have a social conscience about blacks as well—an awakening not confined strictly to the North. In striking contrast to their husbands and brothers, women in many southern denominations voiced concern about the subordination of blacks. Additionally, the rise of labor as a political force initially muted racial antagonism and accentuated shared economic grievances. Joining the cry was a talented black middle class. People like Emory Jackson were no longer content with leftovers. They would be joined by a handful of white editors and teachers who, imbued with the new currents of modernism, began to chip away at old assumptions.

World War II galvanized these forces. Just as the Civil War left in its wake the Thirteenth, Fourteenth, and Fifteenth amendments, and World War I, the Eighteenth and Nineteenth, so World War II yielded its own surge of moral commitment. The possibilities inherent in that flood were nowhere clearer than in the demands of returning black servicemen. Having sacrificed to end one form of racism in Nazi Germany, they were not content to be victims of another at home. In the late spring of 1945, Sergeant Harry H. Smith wrote from "somewhere in Germany" a letter to Raymond Ross Paty, then president of the University of Alabama. Smith had learned that the university was opposing the extension of compulsory military training because "it might jeopardize our relations with other countries at the coming peace conference by placing us on an all-out military basis." Sergeant Smith asked why the university had not gone on record against "Jim Crow" laws since "such laws have jeopardized the working Democracy in our whole nation." "In the words of my Tec. Sgt.," Smith warned, " 'You had better wake up' or even defeated Germany will pass you up by being a working Democracy."[7]

Sergeant Smith's warning and forces for change notwithstanding, the University of Alabama was not about to lead the advance. Despite the capable leadership of President Paty, the university had another agenda. Besides, Paty's departure to be Chancellor of the University System of Georgia in 1946 returned the university to a leadership more in keeping with its past, a past best summarized in a letter to Paty from Lee Bidgood, then Dean of the School of Commerce and later an interim president. In reference to Sergeant Smith's inquiry, Bidgood recommended that Paty make no response. He warned, "The next move may be for this man to apply for admission under veterans' training. If he is a white man, he could be as dangerous to us as if he is a Negro. Even if he is white, unless he is a bona fide resident of Alabama, we do not need to admit him. There are enough racial agitators on the campus already."[8]

In fact there were no racial agitators on campus. What passed for agitation was little more than a general sentiment that black demands for equality should be met within the confines of segregation. Paty shared this sentiment and left one legacy to the files—a suggestion that blacks be educated on campus, albeit on a segregated basis, by expanding selected graduate programs. (Though the *McLaurin* decision in 1950 forbade the segregation of blacks within an institution, Paty's suggestion is interesting, for it posed the prospect of blacks on campus when segregation finally collapsed. It could have made for a true test of gradualism.) Paty's memorandum went to the edge, though it did not go beyond what many thoughtful southern whites were willing to accept in 1943.[9]

By 1944, however, constructive talk, even of limited scope, had turned to plans of evasion. The state never had made provisions for out-of-state grants, and despite their lack of constitutional standing as a means of evasion, efforts to provide money for blacks willing to go out of state now proved near frantic. When a young black from Homewood, Alabama, requested assistance, the Director of the Division of Negro Education in the state Department of Education responded that no funds were available. He then requested immediate action by the presidents of white institutions and members of the legislature to find funds. He reported that Governor Chauncey Sparks had requested

$12,000 from the General Education Board but that counsel for the board questioned "the advisability of such an organization as the General Education Board seemingly scheming with the state in circumventing the *Gaines* ruling of the Supreme Court in 1938." Accordingly, the General Education Board refused the governor's request.[10] Still, circumvention was not out of the question, and by 1945, the legislature provided the money.

Adams and the State Bar

Paty's talk of accommodation and efforts to bribe blacks into going out of state did not represent the reality of university policy. That policy was shaped day to day by decisions in the Office of Admissions. Ralph Adams was an administrator's administrator. His whole life had been spent making order of bewildering files, a life eventually rewarded with the title Dean of Administration. As the chief admissions officer he followed no consistent policy in dealing with black applicants, except rejection. Some inquiries were simply ignored. To a young man asking whether "colored veterans" would be accepted, Adams wrote at the bottom, "No catalogue extended; nothing written." And to another applicant he replied curtly, "Since the institution you attended is maintained by the State for Negroes the assumption is that you are a Negro. The University of Alabama does not admit Negroes."[11]

Adams knew how to curry favor with those in authority by involving them in his efforts. When a student from predominantly black Talladega College applied, Adams asked Brewer Dixon, a university trustee from the city of Talladega, to check her out. "It would seem reasonable," Dixon replied, "to assume that one of the intelligence of this applicant must know that the Alabama Law School does not admit negro students and therefore her making application is perhaps a part of some plan to make a test case of our segregation laws." Adams also counseled individual deans concerning their replies to black applicants. When the historian and Dean of the Graduate School, Albert Burton Moore, got an inquiry from Private H.W. McElreath of the Tuskegee Army Air Field, Moore wrote Dean Adams saying, "I strongly suspect that this person is a Negro." Whereupon Adams and Moore agreed to respond with a simple statement that the University of Alabama offered no courses in the area of Private McElreath's interest.[12]

From 1944 on, many of the applicants were veterans. One veteran, writing from Ward Seven East of the Army Regional Hospital in Oakland, California, knew his rights and forced the university into a serious round of self-examination. Captain Nathaniel S. Colley, a native of Snow Hill, not far from Selma, and a 1941 honor graduate of Tuskegee, applied for law school in the spring of 1946. "I am aware of the fact," he wrote, "that the law in Alabama makes my attendance at the State University (which I help to support) illegal." He added that it was a denial of his constitutional rights "since none of the fictional 'separate-but-equal' facilities in Alabama provides legal training."

Moreover, Colley refused to "accept a grant-in-aid for attendance elsewhere," nor did he "plan to attend some makeshift segregated school." Scoring a direct hit with *Gaines* and anticipating *Sweatt*, Colley concluded that "the Federal courts should have the opportunity to pass upon my specific right to attend the University at Tuscaloosa."[13]

To this threat, Dean Adams huffed with a lengthy explanation of why he did not have to be told the implication of the *Gaines* ruling and added a proclamation of his own. "While this may be gratuitous," he lectured Colley, "I am adding that we at the University of Alabama are convinced that relationships between the races, in this section of the country at least, are not likely to be improved by pressure on behalf of members of the colored race in an effort to gain admission to institutions maintained by the state for members of the white race."[14] This assertion became a form-letter response to black applicants, but the Colley application could not be dismissed with simple arrogance. Dean William M. Hepburn of the Law School knew it and sought advice from a friend in the Regents Office of the University System of Georgia. His correspondent made it clear that "from a strictly legal standpoint there is no answer to the demand made." Nonetheless, he continued, Georgia was prepared to delay through all routes of appeal, "and if we had not by that time established adequate [separate] facilities we would not know what to do."[15]

Dean Adams sought advice elsewhere. In a letter to the president of the Alabama State Bar Association, he inquired about an appropriate response to Colley. The initial reply was not much help, simply referring Adams to legislative authorization for out-of-state grants. In a second letter, Adams demonstrated at length his knowledge of the law and, in a gratuitous aside, blasted those calling for token admission of blacks. "You and I know, of course, that the negro in the graduate school is just as black as the negro in the freshman class, and I do not believe that we Southerners are going to permit our children to eat and sleep with negroes, which will inevitably follow their admission to our classrooms."[16] The president of the state bar, a resident of Grove Hill, was more thoughtful. In a follow-up letter to Adams he advised, "Confidentially, I have very serious doubts, after having given further thought to the matter, that the Supreme Court will regard our legislation on the subject as being adequate; I rather think that if this applicant pursues his purpose to conclusion, he may get a court order requiring that he be admitted."[17]

With Paty's departure in December, Dean Adams became acting president for the 1947 calendar year. In his new capacity, he asked the new president of the state bar, Logan Martin, to influence legislation permitting blacks who attended approved out-of-state law schools to be admitted to the bar without examination. Because graduates of the university's law school already had that privilege, Adams believed his proposed amendment "would strengthen the State's position that it does furnish comparable facilities and would lighten our task in trying to persuade these [black] students to obtain their professional work elsewhere." Martin replied that such a plan would not meet the test of *Gaines*. Moreover, he was sure that the NAACP would not accept the proposed change "as a sufficient compliance with its demands."[18]

Six years later, when the university faced squarely the suit brought by Myers and Lucy, an inquiry was again made of the bar association to see if the Martin committee had ever filed a report. John B. Scott, then president of the bar, replied that association records disclosed a meeting held on August 21, 1947, but nothing more. "It is quite possible," Scott guessed, "that when the situation [Negro applicants in 1947] eased itself the committee did not proceed further with its work." Scott added, "As you know, we have had to fight a rear guard action as to this question. It looks like that for the time being this is a matter for the diplomatic department of the university as our legal position is none too secure."[19]

In the meantime Nathaniel Colley graduated from Yale's law school.

Gallalee and the closed window of opportunity

The futility of legal resistance became the persistent and consistent advice of the state's reputable legal community. The diplomatic alternative would have been viable only if the university had been willing to budge from its categorical rejection of black applicants. Unfortunately, Dean Adam's intransigence was matched by the man who replaced him in the presidential office. John Moran Gallalee, a mechanical engineer, came to Tuscaloosa from Virginia in 1912. Gallalee's attention to detail was exceeded only by his lack of humor. He got the sobriquet "shoot-em-in-the-leg Gallalee" when he ordered university police to "shoot in the leg" any student who attempted to destroy a fraternity-owned Confederate flag. The letters "G.G.G." (Gallalee's Gotta Go) appeared all over campus.[20] His stiff relationship with students acquired statewide notoriety and prompted a letter from a prominent member of the Birmingham bar who feared that Gallalee would go too far in punishing students involved in a panty raid. His relationship with the faculty was no better. He launched a building program and paid for it in part by diverting a million dollars intended by the legislature for faculty salaries. Only by promising not to redirect money in the future did Gallalee forestall an investigation by the Alabama chapter of the American Association of University Professors.[21]

Gallalee's administration was beset by dramatic swings in enrollment after the end of World War II. The numbers ballooned to 10,000 in 1948 and dropped to 6,500 by 1953. Despite these swings, the university's assets increased by 50 percent, doctoral programs were begun in eight departments, a School of Nursing was added in Tuscaloosa, and a School of Dentistry was established in Birmingham. These credits notwithstanding, Gallalee was no visionary. He stuck his finger to the wind of postwar racial thought and detected no change in direction for higher education. Asked about admitting blacks to professional programs, he responded, "It is my opinion that admitting a Negro to one of our white institutions would bring about a disturbing reaction from many worthwhile people of Alabama. In view of the problems of housing, feeding, social functions, and many other problems, we sincerely hope that we will not be faced with an issue of this type."[22]

In the face of these circumstances, Pollie Myers and Autherine Lucy made their decision. Both were encouraged by signs that not all students at Alabama shared the views of their elders. A 1948 student opinion survey, published in the *New York Times,* revealed that a majority (54 percent) favored abolition of the poll tax; 57 percent opposed arbitrary restrictions of black suffrage; 84 percent categorically opposed lynching; 68 percent favored a federal anti-lynching law; and half who expressed an opinion favored the establishment of a Fair Employment Practices Commission. Significant minorities favored admission of Negroes to the medical, law, and graduate schools (38, 35, and 33 percent respectively). Exactly half said they would not object to Negroes in the undergraduate school, although 64 percent of those holding this view favored some form of internal segregation.[23]

Morrison Williams, an aspiring student politico and son of Aubrey Williams, one of the state's most outspoken liberals, called for the admission of blacks. Two years later in 1950, Sam Harvey of Guntersville, editor of the student newspaper, predicted the eventual admission of blacks and failed to see anything "so terrible" about the prospect, for which he was called a "damned nigger loving Yankee." Harvey observed that people rode buses and shopped in stores with blacks, and he doubted "if violent proponents of segregation leave the room when a Negro janitor comes in to sweep." "Like it or not," Harvey concluded, "we might as well get ready. They're on the way." The administration observed all this with gimlet-eyed apprehension. The Dean of Students sent Gallalee a clipping from the Auburn student paper in which similar sentiments had been expressed. The dean attached a note saying, "Cold comfort, perhaps, but Auburn too."[24]

As president of the NAACP Youth Council, Pollie Myers took all this in. But the signs of change in the student population, though helpful, probably had less to do in forming Myers's decision than the words and deeds of Roy Wilkins and Emory Jackson. In 1951 Wilkins predicted in a Birmingham speech "that Negroes will be attending the University of Alabama and Alabama Polytechnic Institute within two years."[25] But Emory Jackson would make the big difference in her life.

Emory Jackson

No leader in Alabama made more important contributions to the awakening of blacks in the forties and fifties than the editor of the *Birmingham World.* Emory Jackson was not a popular leader. He was too aggressive and occasionally too full of himself. He craved recognition but was too abrasive to get it even when he deserved it. Still, Jackson pushed and prodded the black community into an awareness of its future. There was more fact than boast in a despairing letter he wrote to his sister in 1965. "Sometimes I wonder," he sighed, "whether the worry and wear of trying to work on a Negro newspaper is worth it. I could have made more money with less of the heart-breaking problems. If anything shows up, I am letting this job go."[26]

Born September 8, 1908, in Buena Vista, Georgia, Jackson moved with his family to Birmingham in 1919. He attended Morehouse College in Atlanta where in 1931 he became the first president of the newly organized Morehouse Student Body. He also worked on the *Maroon Tiger*, where C.A. Scott, editor and general manager of the *Atlanta Daily World*, had also worked as a student. That contact put Jackson in the newspaper business, but not immediately. Graduating from Morehouse in 1932, he first taught at Carver High School in Dothan, Alabama, then at Westfield High in Jefferson County. In 1934, Jackson became managing editor of the *Birmingham World*. He was an active member of Sardis Baptist Church, director of the Social Action Program of Omega Psi Alpha fraternity, and stalwart in the Alabama Branch of the NAACP.

Jackson looked like an editor. He rounded off his large frame by conducting a lot of business over food. He liked action and being in the presence of important people. When J. Edgar Hoover and the FBI went after Jackson in April 1943 for publishing a syndicated article critical of Hoover's racial policies at the FBI, Jackson seemed more interested in wangling a personal interview with Mr. Hoover than in defending his paper against the director's criticism.[27] The incident typified Jackson's style. Seeing little to be gained in a First Amendment war against Hoover, he used the occasion to advance his own interests. Besides, Jackson had his own agenda back home.

Among other things, he wanted integrated education in Alabama, but he believed that such a goal had to occur within a larger framework of black militancy. His own courage was never in question. He was escorted out of the Dixiecrat Convention, which Birmingham hosted in 1948, and told to run under threat of being shot. In 1951 T. Eugene (Bull) Connor menaced Jackson for having reported to the NAACP's Atlanta convention Birmingham's policy of deliberate police brutality. In 1953 he had a pistol drawn on him while attempting to register voters in Henry County, and in 1957 he was roughed up in his office because of his outspoken position in the Teachers Equal Salary fight.[28]

Jackson expected much of others. Like his conservative mentor, C.A. Scott, he supported the Republicans and Eisenhower in 1952, but he acknowledged that the GOP had had diminishing appeal for blacks since 1922 when the Alabama party excluded Negroes. Even though the GOP now tried to lure blacks back, Jackson believed Eisenhower's "lukewarm stand on civil rights" drove many away. When Harlan Hobart Grooms, a prominent Republican in Birmingham with prospects for a federal judgeship, invited blacks to an Eisenhower rally in the late summer of 1952, Jackson warned the black elite that "the masses of Negro followers became suspicious of those leaders who only show up as platform guests . . . or ride in gaudy motorcades."[29]

Conservative black leaders drew Jackson's special ire. When W.E. Short-ridge announced plans to use a black fraternity to conduct the NAACP's membership drive, Jackson flew into a rage, ostensibly because it removed any one individual from accountability, but more likely because the proposed fraternity, Alpha Phi Alpha, was a rival of Jackson's own, Omega Psi Phi. Even so, Jackson threatened to resign as secretary of the NAACP and sug-gested that Shortridge be investigated by the NAACP for financial involve-ment in a new, segregated golf course. Jackson reminded his readers that support of any segregated facility was against NAACP policy.[30]

Jackson's blurring of personal motives with philosophical positions dimin-ished neither and intensified both. He was always on the lookout for what Pollie Myers had called Judases in the black community, and though he lent his support to interracial groups, he was skeptical of their success.[31] Jackson was not much more optimistic about promises white liberals made to work with blacks. Noting that only two of twenty-five interracial groups organized under the aegis of the Southern Regional Council in Alabama showed sem-blance of life, Jackson lamented, "In Alabama the discouraging observation is clear that the two dominant racial groups are not working together."[32]

Myers absorbed all that Jackson had to say and seemed eager to please him. When he gave her the title of society editor, she felt uncomfortable and perhaps a little embarrassed because she lacked training in newspaper work. She determined to do something about it. At the same time, the sports editor of the *World* encouraged her to get formal schooling in journalism, but there were no journalism programs in the state's black colleges. So in the late spring and early summer of 1952 Myers's personal ambition merged with mounting frustration over segregation. The time for decision had come.

Arthur Shores

Though lawyers for the university later tried to prove that Myers's and Lucy's decision was made by committee and amounted to nothing more than a legal challenge by the NAACP, the evidence suggests something less. It will be recalled that Pollie worked part-time for the regional director of the NAACP, Ruby Hurley. Pollie had confided her personal ambition to Hurley, who put the support machinery in motion. Hurley later recalled, somewhat defen-sively, that the NAACP felt it could not be too selective about potential candidates in Alabama.[33] There were few who wished to run the gauntlet of threat and intimidation, and the brightest students saw no need to compro-mise a certain future that included opportunities to study out of state. But Pollie was determined, so Hurley decided to support her and suggested only that she get a friend to go with her for mutual support.

17

The next step was to get a lawyer. Arthur Davis Shores occupied center stage in the legal battle to end segregation in Alabama. A 1927 graduate of Talladega College, Shores spent ten years as principal of a school in Bessemer. During summers he attended law school at the University of Kansas ("because it was cheaper"), and in 1937 began a full-time law practice. He looked something like a compact, dark-skinned Errol Flynn. A pencil mustache highlighted the crisp details of a wardrobe that always appeared to have been pressed against a straight edge. His manners were more relaxed than his appearance, but he displayed a quiet tenacity in the legal arena. He forced his way into the all-white Democratic party. He was forever turning adversity into advantage. A black police informer was hired to take a punch at him to break up a trial Shores had instigated against the Birmingham police for brutality. The unsuccessful assailant later used his inside knowledge at the police department to turn criminal cases Shores's way, precisely the kind of boost he needed for his fledgling practice.[34]

Shores was supremely confident because he knew the law was on his side. He watched in bemused satisfaction as white lawyers squirmed in legal contortions trying to preserve tradition. His office was in the Masonic Building, a short step from the offices of the NAACP; so naturally Ruby Hurley told him about Myers. He had every reason to think that integrating the University of Alabama would be considerably less dangerous than efforts in behalf of black voting rights. With the law at his back, Shores expected legal resistance, but little more. His first task was developing a strategy to deal with the question of NAACP involvement. He knew that the public would doubt the sincerity of Myers's and Lucy's personal goals if they had NAACP support. Thus, to protect what white lawyers would obviously impugn—the good faith of their applications—a small lie was created and adhered to strictly throughout subsequent litigation. The falsehood maintained that Myers and Lucy got together at Myers's house, prepared their requests for applications on September 4, 1952, and mailed them the same day; that Myers and Lucy contacted Shores on September 24, only after they had been denied admission.

It was not a necessary lie. The court would not inquire into what motivated an application even if white lawyers sought to establish the absence of good faith. But in 1952, one could not escape the feeling that lies were necessary. After all, a system that cheated the most fundamental principles of democratic government encouraged, indeed expected, blacks to lie to whites and vice versa. The only legacy of this particular evasion is the continuing need to perpetuate its fiction and, accordingly, failed memories as to who helped the two women apply and how that assistance was given.

Opening skirmish

The truth seems to be that the letters of inquiry were prepared in the Masonic Temple Building.[35] Both were simple, two-sentence requests but contained a blatant error. Whoever typed the letters (Myers claimed to have done so) inadvertently put Myers's name at the bottom of the inquiry about library science and Lucy's on the request about journalism. They were signed accordingly. Though an obvious mix-up, it would be more evidence to doubt their sincerity. Nothing in the letters identified them by race.

Myers's and Lucy's mix-up, however, was nothing compared with the university's response. The day after they mailed their inquiries, September 5, the university sent application forms. Three days later the Assistant Dean of Women, Miss Ola Grace Baker, sent a letter to each girl promising modern and comfortable dorms and balanced diets of good wholesome food for only $216 a semester. To reserve a room Myers and Lucy were asked to send a deposit of five dollars each. Promptly done, the university acknowledged receipt of the deposits on September 10 and assigned them to Adams-Parker Dormitory. If they were surprised at how easily the process was going, it was nothing to compare with the letter from President Gallalee on September 13. It was a form letter, to be sure, but it welcomed them to the campus on the advice of the Dean of Admissions and promised that they would find the university "a delightful place to live and work."

The university did not realize the impact of its loose, if not open, admissions policy until September 19, when the Admissions Office received the applications of Myers and Lucy by certified mail. (On that same day Lucy resigned her job at the Britling Cafeteria in order to attend the university.) A precise man by nature, the Dean of Admissions, William F. Adams (no relation to Ralph), went straight to President Gallalee's office to explain his mistake. Gallalee saw no reason for alarm. All previous applicants had been dissuaded, so he got on the phone to see what could be done. The approach called for contacting white lawyers or business men with good connections to the black community and getting them to use their influence. In this case, the contact was Judge Clarence Allgood.

Like Shores, Judge Allgood had taken an irregular, though not unusual, route to the legal profession. In 1938 he became a referee in bankruptcy and set up what became known as debtors' court. That was where he first met Shores, many of whose clients were debtors. Allgood later got a law degree from the Birmingham School of Law. Allgood was a staunch supporter of Lister Hill and allied himself with the progressive wing of the state Democratic party. As a result, his ties to the black community were considered good.[36]

Gallalee reached Allgood through intermediaries. Allgood did not get to Shores immediately but promised to work on the case. In the meantime, Gallalee learned that the two women were to appear in the Admissions Office at nine the next morning, September 20. Shortly before they arrived, Gallalee called Adams into his office and told the dean to return the girls' dorm deposits. Gallalee wanted it done in cash. Before the conversation ended, a secretary appeared to say the women had arrived. A quick search turned up no cash, but Adams had two five-dollar bills in his wallet. Thus, with two bills bearing the likeness of Abraham Lincoln, Adams prepared to meet this latest test of the university's segregation policy.[37]

The Reverend J.L. Ware, president of the influential Baptist Ministers Conference of Greater Birmingham, accompanied Myers and Lucy. Ware was along to mediate if trouble arose. What happened next is a matter of dispute. Myers said that after the initial greetings Adams confessed that "an error has been made." When she asked if their transcripts had been accepted, Adams said, "No it isn't that, it is just that we can't accept you." Myers asked if the reason was her color, to which Adams responded by asking whether they had applied to Alabama State or Tuskegee. Myers protested that to go to the University of Alabama "would be a distinction," whereupon Adams simply repeated that "an error has been made."[38]

Myers also alleged that Adams said the laws of Alabama prohibited their enrollment. In a sworn affidavit, Adams denied having confessed an error or having said anything about the laws of Alabama. Who did or did not say what would never amount to much. The question of why or how Myers and Lucy had applied was immaterial. Both sides postured and prevaricated without need. The incidents of that September morning provided fodder for polemicists but did not affect the course of events.

Adams did offer to return the deposits, which the girls refused, ostensibly because their mothers would not believe the money had been returned—certainly one of the stranger inventions in that surreal encounter. When asked to accept in their behalf, Ware too refused, saying he had nothing to do with it. With the little curtain-raiser in Tuscaloosa thus enacted, the stage was set for Arthur Shores. On September 24, one day before his forty-eighth birthday, Shores wrote President Gallalee saying that he had been retained by Misses Polly [sic] Anne Myers and Autherine J. Lucy. Shores appealed to Gallalee "as President of the University" to intervene and grant them "admission to the University of Alabama."

The strategy called for appealing the decision all the way to Governor Gordon Persons. Failing there, the case would go to court. Myers and Lucy were to wait. Whatever career ambitions had spurred their original desire to apply would now be subservient to "the cause." They were content with that result, for having given themselves over to a cause, they could attain support, perhaps even a measure of notoriety, at least until the cause itself was exhausted. With the issue entrusted to lawyers and officialdom, the ugly denial of two attractive young women could be sanitized in professional courtesy. In a letter to Shores, Gallalee reported that according to the agreement reached by Shores and Judge Allgood, he was returning the five-dollar deposits. Gallalee went on to express his "appreciation for the consideration that you have shown in this matter. I intend to convey to you personally at an appropriate time our gratitude for your cooperation."[39]

Having joined the issue, the parties would wait. How long was uncertain. Ever optimistic, the *Birmingham World* looked for a speedy resolution. Noting that administrative remedies would be exhausted before going to court,

Emory Jackson observed, "Should this formula be followed and the Court follow other recent procedure, it means that this term Negro students will be studying in the University of Alabama."[40] Such optimism underestimated the determination to "keep 'Bama white." The *Birmingham News* kept the issue in perspective for white Alabamians by mentioning the attempted enrollment on page one, bottom right, of the September 21 edition; then dropping all discussion. More important to its readers was the football season. On the evening before Myers and Lucy arrived to register, the Crimson Tide had beaten Mississippi Southern 20 to 6 before a crowd of 14,000 in Montgomery's Crampton Bowl. It was a good start for a freshman quarterback from Montgomery named Bart Starr.

But even football could not completely submerge the question. President Gallalee, who had ordered Coach "Red" Drew to take his team off the Orange Bowl field if Syracuse University played one of its blacks in 1952 and who kept a clipping file the better to admire the defiant words of South Carolina's Governor James Byrnes, was now face to face with the dread issue.[41] Gallalee's brief five-year presidency had spanned most of the gestation period for the impending desegregation. The institutional response was intransigent from the beginning, though voices of moderation and accommodation were never completely silenced. Students and faculty had shown signs of acceptance or, at the very least, acquiescence. The legal community had read correctly the changing mood of the federal courts and counseled the futility of resistance.

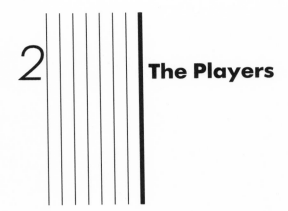

2 | The Players

Spring 1953 promised a new era for the University of Alabama. As usual, the azaleas around the president's mansion bloomed crimson and white, while towering oaks feathered green. The board of trustees announced that the university's next president would be the most distinguished educator the state had ever produced. Oliver Cromwell Carmichael, president of the Carnegie Foundation for the Advancement of Teaching and chairman of the Board of Trustees for the New York University System, was at age sixty-one returning to his alma mater. The student newspaper happily predicted "an era of recognition of the University's true worth to the state and to the South."[1]

Before Carmichael, the university had been served by a series of capable, and occasionally interesting, presidents. But they performed in another time. When Lee Bidgood stepped in as interim president between the retirement of John Gallalee in the spring and Carmichael's assumption of duties in September, the former Dean of Commerce made one important decision. He replaced the solid door which led to the president's office with one that had a glass pane, thereby ending years of frustration at not knowing whether the president or his secretary were in or out. Gallalee, it seems, had gone and come from his office as might a faculty member, showing little regard for whether the office was attended in his absence. Bidgood made one nod to privacy. He installed a shade that could be pulled.[2]

Such a casual approach to administration held over from a time when presidents operated like deans and led the faculty by force of personality. George Hutcheson Denny, who presided over the campus from 1912–36 and again on an interim basis in 1941, had used the personal touch in building the modern university. Gallalee, an engineer, was not much disposed to tinker with Denny's formula. However, the new man was. Carmichael was at the pinnacle of his career and spent late May and early June 1953 dispensing accumulated wisdom at commencement exercises. At the University of Texas he called for an end to isolationism in education, more attention to international studies, and an "inculcation of spiritual and moral values." In St. Louis he attacked "over-zealousness in rooting out Communist infiltration." Warning about McCarthyism, he declared, "It would be ironical indeed if in our efforts to root out Communist ideology we should by intimidation uproot a basic American tradition and put in its place a form of regimentation so despised in totalitarian regimes."[3]

Newsweek applauded his "Cromwellian tenacity" in attacking "the increasing emphasis on facts in education." The *New York Herald-Tribune* praised his belief that "the average college graduate" should appreciate "the three great streams of culture—Roman, Greek, Hebrew—that undergird our own" and noted his forward look to a " 'science of society' in which the humanist will have the aid of the social scientist."[4]

Cultural assumptions

Carmichael came honestly by his convictions about society and traditional values. Five years before his return to the university he attended a family reunion at a church camp not far from Goodwater, the small community in east central Alabama where he was born on October 3, 1891. The family was justifiably proud. Each of seven brothers had doctorates in medicine, theology, or education. Each had tilled the soil of Clay County and had his horizon shaped by a Calvinist dedication to duty and learning.[5] Carmichael's personal road to and from Tuscaloosa had been long but his progress enviably successful. In 1907 he traveled the short distance from Goodwater to Anniston where he enrolled at Alabama Presbyterian College. He helped pay for his education by working in the dormitories and milking the school's cows for the dining-hall tables. Two years later, Carmichael transferred to the university in Tuscaloosa. He excelled in public speaking and languages. He won a gold watch in the Senior Oratorical Contest, an accomplishment that also meant giving the commencement oration for the class of 1911. During his senior year, he received a teaching fellowship in romance languages, which was renewed an additional year so that he might work on a master's degree. While pursuing graduate study, he got his first taste of administration as secretary to the faculty. Mature beyond his years and a prince among students, Carmichael became acting professor of romance languages at the normal school in Florence, Alabama, even as the university pushed his candidacy to become its third Rhodes scholar.

Carmichael entered Wadham College, Oxford, in 1913 and began a six-year adventure. He was in Germany when war erupted in 1914 and made his way out on one of the first trains carrying Americans. He volunteered for Herbert Hoover's Belgium Relief Commission, and worked with the Commission in Antwerp, behind enemy lines. It was young Carmichael who smuggled out Cardinal Mericier's letter from the Belgian people—a letter that did much to counter German propaganda about conditions in occupied Belgium. Working for the Relief Commission, Carmichael made friends with the same ease that had characterized his relationship with the faculty in Tuscaloosa. One friend for life was Herbert Hoover. Carmichael left the Commission to join the British Army YMCA and sailed for Bombay on August 14, 1915. India impressed him for the first time with the differences among peoples, an impression that confirmed his Victorian appreciation for a society tiered by classes and races. He transferred to East Africa in 1916, and there, too, imbibed the British sense that certain lands, even on the dark continent, were white man's country. Ever the student, Carmichael added Swahili and Urdu to a growing list of language proficiencies that already included German, French, Spanish, Italian, Dutch, Latin, and Greek. In the spring of 1917 Wadham College awarded him the B.S. degree with diploma in anthropology.

One success followed another. Princeton awarded Carmichael an appointment as Charlotte Elizabeth Proctor Fellow in languages, but with the U.S. entry into World War I, he resigned the appointment to sign up for officer's training at Fort Oglethorpe, Georgia. Transferred to Camp Jackson in Columbia, South Carolina, he met Mae Crabtree of Atlanta. Eight months later, on July 13, 1918, they married. Two weeks later, he was on his way to France. Because of his foreign language competency he was assigned to the Intelligence Section of Division Staff. Sent to the Swiss frontier, he eventually wound up at Verdun. He mustered out at Fort Dix, New Jersey, with more education than a lad from Goodwater could have thought possible. He also returned with a nickname, "Mike," which he picked up in bridge games during idle hours.

The career ladder

Failure to complete an earned doctorate would be Carmichael's only frustration. He reapplied for the Proctor Fellowship but was turned down because married students were ineligible. He took a series of appointments in the high schools of Birmingham, teaching French and Spanish, and even ran for a seat in Congress, only to lose. But if he did not win election or the doctorate he so earnestly sought, he did win the hearts of his students, especially the girls. One of those charmed by his easy manners and good looks was Virginia Durr, who later figured prominently in the state's small liberal circle. "There wasn't a girl in his class who failed to fall for him. He was very handsome, with reddish blond hair and blue eyes. He had to be extremely proper and stiff and dignified or the girls would have mobbed him."[6] In 1922, he became Dean and Assistant to the President of the Alabama State College for Women at Montevallo. He tried one more time, in 1923, to get back to graduate study at Princeton, but the college's board of trustees refused to grant a leave. Two years later, in January 1926, the same board named him president. Not only had he proved himself by spearheading the "Million Dollar Drive" for Montevallo, he was immensely popular with the students; the women of

Montevallo took up Virginia Durr's earlier infatuation. Carmichael now began a series of executive appointments, each lasting the better part of a decade and marked by solid accomplishments.

In 1935, he became Dean of the Graduate School and Senior College at Vanderbilt. Chancellor Kirkland had noted his work on regional and national education committees and thought highly of him. The decision to leave Montevallo was not easy. Not only was he thoroughly admired in that community, he also had his eyes set on succeeding Denny at his alma mater. Still, the opportunity was too great to pass. By 1936, he moved up to Vice Chancellor, and on July 1, 1937, at age forty-six, he assumed duties as the third Chancellor of Vanderbilt. Again, his administration was one of accomplishment. Among other things, he started a law school, raised two million dollars for a library that bears his name, and, to the chagrin of some alumni, ended preferential treatment for football players. In the process he acquired still more influential friends, including John D. Rockefeller, Jr.

In 1946, Carmichael became president of the Carnegie Foundation for the Advancement of Teaching. The position appealed to him because it allowed him to experiment with more far-reaching ideas about higher education. He wanted to approximate the European model in graduate education, encourage programs in American Studies, and promote scholarship in international relations. While thoroughly modern, Carmichael stressed the teaching of values, values which in practical application resembled the Victorian standards of his youth. Living in New York completed his cosmopolitan journey. He got to know and like Dwight Eisenhower, then president of Columbia University, and renewed his friendship with Herbert Hoover. He also got to know another prominent Republican, Thomas E. Dewey. Recognizing Carmichael's talent for planning, Governor Dewey made him the first chairman of the Board of Trustees for the New York State University System. Carmichael responded with blueprints to democratize higher education in the Empire State, an effort that continues to bear fruit.

Why Carmichael returned to his alma mater is not clear. As early as the thirties, he showed an interest in succeeding Denny, but that was before Vanderbilt and his success in New York. Perhaps he thought of Alabama as a place to retire, a smaller garden in which to tend some of the educational flowers he had developed elsewhere. Or perhaps he could be taken at his word: "I want to repay my debt to the state for bringing me up."[7] Knowing that President Gallalee's retirement was imminent, the Board of Trustees approached Carmichael in 1952 through its executive committee. Carmichael turned them aside, saying that his work with the foundation was not finished. Within the year, he changed his mind. Moreover, he was now eligible for retirement from the foundation. Carmichael arranged a meeting with the executive committee and asked if his age would present a problem. They said no.[8] With that, the university had a president, and the president had a set of problems he could not have imagined.

The board

The board of trustees was a self-perpetuating body, whose prerogative was jealously guarded against attempted incursions by governors and legislators such as plagued the board of the Alabama Polytechnic Institute at Auburn. Chosen by congressional district, the ten-member board was composed of a textile manufacturer, a banker, a doctor, a real estate developer, and six lawyers—soon the banker and the doctor would be replaced by a corporate executive and another lawyer. None, with the exception of Thomas D. Russell, head of Russell Mills and chairman of the state chamber of commerce, would have been considered among the state's wealthiest or most powerful businessmen. Hill Ferguson had accumulated a modest fortune as vice president of a realty company but was better known as a tireless promoter of community development in Birmingham. Robert Eugene Steiner, Jr., head of a highly regarded law firm in Montgomery, followed his father on the board, as did Brewer Dixon, a tenacious and well-heeled country lawyer from Talladega. Gessner McCorvey, an attorney from Mobile, was the state's chief Dixiecrat in 1948. An affable man of the old school, his father, Thomas Chalmers McCorvey, held the first chair of history and moral philosophy and taught at the university for fifty years. Gessner grew up on campus.

Thomas S. Lawson of Greensboro sat on the state supreme court, having first drawn notice as a prosecutor in the Scottsboro case. Though his role in that tragedy should not be exaggerated, it was Lawson who effectively ended the trial phase by nol-prossing the cases against the final four boys. Also involved in hiring Carmichael was William H. Key, an attorney from Russellville. During Carmichael's tenure, William Henry Mitchell, a lawyer from Florence, and Gordon Palmer would be replaced on their deaths by John A. Caddell, a Decatur attorney, and Ernest Williams, university treasurer and a local bank director. Eris Paul, a circuit judge from Elba in southeast Alabama, filled a vacancy already existing at the time of Carmichael's arrival.

The appointment of Paul revealed much about the character of the board. Membership had little to do with representing constituent interests in the state. The main criterion was loyalty to the university. When the death of Dr. Paul Salter occasioned an opening in the Third District, Gessner McCorvey discussed the merits of Wallace D. Malone, "who is certainly one of the best known, and most prominent, citizens of Southeast Alabama. I imagine that Wallace is probably the best fixed man, financially, in that section of the State." However, McCorvey sensed a drawback. "He is an Alabama alumnus, but someone told me that Wallace sent his son to another University."[8] In fact, his son, who later graduated from Alabama, was attending Auburn. Still, Malone should have been McCorvey's man. Both were staunch segregationists and bolting Democrats in national elections. They were among Alabama's "Big Mules," a term denoting the planters of the state's Black Belt and their industrialist/banking allies in Birmingham. They shared the planters' orthodoxy on Negro inferiority and the industrialists' interest in cheap, passive labor. So why did McCorvey abandon a wealthy and influential ally?

In particular, why did the board turn to a little known, recently elected circuit judge from the small town of Elba in Coffee County? It helped that Paul had been active in the alumni association, much more so than Malone. A more important consideration was, for once, politics. Paul had supported Big Jim Folsom, and Folsom had just won election to his second term as governor in spite of and because of Big Mule opposition. McCorvey had done his own fast-stepping to stay on as chairman of the state Democratic party executive committee, and, he reasoned, it would not hurt to have someone on the board with political ties to Big Jim. Bob Steiner also supported Folsom, "the little man's big friend," but everybody figured his support owed to his law firm's business with the state. Paul, on the other hand, had worked for Folsom when the governor got his start in Elba.

So Paul became the board's choice. Their reasoning was simple if a bit gothic. "Judge Paul voted for Governor Folsom on the occasions when Folsom was a candidate and Folsom and his connections in Coffee County have supported Judge Paul. . . . This support has not been enthusiastic on the part of either for the other, but has been because of local conditions and family ties."[9] This half-nod to expedience constituted an exception to the trustees' general rule of keeping the board out of politics, and certainly out of Folsom-style politics. In the process, however, the board acquired a member whose judgment would be valued for years to come.

Had there been no integration crisis, board meetings would have been occupied with routine considerations of honorary degrees, real estate transactions, construction authorizations, which weekend the Alabama-Auburn football game should be played, and other matters pertinent to a closed corporation. More happily, correspondence between the board and its president could have focused exclusively, as it did in quieter times, on football tickets. But trouble in paradise was not to be avoided, and in the crisis that followed, five members emerged to set the tone for the board and the university: McCorvey, Steiner, Lawson, Dixon, and, chief among them, Ferguson.

Ferguson and the new president

William Hill Ferguson was a child of the nineteenth century. Born in Montgomery on Janury 6, 1877, he moved with his family to Birmingham ten years later where his alcoholic father, through the patronage of a friend, became receiver for a bankrupt railroad and later opened a law office. His father never made much money, but his mother was frugal and used an $8,000 inheritance to supplement the income. The Fergusons lived near many of early Birmingham's important people. Robert Jemison, Sr., was a boyhood friend who became a lifelong associate in business. Jemison and Ferguson made the relationship even closer when they married sisters, the daughters of another pioneer Birmingham family.[10] Ferguson followed his older brother, Burr, to the university in 1891. He quarterbacked the football team in 1896 and was one of eight scholar-athletes to receive Phi Beta Kappa keys retroactively when the university got a chapter in 1913. He helped form the glee club, enjoyed fraternity life (Sigma Nu), and was proud to be a member of the Alabama Corps of Cadets, whose members wore Confederate gray and lived in "Woods Barracks." After receiving his A.B. in 1896, he stayed another year in law school. During that time, he earned money as secretary to President Richard C. Jones and as manager of the yearbook, *The Corolla*. He also promoted a Chautauqua series that featured Thomas Dixon, best-selling author of *The Clansman*.

In Ferguson's day some 400 students clustered around the sixteen buildings making up the campus, a safe haven from the modernist era that brimmed uncertainly. They were a self-conscious elite, extending their adolescence through military training that smacked of the old order and sports like football that tokened the new. They stare back stiffly from photographs that reveal them in all their high-collared, scrubbed, and parted-in-the-middle innocence. Most appreciated their family's sacrifice in sending them to school but not to the point of dulling their enthusiasm for the new century. They could rest assured that whatever new, personal freedoms lay ahead would be an earned dividend, would come to them and their class for having made the sacrifice. They were assured of remaining a protected class so long as they maintained the discipline of their rank.

In Hill Ferguson the university got loyalty. As president of the Alumni Society from 1905 through 1907, he devised and spearheaded the "Greater University of Alabama Campaign." Under the slogan "Quarter of a Million Dollars—University Improvement in the Next Five Years," the campaign targeted the legislature for half a million dollars in capital improvements. As a thirtieth birthday present, Ferguson got a $550,000 grant above the $36,000 annual appropriation. With that money and architectural plans commissioned by Ferguson's committee, the twentieth-century university began to take shape. Not content with buildings, Ferguson wanted to fill them with students. To that end he pushed for the election of Denny in 1912. Denny shared Ferguson's love of football and by the mid-twenties the university had a national reputation. In the process, Denny filled not only the stadium but classrooms. By the thirties, the university had become a bustling institution of 4,000 students; though one where, in W.J. Cash's mind, the academic department had been reduced to "an apanage of fraternity row and a hired football team."[11]

In 1919, Ferguson was honored with election to the board of trustees, a post he held for forty years. By the time Carmichael arrived, Ferguson was said to "own the board." In fact, he did not need to own it. All but the younger members were of like mind, and they deferred appropriately to their elders. Ferguson would try to own Carmichael. A letter in late July 1953, a full month before Carmichael took office, warned of a meeting being organized by the "Fund for Development of Education." Bob Steiner had called from Montgomery to say that Governor Byrnes of South Carolina had alerted Governor Persons to the danger of southern universities participating in a meeting called to discuss segregation, especially before the Supreme Court made its ruling.[12] Carmichael received the letter at Biltmore Forest, the Vanderbilt estate near Asheville, North Carolina, where the Carmichaels maintained a retreat. Ferguson's unease over his president's connections with the "Foundations" was palpable, and, in fact, Carmichael sat on the Advisory Committee of the Ford Foundation's Fund for the Advancement of Education. On the same day Ferguson sent his warning, the Fund invited Carmichael to the meeting Governor Byrnes feared, and rightly so, for the report to be considered was Harry Ashmore's study on biracial education in the South, a study that was published as *The Negro and the Schools* and joined the chorus for change in the South. Carmichael begged off the meeting but wrote to say that he had read some of the material and felt "sure if Governor Byrnes were to read all of it carefully he could find no objection to your approach, though I suspect that he would still hope that no publicity of such a study would be given until after the Supreme Court decision."[13]

Carmichael was happy to avoid both Ferguson and his friends at the Fund. He had no deep conviction on the race question, and tried to duck forums where he might be required to take a public stand. He was a segregationist by inclination and would have preferred desegregation to come later rather than sooner, but he also knew the drift of his friends in the North and did not wish to alienate their affections. The result was personal ambivalence and public ambiguity, a deadly combination in his new circumstances.

A cautious leader

For the moment, however, Carmichael could come to Tuscaloosa with applause ringing in his ears. Late summer heat still hung close when the Carmichaels arrived, but faculty receptions in September could be tolerably comfortable, requiring only that they acknowledge the congratulations while the punch-and-cookies banter drifted to football. In expressing gratitude to the Carmichaels, the university set a precedent by having the mansion furnished. Apart from getting her piano into the mansion, Mrs. Carmichael said she left all details to the decorators. To reporters' questions about her short coiffure, she patiently said that she had had it cut in 1926, the year her second son was born, and had worn it that way since. She found the shorter skirts unbecoming. And so it went, from one society page to the next congratulatory editorial.

The air of calm continued through the fall of 1953. The trustees had effectively taken over the Lucy case in July when they hired the firm of Burr, McKamy, Moore and Tate. Senior partner Borden Burr had been the other credible candidate when Ferguson was appointed to the board in 1919. But members of the board felt they already had too many lawyers and elected Ferguson for his business and promotional acumen. Burr's firm now did much of the university's work, and at the request of the board, they prepared a brief for Carmichael "with reference to the present status of the case."[14] If Carmichael sensed danger in board control of this sensitive issue, it was not apparent in 1953.

Carmichael did give the race question some consideration. In December, Governor Persons solicited opinions from leaders throughout the state on the question of desegregation. "Alabama *must* be prepared," he said. "Unfortunately, we do not know exactly what to prepare *for.* This makes it all the more necessary that we, in the meantime, explore every possible solution so that we can be prepared for any eventuality." Carmichael's reply was nonspecific and safe. He called for the establishment of a large committee, perhaps 100 of Alabama's leading citizens representing all constituencies, to permit a maximum of discussion and an opportunity for those who wished to be heard.[15] Of course, this was precisely what the governor was already doing.

Carmichael also had received for comment a copy of the Ashmore report. He made a number of editorial suggestions showing the North's transgressions in racial matters. "The benefit of the doubt ought to be given the South," he thought. "Would it not be true to say that the worst race riots in all our history has [*sic*] occurred in recent years in the non-South." Ashmore incorporated the fact about riots but never acknowledged Carmichael's reading of the manuscript, perhaps because Carmichael's commentary was so limited and defensive in tone. Carmichael later refused to review the book for the *Virginia Quarterly Review,* saying that while he had read the manuscript he would not have time to prepare a suitable essay.[16]

Carmichael pursued one constructive approach to the problem quietly and diplomatically. In a conversation with President Eisenhower, he tried to persuade him that a major speech on human rights, if made in the South following the Supreme Court's ruling, would have good effect. "The more I have considered it," he wrote in follow-up, "the more it has seemed to me that the announcement of the Court's decision on segregation might prove a unique occasion and opportunity for a profoundly important statement on human rights. Such a statement made in the South would probably be more effective than if made elsewhere." Not surprisingly, Carmichael suggested the university's spring commencement as the venue for such a proclamation if the Court reached its decision before May 30.[17]

The White House response encouraged Carmichael to write a lengthy memorandum suggesting what the President might say. Again stressing northern transgressions and appealing for an appreciation of the South's unique situation with respect to its large black population, Carmichael nonetheless urged the President to come down strong on the nation's common heritage of human rights and the South's own progress over the past ninety years. He wanted Eisenhower to emphasize national "idealism" as a beacon to the rest of the world. From the international perspective, the President could say that the Supreme Court's decision "has obviously not pleased every segment of the American people, but there is no doubt that the Court was mindful of the profound issues involved and that it sought to achieve what every American hopes and prays for, a step forward in America's effort to strengthen the basis for peace and security in the world."[18] Carmichael's draft was an artful apology for a Supreme Court decision not yet written, and a brilliant suggestion for presidential leadership. For all his personal indecision, Carmichael sensed that a Court decree, backed by a clear signal from the President, would facilitate change. Such a statement made in the very heart of Dixie could prove all the more effective. But temperamentally Eisenhower was disposed to the "deliberate speed" doctrine that came to mark his administration and turned aside not only the invitation for early June but renewed appeals in September and again the following January.[19]

Grover Hall, Jr., of the *Montgomery Advertiser* was the first to question publicly Carmichael's own lack of leadership. When Carmichael observed that the Supreme Court's decision foreshadowed "many problems the nature of which is not yet clear," Hall retorted with his customary plain talk. "Contrary to Dr. Carmichael's statement, the 'nature' of the decision is as clear as Denny Chimes [the carillon tower that stands at the center of campus]. The court has ruled that it is unlawful for Dr. Carmichael to deny admission to a qualified Negro student to the university campus and dormitories. There aren't any ifs, ands or buts about that."[20] Hall would not be the last to speak loudly where Carmichael felt constrained to whisper. Over the next year, the Capstone's president stumbled into three controversial incidents that drew unwanted attention.

Miami

In 1954, the Commercial Law League invited Carmichael to address its July convention in Miami. A United Press correspondent interviewed Carmichael and reported him as saying that "non-segregation" could be brought about if both sides adopted "a tolerant and intelligent approach to the problem." He praised black and white leaders in the South for a mutually "sympathetic approach to the problem, one not to be treated flamboyantly if it is to succeed." He added that the University of Alabama "will make every effort to comply with the Supreme Court ruling banning segregation in public schools." Carmichael acknowledged the difficulty of bridging "folkways and traditions" but believed that "with mutual understanding and clear thinking the segregation problem can be worked out just as it has been worked out in the armed services."

This time Carmichael's rebuke came from another newspaperman, whose name, if not influence, was equal to that of Grover Hall, Jr. If anyone could have taken the measure of Carmichael, it was John Temple Graves, editor of the *Birmingham Age-Herald* since 1929 and by the fifties senior editorialist with the *Post-Herald*. His father had succeeded Henry Grady as editor of the *Atlanta Constitution* and delivered the now famous oration on the occasion of Grady's death. The Princeton-educated Graves, like Carmichael, was a discreet representative of his class, conscious of family inheritance and confident in his duty to speak about the human condition. The two were alike in at least one other respect: both were idealists. Truth and beauty were not mere abstractions; as essential ideas, they both dissolved and resolved reality. To acknowledge the evils of pellagra, rickets, and brain damage, which Graves often did, was not to condemn the social order that made victims of the classes most vulnerable to privation. In fact, social order itself was an ideal, requiring for its maintenance social stratification, or what Graves once called "the sacred differential."[21]

Thus, when Graves rebuked Carmichael for his Miami remarks, it stung. Expressing dismay, Carmichael sent a personal note to tell his side of the story. He explained that the statements attributed to him had not been part of the speech. In fact, the reporter had pressed him for his views even before he could check in and change clothes at the hotel. During the interview, Carmichael gave the reporter what he thought to be an accurate statement of the board's policy. The reference to the army situation and other comments, Carmichael asserted, were matters the reporter suggested and discussed on his own initiative "with a minimum of comment from me." Carmichael told Graves that he wished no retraction of the story because it would only stir more discussion and do more harm than good. Graves refused to let Carmichael off so easily. "Since the editor of this paper—and many other Southern leaders—take the position you were quoted as taking," he replied, "I did not feel it necessary to check with you before commenting. It was a perfectly respectable point of view even if one I oppose. But so many people have sent me the news story that I do wish there were some way to have the public record set right."[22]

Carmichael was not about to set the record straight. Given the opportunity to speak his mind by Governor Persons, he had ducked. Having tried and failed to get Eisenhower to speak, the university president was not about to take a field abandoned by a five-star general. Yet Graves was right. Having purposefully or inadvertently spoken his views, Carmichael had an opportunity to take a position that reasonable people could defend. Whatever liability would attach to the Miami utterances had already done so. The time was right for speaking out. What could the board have done—fire the state's most celebrated educator a year after his employment? Virginia Durr, by then a well-known Alabama liberal, wrote from Montgomery to congratulate and to encourage Carmichael for his remarks. "I have often thought and said (no doubt unwisely)," she wrote, "that while Southern men are noted for their physical courage, they have nearly all proved to be cowards when it came to the race problem."[23]

People like Durr and Graves, born and educated to the same tastes as Carmichael, stood together when the road of southern liberalism forked in the late 1930s. Seeing the fork, Graves veered right, becoming one of the South's most trenchant critics of the New Deal and anything that smacked of change in the social order. Durr went the other way, from championing abolition of the poll tax to giving aid and comfort to Freedom Riders. Finding their mutual friend, Carmichael, hesitating at the fork, both beckoned, but the ever-cautious Carmichael committed to neither.

The Race Relations Institute and Kern-Sims

Carmichael's default was more than deference to board policy. It mirrored a prevailing mood of escapism, a reaction to what Ashmore called a "cold atavistic wind of fear" sweeping the South.[24] Virginia Durr reached into her considerable bag of southern charm to steel Carmichael to the task ahead. "Whatever your problems, and necessities, and the importance of timing, proper utterance and so on, nothing will convince me that you are not too enlightened a man . . . to believe anything so barbarous as segregation. . . ." "There is a problem with being a hero to a fifteen year old girl," she cooed, "and that is you keep on being one. . . ."[25] Carmichael had other things on his mind in the late summer of 1954. The Carnegie Corporation came through with $150,000 for an institutional self-study.[26] He could now lay the foundation for an American Studies Program and promote the democratization of higher education through educational television and the development of extension centers.

Carmichael did keep his contacts with the foundation and, through them, the continuing debate over desegregation but remained reluctant to do other than make suggestions from the sidelines.[27] Even so, he could not avoid controversy. In the spring of 1955 Birmingham Southern College hosted a conference to explore "Negro Progress in Alabama and Its Effect Upon Race Relations." The aims of the conference were conservative. Still, organizers succeeded in getting permission for integrated seating, and Talladega College's president had the temerity to suggest that Americans demonstrated a "growing realization of the sin of segregation."[28] While the conference offered some promise of improved race relations, hardcore segregationists saw it as an opportunity to enlarge their enemies list. Each participant could expect a mimeographed hate sheet. One letter came to Carmichael almost a year later. Signed by Joe Adams, chairman of the Anniston Citizens' Council, the letter questioned Carmichael's participation in the "Race Relations Institute" and demanded to know his position on segregation. Carmichael replied that prominent lawyers, bankers, and industrialists had played a part in the conference and said that while the university was represented, he himself had not attended. Carmichael closed, "You suggest you would like to know my views on integration of the races. I have never favored it."[29]

Both the Miami incident and Carmichael's reply to the Citizens' Council chairman raise questions about his judgment. Why had he disavowed his own counsel of moderation? Why did he bother with a lengthy reply to a radical segregationist? The probable explanation is that Carmichael had little stomach for confrontation. His success had been built on resolving differences and getting people to talk. Admirably he did not reserve his communicative style for the boardroom. In the dark days of the Lucy crisis, thousands of letters poured in, many of the hate variety. Exceeding virtue, Carmichael set himself and his secretary the task of answering them all. Eventually they gave up, but not before the vast majority had been answered.[30]

Carmichael persisted in treating race as a matter for rational discussion. On May 5, Willis F. Kern, a former math instructor at the Naval Academy and resident of Arlington, Virginia, telegraphed Carmichael to complain that his daughter had been brainwashed on the subject of integration in a class taught by Dr. Verner Sims. Carmichael responded on May 12, defending Sims, who had taught in the university since 1928, and suggesting that Kern visit the university to discuss the matter with Sims directly. As it turned out, Kern was a member of the Birmingham-based American States' Rights Association, a rabid white supremacist organization. Worse still he associated with the Seaboard White Citizens' Council, headquartered in Arlington and run by Frederick John Kasper, a sycophant of Ezra Pound. Kasper also identified with Admiral John Crommelin, the ex-naval air commander, who prated a mixture of white supremacy and military preparedness in quixotic campaigns against Senator Lister Hill. On receiving Carmichael's reply, Kern circulated the correspondence to legislators, trustees, and others he deemed influential. He also sent copies to the segregationist *Alabama* magazine which published the correspondence on May 27 with editorial comment. Carmichael and the board reacted by circulating an official statement that criticized the magazine for publishing a story without investigating the facts.[31]

Unrepentant, the magazine of Alabama's conservative Big Mules struck back with a letter from a university freshman named Leonard Wilson. Headed "Awake! Sirs," the letter commended the magazine for its editorial "relative to the racial bosh that is taught at the University of Alabama." "Despite what the trustees and Dr. Carmichael may say," Wilson continued, "the warning of Mr. Kern is timely. The people need to know more about the affairs of the University and of the slaps that are often made at Southern customs and traditions." The sophomoric closing, "Alabamians awake! Paul Revere has ridden," signaled trouble ahead.[32] When the legislature showed interest, Carmichael's top assistant, Jefferson Bennett, recommended that Carmichael craft a statement on desegregation.[33] A prepared statement would have been preferable to the letter he sent a prominent businessman about the Kern-Sims affair. "I do not believe," Carmichael wrote, "there is any professor on the staff who believes integration of the races or desegregation in the schools and colleges would be desirable any more than you or I would believe it to be. I am convinced, therefore, that no one has been teaching such a view either directly or indirectly."[34] Carmichael's evasions amounted to rope in plenty.

During this period of travail, the board was not much help. If anything, its chairman became an added nuisance. Ferguson kept in constant contact with Carmichael on a variety of issues. From his office at the Jefferson County Courthouse, Ferguson garnered information from his "very active grapevine," especially about matters pertaining to football. But Ferguson could be more than an irritant. In the wake of the Kern-Sims affair, he goaded Carmichael by sending a speech in which Mississippi Senator James Eastland denounced liberal foundations in ways Ferguson found persuasive. Forced to waste time in response, Carmichael walked gingerly through Ferguson's minefield, trying to separate the Carnegie Endowment for International Peace which had supported Alger Hiss and the Carnegie Corporation which sponsored Gunnar Myrdal from the Carnegie Foundation for the Advancement of Teaching which had supported Carmichael.[35] It was a bewildering, if not hopeless, task.

By mid-summer 1955, Carmichael had been at the helm nearly two years, yet he was adrift. He had refused all occasions to set a steady course on desegregation and now felt himself at the mercy of a vast unknown. When the National Planning Association's Committee on Southern Development inquired about establishing a subcommittee on Social Environment for Economic Development, Carmichael resisted, saying, "I think a little later such a committee might be very useful, but at the present time I believe the tensions are so great as to make it unwise to announce such a committee."[36] When the NAACP asked if anything was being done voluntarily to comply with the Court's implementation decree, Carmichael was short. "I wish to advise that no laboratory or discussion programs or courses dealing with desegregation processes are being offered or planned at the University of Alabama."[37]

It is unlikely that Carmichael perceived his own failure to get in front of the desegregation issue. The university's case was in the hands of lawyers hired by the board, who were instructed to delay by all legal means. From his perspective, prudence dictated a low profile. Certainly, there was nothing he could do as his second year drew to a close. His last chance to influence the board and to effect a legal tone of compliance went aglimmering on a hot summer's day in Miami when he failed to own up to words of moderation, either his own or those put in his mouth by a reporter.

The NAACP gears for battle

While alumni applauded Carmichael's appointment, the state's black leadership was guarded. In a report to the Executive Committee of the Alabama State Conference of NAACP Branches, Emory Jackson observed that Carmichael's record was well known by the national office. "From our viewpoint," Jackson concluded, "the record lacks much that the NAACP would desire."[38] Still, activists in the black community found renewed hope in the "University Case." Since 1948, NAACP membership had declined as unease replaced postwar euphoria and the organization parried charges of communist infiltration. The Birmingham branch alone lost 5,000 members, from 6,614 in 1948 to 1,508 in 1951. But morale improved in 1952. The opening of a Southeast Regional Office in Birmingham and the commitment to do battle in Tuscaloosa provided the rallying point. In the long run, however, the Myers and Lucy case spelled disaster for the NAACP in Alabama and revealed the organization's limitations as a force in the civil rights movement. The NAACP proved to be conceptually and organizationally out of touch with movement reality: conceptually because the solution remained for them a legal one and organizationally because their rigid top-down structure worked against mass involvement.

The organizational weakness showed clearly in the effort to raise money for the University case. Lying at the heart of any crusade, fund-raising interweaves the essential threads of organization, persuasion, and commitment. On September 11, W.C. Patton, president of the state NAACP, wrote the national office requesting permission to raise $15,000 for the impending suit.[39] Written before Myers and Lucy had been denied admission, the letter was more evidence that the case had become part of a larger movement. It was equally clear that the effort was local, not national. If anything, the behavior of the national office frustrated the enthusiasm with which the project began. It reduced Patton's request to $8,000, saying that the *McLaurin* case in Oklahoma had cost only $5,000. It stipulated further that the money not include membership fees and that any funds over $8,000 be divided equally between the State Conference and the national office.[40] The directive fit organizational needs but not the times. The embers of freedom, which the NAACP had tended through a dark half-century in race relations, would soon explode in new tinder and race ahead of institutional imperatives. Unable to direct freedom's flame, the NAACP would try to harness its energy by applying conventional rules and regulations. Not until the Myers and Lucy case ran its course would the NAACP have clear evidence that it was in eclipse.

For the moment, however, dickering between state and national offices over who would get what seemed little more than annoyance. Spirits were up. The Birmingham branch instructed its education committee "to shift . . . attention to the 'University Case' " and set up an Education and Scholarship Fund.[41] On January 9, 1953, Emory Jackson circulated a letter to all branches setting a statewide goal of $15,000, despite the requirement that half of each dollar above $8,000 go to New York. Jackson suggested that pastors and churches that had not yet "lifted" an offering designate Sunday, January 25, as "Educational Freedom Day." The Tuskegee branch, which because of its relationship with the Tuskegee Institute did not stand to benefit financially from integrated education, used "baby contests" to raise $500 and earmarked the entire amount "for the prosecution of our University of Alabama case." By July 7, Jackson had raised $4,569.06. Of that amount $2,500 went to attorney Arthur Shores.[42]

Tension did not mark all relations between the national and state offices.

Signs of movement were all around. The State Conference enthusiastically joined the national Fighting Fund for Freedom campaign. Designed to culminate with the 1963 centennial celebration of the Emancipation Proclamation, its goals were voting rights, public accommodations, and education. All branches received manuals with detailed suggestions for mass meetings, down to using choral groups (preferably interracial) and stressing the importance of "no solos." Members who pledged at least twelve dollars a year received FFF pins. It was an inspirational campaign and coincided with local fervor for ending Jim Crow segregation.[43]

Putting lives on hold

If blacks felt optimistic, whites were not correspondingly alarmed. "Separate but equal" was still the law, and legislators promised a spate of legislation to preserve segregation. If laws did not work, there were always delaying tactics. For whites, tomorrow could be made to wait endlessly. Arthur Shores's own "deliberate speed" offered encouragement. Having elected to exhaust the University's appeals process, Shores allowed first Gallalee, then Governor Persons, to bide their time in deciding the fate of the applications. Shores had reason to be slow. Everything had to be cleared in New York. In December 1952, he wrote Thurgood Marshall congratulating him on a recent television appearance and enclosing a draft of the case. At the end of February, Shores wrote again, enclosing a copy of Governor Persons's decision to turn the matter over to the trustees and asking whether they should wait until the board met to file suit. Marshall advised waiting.[44] Shores continued passing information to Marshall. "I have heard from sources close to the administration of the University," he wrote, "that they will probably agree to admit the students without legal action. However, this would be a miracle in Alabama." Shores added that anticipating a negative decision he was "drawing his bill [of particulars]."[45]

Shores had given Governor Persons a June 10 deadline. On Monday, June 1, reporters gathered in Tuscaloosa to get the trustees' decision. Snippets of conversation from the boardroom was all they got. The trustees had no comment. Rufus Bealle, the board's secretary, waited until Saturday, June 6, to relay the decision by letter to Shores. He said the board had deferred "final action pending receipt of a court decision concerning litigation now before the Supreme Court of the United States." Bealle advised that the two women consider Alabama State College or Tuskegee.[46] Shores received the letter on the 8th and copied Marshall on the 9th, saying, "I presume there is nothing for us to do but go into Court now." He added, "Please return my complaint and whatever suggestions and additions that you propose, in order that we might get the complaint filed immediately."[47] Unfortunately, the New York office had misplaced the complaint. Near frantic, Shores reminded Marshall that it had been sent in December. "Please check your files," he pleaded, "as I did not even make a copy. We, too, would like to get this case filed before the [NAACP state] convention." A search failed to turn up the complaint, so Shores labored through another draft which he mailed on June 27. He urged haste "since the State Conference had expected it to have been filed at least a couple of weeks ago."[48] After quick editing, Shores filed it on July 3, 1953, the ninetieth anniversary of Pickett's charge at Gettysburg.

Chief Judge Seybourn H. Lynne of the United States District Court, Northern District of Alabama, received the complaint and upon application of the university enlarged the time for filing to September 11. Shores did not protest other than say that the process of deposition and discovery should proceed since cases such as this were "usually fully tried and won before actually coming to trial."[49] Judge Lynne agreed. The extension gave Shores time to feel out support from the state's white leadership, many of whom could read handwriting when it appeared on the wall. Among them was the state's attorney general, Si Garrett. Though later implicated in the corruption that plagued Phenix City, Alabama, and in the murder of Attorney General Albert Patterson, during Governor Persons's administration Garrett had the reputation of a reformer. Shores liked Garrett. In a letter to Marshall, he noted that the attorney general was "emphatic" in pledging not to cooperate in the university's defense, believing it to be "useless." Saying that Garrett "appears to be in our corner," Shores urged Marshall to meet with the attorney general when he passed through New York from a meeting in Boston. "He has been very cooperative with us in our efforts here," Shores emphasized.[50]

Arthur Shores was not a gullible man. There were plenty of "right-thinking" whites in positions of leadership. Shores's mistake came in thinking these whites could be counted for other than behind-the-scenes help, and desegregation was not about to stay behind the scenes. The best that could be hoped were small bits of encouragement from people like Garrett or, more often, the moderation of leaders like Senator John Pinson of Sumter County who, as a trustee of Tuskegee, argued for increased appropriations to that institution as a way of getting Shores to drop the lawsuit, "at least until the Supreme Court rules on segregation." Pinson told his fellow legislators that "the day of the demagogues has passed. No more Bilbos and Rankins—no more rabble rousers—no more Tom Heflins who thought an emissary of the Pope was hiding behind every bush and rock to do him bodily harm."[51]

Moderates like Garrett and Pinson served the cause in the long run by bowing to the inevitable, but in the short run they simply abetted the process of delay. Reliance on them was like accepting the offer of a ride from a friend who has no car. Even at that, moderation contributed less to delay than the NAACP's own legal maneuvering. On August 17, 1953, newly appointed federal judge Harlan Hobart Grooms swore his oath and immediately received from Judge Lynne the Myers and Lucy case. The NAACP had lodged its complaint against the state of Alabama, but the state constitution prohibited citizens from suing the state. Judge Grooms allowed the NAACP fifteen days to amend and was surprised to learn that rather than amend, the NAACP wished to appeal. A simple amendment naming a university official as defendant would have gotten Myers's and Lucy's case in court, but after a meeting in Atlanta, the NAACP decided to make a test of its right to sue a state agency in pursuit of constitutional rights. Openly disappointed, Shores bowed to the NAACP's desires. The Fifth Circuit dismissed the appeal on June 12, 1954, as "prematurely filed."[52] The NAACP gained nothing and lost eight months of Myers's and Lucy's time.

Another year and a day passed before the court ordered the case for final hearing. During that time plaintiffs and defendants argued over who would be named in the suit. As finally ordered, the case proceeded against Dean William F. Adams, and Grooms set the hearing for Wednesday, June 29, in Birmingham. Two years and nine months, about as much time as it had taken to fight the Korean War, had passed since the university denied Myers and Lucy admission in Dean Adams's office. The case had become larger than the personal desires of the two women, but they were still parties to it, and they clung to the case as if their wishes still mattered. During those years of waiting, adolescence gave way to the necessity of finding jobs. Lucy secured more stable employment. Through the Dixie Teachers Agency, she got a job at Conway Vocational High School in Carthage, Mississippi, a small stop on the road between Philadelphia and Jackson. She taught English and in the second year sponsored the senior class. She spent summers in Birmingham. Myers took a variety of jobs, teaching English as a "supply teacher" in Greene County, Alabama, and working in Birmingham at a cafe called Snack Bar No. 2. She continued to volunteer at the NAACP Regional Office and wrote occasionally for the newspaper. During the summer of 1954, Myers stayed with an uncle in Detroit and enrolled for courses at Wayne State University.

Setting the trial

The year 1954 was transitional. Lawyers were bogged down in procedural issues. The South awaited the Supreme Court's ruling on segregation, and when it came, only a handful of extremists expressed alarm. Even optimistic liberals thought integration would be a generation in coming. The Court itself delayed an implementation decree, preferring to invite southern states to participate in the formulation of its desegregation plan. Threats of reprisal and intimidation were yet to fill the air, and on the very eve of *Brown*, Thurgood Marshall could speak without incident before an audience of 1,500 blacks in Mobile. It was a time for uncertainty but not alarm. The first Citizens' Council (dubbed unofficially White Citizens' Council) held its organizational meeting on July 11, 1954, in Indianola, Mississippi, but that engine of massive resistance would not roll into Alabama until late November. Even then the formation of a council in Selma on November 29 and one in Linden less than a week later signaled no departure from a cautious wait-and-see attitude. At the political level, Governor Persons refused to call a special session of the legislature to plan for resistance, and the election of Jim Folsom for a second term to begin in 1955 discouraged those who counseled an all-out fight.

There is a curious weather pattern that sets up on warm, fall nights in the South. It begins with a stiff westerly breeze that picks up shortly after dark and blows steadily for about an hour. Then, without warning, the wind dies, but only for a moment. Windsocks quiver in anticipation of an easterly shift that will drop buckets of thick, incandescent gray rain. In the meteorology of race relations, the year 1954 was like those few, anxious seconds before the deluge. A nervous calm settled over the South. Then, on May 31, 1955, the Supreme Court handed down its implementation decree, and the heavens opened. The decree symbolized for many a step beyond return. The high court empowered the district courts to order desegregation wherever plaintiffs demonstrated the presence of laws, policies, or customs that promoted segregation. In deference to those same laws, policies, and customs, the Supreme Court urged that the process of desegregation be carried out "with all deliberate speed"—a formula which for blacks sounded all too familiar, echoing as it did the timid cry, voiced a century earlier, of "gradual emancipation immediately begun." The Court's gradualism did not discourage Arthur Shores. He was too busy preparing for the first test of the implementation decree. Because he asked that the relief sought by his clients extend to all similarly situated, the press and the public viewed the case with special interest. Though the class-action feature applied only to the University of Alabama, all parties knew that other institutions in the state would fall like dominos.

Two weeks after the decree, Judge Grooms set the trial date of *Myers and Lucy v. Adams* for June 29, 1955. Both sides quickened the pace of preparation. On the plaintiffs' side Shores checked with New York to iron out final details, an effort that included making sure that Marshall's assistant, Constance Motley, did not refer to Jeff Bennett, Carmichael's assistant, as Jeff Chandler, the movie star, a mistake she had made on her trial notes. At the same time, Emory Jackson drummed up support in the community. He observed that "the two young women [had] stood up well in this fight" but were in difficult financial circumstances. Neither was "substantially employed" for the summer, and while they did not complain, "both recognized their plight." Jackson noted that machinery had already been set up "for promoting this case" and that "a committee of outstanding women" had been asked "to look out for the personal and related welfare of these two women." Equally interested in garnering moral support, Jackson encouraged Negro leaders to turn out for the June 29 trial.[53]

For its part, the university was less interested in obtaining public support than in digging up dirt on the two women. Hill Ferguson wrote Bob Steiner, the Montgomery attorney who acted as the trustees' contact with the university's lawyers, and copied his letter to Carmichael and to Gordon Palmer of the trustees' executive committee. Ferguson recommended hiring the Bodeker Detective Agency to investigate Lucy and Myers. He believed that the cost would be minimal; $25 a day plus expenses, with the total cost not exceeding $250. The next day Carmichael talked over the proposal in Steiner's office. Both liked the idea and hoped they could "develop something in this investigation." Carmichael also talked with Palmer, who suggested only that the contact with the agency be initiated by counsel. Having reached agreement, Ferguson called the attorneys on the same day Judge Grooms ordered the trial.[54]

Before June 1955, the university made little effort to learn about Myers and Lucy. Of course, massive resisters were busy sifting rumors and facts until the two became indistinguishable, but the university felt no need of exertion. The board simply hired lawyers. Frontis H. Moore, a senior partner in the Burr firm, and Andrew J. Thomas, its best litigator, took charge. This minimal exertion seemed sufficient, especially during the procedural phase when the trustees saw what they considered incompetence from the opposition's legal counsel. Andrew Thomas wrote about deficiencies in the plaintiffs' case which was presented "by a Negro woman [Motley] whom I presume was out of Thurgood Marshall's office." Thomas belittled the plaintiffs' lawyers for endlessly substituting defendants and scoffed at their general strategy, though he proved sufficiently cautious to make "no prediction" as to outcome.[55]

The trial

While Shores prepared interrogatories and divided courtroom responsibilities with Constance Motley, Thomas and Moore worked out their strategy, based in part on information received from the detectives. The trial had been set for one of the larger courtrooms in Birmingham's Federal Building, and to the delight of Emory Jackson, blacks packed the room four to one. Judge Grooms, a native of Kentucky and 1926 graduate of the University of Kentucky Law School, brought his border-state Republicanism to Alabama. He chaired the state GOP Finance Committee and organized Eisenhower's 1952 tour of the state. His appointment at age fifty-seven to the federal bench came as no surprise. Black attorneys correctly assessed him as "no crusader,"[56] but then none was needed. Grooms could be counted on to do his duty, if conservatively. It also happened that Grooms's son was attending the university's law school, where he was a member of Jasons, the top honor society for men. His daughter had graduated from the university in 1954.

The trial convened at 9:30 the morning of the twenty-ninth. After two years and nine months, there should have been more to it. Shores and Motley called ten witnesses; five connected with the university and five, including Myers and Lucy, representing plaintiffs. Shores examined each in a perfunctory manner: What is your name? How long have you been connected with the university, and in what capacity? Are you familiar with the fact that by law and custom Negro and white students in Alabama have been and presently are educated in separate public educational facilities in the state? At this point, the defendant's lawyers would object, saying that the witness could speak only of the university, whereupon Shores would ask about the policy of the university, only to have the trustee or official reply in studied ignorance. It was Kafkesque to have Hill Ferguson say that to his "knowledge" Myers and Lucy had not applied, though he did "have evidence that they did." Counsel for the university, observed a black reporter, "argued that there was no evidence of racial denial, suggested that the wrong parties were being asked for relief, challenged the class action feature, and contended that the University had a 'no policy' policy with regards to racial applicants."[57]

Andrew Thomas, a physically imposing man whose voice commanded attention, concentrated on the motives of the two women. He doubted their purpose, evidenced by the mix-up that found Myers asking for a program in library science and Lucy for journalism. He suggested that Shores and the Legal Defense Fund set up the case to further the NAACP's own objectives, an allegation Shores heatedly denied. Not surprisingly, Thomas also went after Myers's personal reputation, carefully going over the dates of her marriage and her application to the university, which revealed that she had been pregnant, though not married, when she applied. Shores and Motley quietly endured Thomas's high fustian, a performance the segregationist magazine *Alabama* itself described as "overborne."[58] Of course, the court made it clear that motives were irrelevant and immaterial, but Thomas pursued them nonetheless and filled the air with objections phrased in Latin whenever Shores examined witnesses. In an earlier time and before sympathetic courts, such labyrinthine locutions would have succeeded. Whether anyone for the defense thought such tactics would succeed this time is not known. The attorneys prepared as if preparations could make a difference, but even they were cautious enough not to predict victory. Perhaps Hill Ferguson thought they could win. Certainly a society that had developed an elaborate mythology to justify segregation would have difficulty in perceiving the truth of arguments against the prevailing system, no matter how constitutional. In matters of race, the southern courtroom long since had subjected reality to the Procrustean blade and rack.

Only Arthur Shores was right. If the law were applied, the discovery process itself would settle the case. The facts would speak. Not only had the university denied admission to Myers and Lucy (or as the university argued it, not had their applications acted upon), the university had also denied admission to other blacks and had made such denials a continuing practice. Shores entered into evidence the application of Wilbur H. Hollins, a young man who had applied to law school in 1950 without success. At the time of the trial, Hollins managed a low-cost housing project and associated with Shores in the Hollins and Shores Realty Company. To demonstrate a continuing policy, Shores had his secretary, Agnes Studemeyer, apply on May 31, 1955. She asked President Carmichael whether the "policy" regarding "the admission of Negroes" had changed. Carmichael advised "that the admissions requirements of the University of Alabama have not been changed in recent months." Carmichael's reply had been reviewed by counsel; the major contribution being the substitution of the word "requirement" for "policy."[59] The university paid for fine and, in the end, irrelevant distinctions.

For all the university's money and effort, Judge Grooms did not allow even a face-saving interval between closing arguments and his order of injunctive relief for the two women. He observed that "on the occasions that the applicants were considered by Dr. Gallalee and the Board of Trustees, neither Dr. Gallalee nor the Board took any steps to deny the applications on grounds other than that of race and color of the applicants." He further noted that while the evidence failed to disclose "a written policy," the university consistently advised black applicants to use the machinery provided for studying out of state or suggested they apply to Tuskegee or Alabama State. Accordingly, the university, having strained at the gnat of whether "no action" could be construed as "no policy," was now ordered to swallow the camel of desegregation. Forty-eight hours later, Grooms extended the relief to all "similarly situated." Both Shores and Frontis Moore expressed surprise at how quickly the class-action ruling came. The university had no comment. Governor Folsom, always ready with a quip, was suddenly unavailable. Reporters finally got to Austin Meadows, the state Superintendent of Education, who could only hope that Negroes would "abide the will of the majority of the people in Alabama which is overwhelmingly in favor of continuing segregated education."[60]

Reaction and preparation

Meadows merely repeated what had become cant in the white mythology of segregation: that a majority of blacks did not want integration any more than did whites. Having never had to consider black attitudes, Meadows could not know the exuberance with which the black community greeted Judge Grooms's decision. Emory Jackson spent the Fourth of July in urgent yet happy correspondence. He wrote the president of the state NAACP Conference urging large numbers of blacks to apply to state-supported institutions of higher education and technical training. He noted that 103 black students were using Jim Crow scholarships to study out of state and feared losing the offensive if many of these could not be persuaded to stay home.[61] Jackson also wrote Roy Wilkins about raising more money to ease the financial burden for Myers and Lucy. Jackson told Wilkins that the national office had promised to put the local group "in touch with a source from which the money could be applied for to be used as scholarship aid." Wilkins replied that he was unaware of "any definite pledge" but promised to help and referred Jackson's request to Herbert L. Wright, youth secretary for the NAACP. Wright in turn asked Jackson to gather the scholastic records of the two women and letters of recommendation "attesting to the integrity and ability of the young ladies by former teachers, ministers and so on."[62] On July 15 Jackson called to say that both women had C plus averages and together would need $1,055 a year in scholarship aid. He promised letters of recommendation immediately.[63]

Lucy received recommendations from her minister, the principal of the school where she taught, and from Myers. Whatever letters Myers got have disappeared, but her letter in behalf of Lucy strained for eloquence and revealed much about its author:[64]

> It is with profound pleasure that I write this letter recommending Miss Autherine J. Lucy for an educational scholarship to attend the University of Alabama this fall.
>
> To put it bluntly, I am a staunch friend of Juanita, I call her, and have been since our college days at Miles College here in Birmingham. I can unhesitatingly say that her scholarship at Miles College is good. But, most of all, her moral character, diplomacy, neatness, and pleasing personality are far above negative interrogation.
>
> Her enthusiasm to get a job well done, her courage to tackle the difficult, her patience, her unselfishness are too above question, as revealed by her stand and legal fight to open the doors of the heretofore all-white institution, the University of Alabama, to qualified Negroes.
>
> Certainly she would welcome a scholarship from you, and, of course, her modesty will not permit her to say that her financial means are just ample for a comfortable living.
>
> Thanking you in advance for any consideration given to Miss Autherine J. Lucy, I am. . . .

Herbert Wright wrote back regretting the inability to use Myers's letter. "I am sure," he explained, "that you were unaware that it is rather improper for one applicant to write a letter of recommendation for another."[65]

The national office made arrangements with the Jessie Smith Noyes Foundation to provide scholarship money.[66] Myers and Lucy, however, needed more than tuition and fees. Because they were unable to attend in the fall, additional resources had to be found to carry them through. Lucy had given up her job in Mississippi, and A.G. Gaston, owner of the Booker T. Washington Insurance Company, provided secretarial employment. Myers continued her odd jobs. Jackson renewed pleas to supplement their small incomes and set up a Personal Needs Fund (later called the Committee on Special Needs) in the fall of 1955. Lucinda Brown Robey, state youth director, coordinated this fund, and Jackson instructed the secretary of the Alabama Scholarship and Education Fund to funnel all money that "persons or organizations" wish to contribute for Myers's and Lucy's personal needs through Mrs. Robey. Jackson reported that $211.35 had been spent already for Lucy and $366.98 for Myers. He believed $700 more would be needed immediately and that other indebtedness would no doubt be incurred "which should be our responsibility to assume."[67]

Myers consistently made greater demands on the committee. In early January 1956, Jackson reported the two women as listing "about $700 in items of personal things they needed" and accounted for an earlier expenditure of $578.24. He further noted that "the things listed by Mrs. Hudson [Myers] did not show up on the original list" and would have to be included. "The other applicant [Lucy]," he noted, "seems to be firm and has made no unusual request."[68] Myers's father came to regard the gifts as a bribe. He remembered a sewing machine, luggage, a washing machine, and a piano, totaling some two to three thousand dollars which he eventually had to pay off.[69] Bitterness may have clouded his memory, but it should not be forgotten that the sudden appearance of gifts in a world of material want justified suspicion.

Stall, stall, stall

While the NAACP worked to get the women enrolled, the university did all it could to impede Judge Grooms's order. Hill Ferguson captured the spirit of the effort in a letter to Carmichael. At a Birmingham Rotary Club luncheon, Ferguson had asked Edward Lund, president of Alabama College and soon-to-be president of Kenyon College, what defense "he and his associates were setting up against the black clouds that were threatening." Lund replied, "STALL, STALL, STALL." He added that applicants would be required to identify themselves by race and "said he would like to require a Wassermann test [for syphilis], but probably would not!!!"[70] If humor were intended, it was lost on Ferguson.

Ferguson had a special friend, Walter Brower, whom he often sought for advice. At one time Brower served as an arbitrator for the International Ladies Garment Workers Union, but association with a liberal union had no influence on Brower, who also was an active Klansman and took his cues from the White Citizens' Council. In the late fifties, Governor John Patterson appointed Brower to chair the Jefferson County Board of Registrars. In that capacity, he purged both whites and blacks from the voter roles along lines recommended by the Citizens' Council.[71] Brower was an embarrassment, but Ferguson considered him a genius. Ferguson was intent on getting Brower's views before the board and called a meeting of the executive committee for July 16 on campus. Brower prepared a document for the meeting. Titled "The Jury Is the 'Saving Clause,' or Rather Our 'Last Line of Defense' in Our Present Crisis," Brower argued that the university should maneuver itself into "criminal contempt" of Judge Grooms's order, thereby entitling the university to a jury trial. *"All this means,"* Brower wrote, *"that our customs, our traditions and our way of life itself is under the protection of the juries of the South."*[72] The recipients of this unsolicited advice, Bob Steiner, Gordon Palmer, Carmichael, and Rufus Bealle, sat in stunned silence as the quadrangle boiled in 90° heat. Once again Ferguson had seized the initiative, and once again nothing constructive could come of it. Immediately following the meeting, Steiner asked Andrew Thomas if there were anything in Brower's suggestion that he might have missed. Thomas said no.[73]

Undeterred, Ferguson called a secret meeting of the board, saying, "I am told there is at least one drastic defense we might invoke, but this is not possible without the hearty concurrence of the Board." McCorvey cheerfully replied, "I cannot possibly imagine any defense that would be too drastic to have my hearty concurrence. . . . The more drastic our action, the better I will like it."[74] No record survives of what Ferguson proposed, but a few days later Carmichael asked the Dean of the Law School about Brower's suggestion that "inaction" on an application would be legal because the court had enjoined only "actions" that denied admission. Dean Harrison was contemptuous. "If such a simple device as that suggested by Mr. Brower could be effective in nullifying an injunction of this type, the segregation cases would be meaningless." Carmichael passed Harrison's comment back to Ferguson.[75]

In the meantime the board became frustrated with its hired attorneys. In February, Ferguson agreed with Bob Steiner that the law firm had done "a lot of defensive work on this matter, and their 'delaying action' has been very successful."[76] That was before Grooms's court order. McCorvey now proposed hiring additional counsel and recommended Joseph F. Johnston, who had drafted Alabama's pupil placement law, a measure designed to circumvent desegregation in elementary and secondary schools. For different reasons, Rufus Bealle suggested that Carmichael follow North Carolina's example and associate the university with the state's attorney general. Such a move "would allow the University [by which he meant the administration, not the trustees] to share the responsibility in our case to some extent."[77] Bealle's suggestion measured how far removed from decision-making Carmichael and his administration had become. The board was calling the shots through its own counsel.

Displeasure with Moore and Thomas was only a passing phase, excited, no doubt, by the hopelessness of the case. Still, trustees could take comfort in Judge Grooms's decision to stay his injunction while the university appealed to the Fifth Circuit. The conservative Grooms reasoned that his order "would be more palatable and consequently more acceptable if it was buttressed by a decision of the appellate court."[78] Shores tried desperately to reverse the stay so that enrollment could proceed for the fall semester. He applied to Judge Richard T. Rives of the Fifth Circuit to remove the stay but was refused. Bob Steiner reported that Rives held the opinion that "under the Supreme Court's decision to proceed 'with deliberate speed,' it was contemplated that the trial judge [Grooms] give the University time to work out the problem pending appeal."[79] Shores turned to Justice Hugo Black, but Black refused to act without the full court, which meant the earliest notice would be October 3.[80] Finally, on October 10, four days after the end of fall registration, the Supreme Court reversed Grooms and reinstated the injunction. Judge Grooms described the high court's action as "crisp."[81] Shores's persistence forced Carmichael to confer with Andrew Thomas and Bob Steiner to plan a response. They agreed "that since October 6 [was] the last day for registration for the present semester, we [the University] could take the position that that applies to the plaintiffs as well as all others."[82] Policies that might have bent under normal circumstances now became inflexible. It was the last barrier the university could legally erect.

Brower again

From July through December 1955, university officials bought time, time they used to develop novel legal approaches to resistance and to find information detrimental to the reputations and qualifications of the two applicants.[83] The strategy reached a frustrating anticlimax on Saturday, November 12. Ferguson had called another meeting of the board for 10 a.m. at the University Center in Birmingham and hoped for a good turnout because the football team played archrival Georgia Tech at Legion Field that afternoon. Despite the lure, less than half the members showed up. Present were Ferguson, Eris Paul, Thomas Russell, Brewer Dixon, and Thomas Lawson. Also attending were Carmichael, Dean of Administration James H. "Foots" Newman, and Rufus Bealle.

Ferguson appeared with Walter Brower in tow. Without delay, he introduced Brower, who focused most of his remarks on a resolution that called for substituting an agent for Dean Adams as final authority for admissions. The agent would be a public figure who could be charged with criminal contempt without incurring lasting liability or stigma. It was Brower's old scheme for getting a jury trial. After embarrassed discussion about the absence of a quorum, the resolution, along with another calling for alumni endorsements for applicants, was tabled. But Ferguson would not stop. He put the meeting in executive session and proceeded with the discussion he was bent on having.[84] The question of alumni endorsements had been discussed since the July 1 court ruling and even invited reasoned response such as Gordon Palmer's observation that its legality was doubtful because the university had no black alumni.[85] However, Brower's suggestion that the university invite a criminal citation raised eyebrows, especially when he acknowledged that winning by juried decision might have the effect of asking "a juror to violate his oath." He argued further that the board had the same right, indeed duty, to devise plans for resisting the Supreme Court's 1954 ruling as did the state legislature in passing laws to perpetuate segregation. Moreover, because the Supreme Court's order did not mandate any action, the university could resist simply by doing nothing. Thus, even if charged with criminal contempt, a crime would not have been committed because no "action" would have been taken.[86]

The lawyers in the room were thunderstruck. Judge Lawson finally stammered, "If purely a restraining order [that is, no action mandated], what harm would there have been in allowing Judge Grooms's order [of July 1] to stand." After some hesitation, Brewer Dixon at last said the obvious. As a lawyer and "officer of the court," he was "embarrassed." "Shall we take defiance to the Supreme Court?" he queried. His loyalty to the traditions of the university and his obligations as a lawyer made it a "difficult decision"; nonetheless, he concluded, "I am bound by the Supreme Court's decision." As an afterthought, Dixon regretted that the university's lawyers "did not see fit to be here."

Unmoved, Brower said that the university was the "capstone of education" in Alabama and that if Negroes were admitted nothing would stop them from doing whatever they pleased. He attacked the board for its timidity, saying that the legislature and Joe Johnston's law firm did not hesitate to find ways of circumventing the Court's decree with the Pupil Placement bill. Reiterating his position on restraining versus mandatory orders, Brower said the worst that could happen, even if Dean Adams remained the admissions officer, would be a fine, an assessment for damages, and court costs—a "slight expense," he concluded, compared with the gain. It was Dixon's turn to be unmoved. "Wouldn't it be obvious," he demurred, "that this would be a direct challenge to the Supreme Court's order?"

"Assuming it is obvious," Brower shot back, "didn't our own Legislature and the state of Georgia obviously attempt to get around it?" At this point Judge Lawson patiently explained the "difference" between a state legislature and a board of trustees. Finally, in an effort to close discussion, Lawson said, *"If the Court of Appeals* holds it [Grooms's ruling] is not a *mandatory order,* then I will go along with you that we may be able to take these steps." Of course, Judge Lawson knew well enough that Brower's tortured construction of the law bordered on dementia. Carmichael, who sat stoically, jotting occasional notes, later agreed with Rufus Bealle that, barring objection from the trustees, no official record of the meeting should be kept. Ferguson, however, would not be silenced. In a letter to Bealle, he said, "I enclose copies of the two resolutions which I submitted to the board November 12th, with the notation that they were fully discussed and tabled. I want these on the record to show that I did try to set up some defense mechanism against the threatened admission of Negroes to the University."[87]

Again, Ferguson had not won, but as before, he set the agenda. There were good lawyers on the board, and men, like Gordon Palmer, of uncommonly good judgment; but they were paralyzed. Above all, they did not want to be found guilty of doing something (desegregation) they did not have to do, and Ferguson kept telling them they did not have to do it. In the struggle for their souls, Ferguson offered a Faustian contract for the here and now. So "duty" was put off to yet another day, awaiting still another federal court to inscribe it on the wall. The board even put off taking action on the "morals" issue that Brower and Ferguson knew would eliminate one of the applicants.

Coda

If only the university had bowed to the inevitable in September 1952, or more probably in June of 1953; if the NAACP had amended its complaint rather than appeal the principle of suing a state agency; if the university had not practiced dilatory motions while awaiting the Supreme Court's implementation decree; or if the university had refused to indulge its futile appeals after July 1, 1955, then desegregation of the University of Alabama might have occurred sometime, anytime, before the winter of 1956. And if desegregation had occurred at any time before the winter of 1956, then the chances of success would have improved measurably. Whatever the possibilities, each passing school term diminished the chances for success, until, finally, the worst possible time arrived.

Ironically, Hugo Black drove the final nail in the coffin of peaceful desegregation when he turned the request to remove Judge Grooms's stay over to his brethren on the Court. That action allowed the October 6 deadline for fall registration to pass by four days: four days that allowed the state of Alabama four months to make the final move from passive to massive resistance. To be sure, there was only a slim chance remaining to avoid a crisis, but there was a chance. Any hope for peace, however, was dashed on December 1, when Rosa Parks refused to move to the back of a Montgomery bus. The civil rights movement with its familiar pattern of non-violent protest in the face of massive resistance now raced ahead of the NAACP and its preference for organization and litigation. In the white community, the boycott galvanized sentiment. The Citizens' Councils, already entrenched in western and central Alabama, now made inroads in the hill country to the north and the wiregrass region to the southeast.

The campus itself was a miserable place in the fall of 1955. The football game against Georgia Tech, played the afternoon of the secret November 12 board meeting, marked the eleventh consecutive loss for the Crimson Tide dating back to the 1954 Tech game. The team would lose the rest of its games that season before sending senior quarterback Bart Starr to the Green Bay Packers as a seventeenth-round draft choice. To make matters worse, intra-state rival Auburn was nationally ranked with such players as half-back (later governor) Fob James, whose 7.7 yards per carry led the Southeastern Conference. For a school like Alabama, with a storied Rose Bowl tradition, the world was truly upside down.

In the waning days of 1955 the real contest, however, was not between archrivals on a playing field but between whites and blacks who viewed each other with increasing suspicion and alarm. The genial acceptance of subordination by blacks that outwardly marked race relations between the wars had given way to a new relationship in which blacks now seemed out of place. For Ferguson and others of his class, equality for blacks was incomprehensible. Having come to maturity in a generation of radical racial thought, they had no language and no perceptual capacity to see blacks as anything but inferior. Ferguson fortified and protected his world by hiring lawyers and private investigators to keep his beloved university white. There were other whites, however, whose lives were less protected, their contacts with blacks more immediate and more fearful. They secured their world by acts of intimidation and violence.

In the beginning, the Myers and Lucy case inspired Alabama blacks to greater exertion. By the end of 1955, however, inspiration had turned to grim uncertainty. In the Library of Congress are boxes filled with sheafs of annual reports from local branches of the NAACP. These reports contain lists of officers and members, as well as accounts of fund-raising events, membership drives, community projects, and special activities. One can spend tedious hours leafing through these reports until, suddenly, the truth of 1955 leaps from a page. It comes from a report by Roy L. Thompson of Chambers County, who sent his list of new members personally to Roy Wilkins but confessed, "We have some that did not wish to reveal their names, so we just numbered them." Or it comes from A.P. Benderson, Sr., who filed his report in January 1956. Memberships and contributions were down and nothing was happening. Then at the bottom, Benderson wrote the unlettered truth: "The Tuscaloosa Branch have not been able to put over any thing fully, due to the fact, the people seem to be affraid."

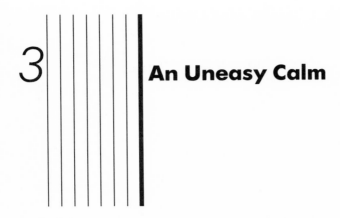

3 An Uneasy Calm

Except that February comes hard on its heels, January is the unhappiest of months. It is an ideal time for penance. Praying and fasting come easier in January because there are fewer alternatives. But January 1956 called for something more down South, especially for blacks. On the night of the 27th, a young black minister who found himself thrust to the front of the Montgomery bus boycott sat alone in his kitchen in total despair. He prayed for deliverance. Epiphany came instead. What else could God give? There was no escaping the white man's system, no Red Sea beyond which lay a promised land. So epiphany would have to do, some divine manifestation that could bring inner peace for those about to enter the storm. What Martin Luther King, Jr., experienced that evening, all blacks similarly situated would have to find.[1] Epiphany was becoming a necessary class action.

The two women who knocked at the schoolhouse door in Tuscaloosa stood on the brink of their own searing experience. Only weeks before, they had had the good fortune of meeting the young minister who was emerging as the symbolic leader of Montgomery's bus boycott. Their meeting was not by chance. Emory Jackson had watched developments in Montgomery closely since the December 5 meeting that gave birth to the Montgomery Improvement Association, forerunner of the Southern Christian Leadership Conference, with King as its president. (Jackson's reports remain among the best evidence of what took place in Montgomery during the crucial months of December and January, 1955–56.)[2] On one of his visits, Lucy and Myers went along. Myers had been invited to address the Dexter Avenue congregation where she was introduced by her nervous and trembling friend.[3] Lucy remembers shaking hands with Dexter's handsome, young Dr. King—a polite exchange between people who, though born the same year, had and would travel different paths; yet within a short time both their names, first Lucy's, would become national symbols.

Emory Jackson's attention to developments in Montgomery overshadowed, for the moment, the approaching denouement in Tuscaloosa. For blacks Tuscaloosa and Montgomery had become separate engagements on the same battlefield. On January 26, the night before Martin Luther King's kitchen visitation, a cross went up in flames on the campus in Tuscaloosa, a visible night sign of white emotions. A visit to the university that day by Lucy and Myers had provoked the burning. Arthur Shores and Emory Jackson had accompanied them to arrange for registration. Apart from waiting forty-five minutes for Dean Adams's arrival, everything seemed to go smoothly. Discussions ranged from transfer credits to fees and room assignments. Though cordial, the visit accomplished little. The university had constructed no plan for registering its first black students; therefore, little more could be said than when and where to appear. About transfer credits, Adams said the applicants would be notified upon receipt of all transcripts. As for room assignments, he was evasive. Still, there was no reason to doubt that after three years of waiting, Myers and Lucy were less than a week away from registering. As Myers said to a reporter, "We are looking forward to enrolling and claiming this as our alma mater."[4]

Little could the two women know that Dean Adams was part of a conspiracy of silence. The board of trustees continued maneuvering for a miracle, moves that demanded silence. On January 9, Hill Ferguson called a meeting of the board at what he termed "the urgent request of President Carmichael and with the approval of the Executive Committee."[5] Carmichael was justifiably anxious. Without specific board approval, he felt powerless to discuss preparations, even with his faculty, and the board seemed in no hurry. His sense of urgency heightened with reports that the Supreme Court would not stay the injunction while the university appealed the Fifth Circuit's December 30 ruling—a fact the high court confirmed informally on January 16. The trustees were to meet for lunch on Saturday the 21st at the Union Building and to stay in session as long as necessary. The plans laid that afternoon were as cold as the weather outside. They instructed Dean Adams to refer Lucy's completed application to the board (her transcript from Selma University and Myers's from Wayne University in Detroit were still being rounded up). They further instructed Adams "to deny admission to Polly Ann [*sic*] Myers Hudson on the ground that the evidence before the Board indicates that her conduct and marital record have been such that she does not meet the admissions standards of the University."[6] Such was the plan. The only thing missing on the 21st was final and formal board approval. Strategically the vote would not be taken until the 29th.

Dean Adams knew all this when he met with Lucy and Myers to discuss registration. Behind his pleasantries that day was a trap set to spring. Not that the decision to deny admission to Myers on the grounds stated was unusual. The university used pregnancy out of wedlock before and after the Myers case as grounds for dismissal or denial of admission. Nor was it applied solely to women. One student, who fathered a child out of wedlock, received a letter of dismissal that expressed the hope that he learn from the experience and "resolve to observe the responsibilities of a married adult in our society."[7]

What was unusual was the timing of the notification. Myers would not learn of her exclusion until the afternoon before her scheduled registration; even though a full ten days had passed following the board's decision to deny her application and months had gone by since the November 12 meeting where trustees talked secretly about what they planned to do in her case. Thus, the words that passed in the January 26 conference with Dean Adams had no relevance as they related to Myers's enrollment.

In fact, all January seemed stranger than fiction. Carmichael, who urgently called a board meeting on January 9, met with the University Council (deans, department heads, and administrators) two days later and never discussed plans for the impending desegregation. Instead, the council took up a proposal to schedule the annual Alabama-Auburn game for the Thanksgiving holidays and another to develop a policy on student absences. The next day Carmichael was in Montgomery arguing the disastrous effects for education of the recent defeat of the Goodwyn tax bill. A day later he signed a request for $23,750 to conduct research for a survival plan in the event of nuclear attack. It was one more sign of the times, and just as children were expected to save themselves from thermal blasts by diving under desks or sitting fetally tucked in school hallways, so did university officials approach the dangers of desegregation blind to reality.

The final board meeting before registration on February 1 took place on January 29, a Sunday. The day was unseasonably warm. Inside, the trustees reaffirmed the decision to deny admission to Myers and, momentously, with one dissenting vote, accepted Autherine Lucy's now completed application— the lone dissenter being Ferguson. Governor Folsom attended, but hearing no alarm about possible violence, left with the impression that the trustees wanted to control the situation without interference from his office. For his part, he was content to let them reap whatever they had sown. With few exceptions, these men were not his political friends, and his dislike for them was instinctive.[8] They did entertain some discussion about procedures for registration and ways to limit disturbances should Lucy go through with her plans, but they still hoped that the trap set on January 21, when sprung, would end the affair. On leaving the meeting Ferguson told a reporter, "We'll know more about it on Wednesday," but when pressed for details, the chief resister said only, "It's a very serious matter."[9]

The surprising thing is how completely the trustees maintained silence. Using the legal process as a shield, they made no effort to notify the two women until January 30, the day the injunction became mandatory. Press speculation focused on the possibility that Carmichael and Adams might not approve Myers's and Lucy's credits. Why no one speculated about Myers's becoming a target because of her marital record is not clear. The simplest explanation is perhaps best. Even though the trustees had known and discussed her background for over a year, and probably longer, they had not made an issue of it. Board silence lulled the NAACP and the press into thinking Myers's pregnancy would not be made an issue. Certainly after Dean

Adams's cordial reception of the two women and their advisers on the 26th, there seemed no reason to suspect a last-minute surprise.

Monday, the 30th, passed in quiet preparation. The grooming committee, spearheaded by Mrs. Geneva R. Lee, operator of a beauty-supply business, had gone over the questions of what to wear and how to behave. Myers and Lucy rehearsed these instructions as they went about last-minute details. No talk or preparation, however, could stay the anxiety. For three years the battle to enroll in the University of Alabama had been a legal abstraction compounded by guaranteed delays. Now registration was two days away. Fortunately, doubts expressed that day by the NAACP's national youth secretary never reached Lucy or Myers. "I certainly hope," he confided to Emory Jackson, "that they prove worthy of the efforts being expended for them."[10] It was a burden they did not need. That evening, for a fourth time, flames from a burning cross licked the crisp, night air in Tuscaloosa, this time at the Negro high school, while in Montgomery a bomb exploded at the home of Martin Luther King, just three days following epiphany.

The university now tossed its own bomb in the form of one-sentence letters to Lucy and Myers—each got copies by both regular and registered mail. Lucy's notice of admission arrived Tuesday morning, and a reporter described the moment. "A three-year-long bumpy but legal road leading to the portals of the University," he wrote, "ended abruptly yesterday on a dirt alley in Ensley. Possibly unaware of the destiny-packed contents, a postman made his rounds on the rutted alley-like street, stopping to deliver two envelopes at 1110 Avenue I." This was the white-painted, neat-looking duplex where Lucy lived with her sister Ethel and her husband Ulyses Moore. Inside, Lucy glanced at the letter. She appeared "reserved, calm and confident; perhaps a trifle uncomfortable, too." Her voice matched her demeanor, "precise, straightforward, but with a detached monotone that bordered on what sounded like indifference." She said simply, "I'm very elated." When asked how she thought she might be treated on campus, she replied, "I haven't any mental picture of how I'll be treated. I always try to take care of situations as I get to them." With that, she left with Arthur Shores to go downtown.[11]

At half past noon, Myers still had not received her letter. Later that afternoon there would be tears. At the last hour the university ended her hopes of enrollment as it knew it would do all along. The trustees had two reasons for delivering the blow without warning. The first was so obvious that it could always mask the second, a meaner and, in the end, a more compelling motive. The university had no mandatory obligation to act on the applications until it exhausted its appeal of the Fifth Circuit's ruling, which occurred formally on January 30. While sufficient to explain the trustees' delay, strict adherence to the legal requirement did not explain their darker hopes—desires that were not lost on Lucy and Myers. The university hoped that Lucy, the more introverted of the two, would not attend without her friend. And in fact Lucy did have second thoughts, but Myers, ever true to her forceful nature, said tearfully, "She's going . . . she's going even if I have to go with her."[12]

Thus ended, on the last day of January 1956, three and a half years of costly and life-absorbing legal wrangling. Whatever the price, it did not compare with what the state of Alabama and its university paid for the next six days in February.

Wednesday, February 1

Lucy arrived on campus at 10:15 on a cold, gray Wednesday morning. Rain, which in a few days would swell the Warrior River to overflowing, had started. She was accompanied by Myers, Emory Jackson, Mrs. Geneva Lee, and the Reverend Fred Lee Shuttlesworth, later to become Martin Luther King's man in Birmingham. Shuttlesworth, who made up in determination what some thought he occasionally lacked in judgment, later imbued the struggle to overthrow Bull Connor's Birmingham with courage by example. On this day he relieved the drama in Tuscaloosa with humor. After Lucy entered Dean Adams's office, Shuttlesworth and her escorts sat on a bench in the hall of the Administration Building. They were joined by Emily Barrett, a secretary-reporter for the Alabama *Citizen,* Tuscaloosa's black newspaper, and Landy Thomas, wife of the *Citizen*'s owner. Though a few students muttered grudging or unkind remarks loud enough to be heard, others made brief, friendly conversation. One student asked Shuttlesworth why the NAACP didn't "send down some basketball players about 7'2"?" Without hesitation, he said they could send "some good football players, too."[13] Later, when Dean Newman fretted to Dean Adams about the continued presence of the Negroes in the hall, Adams went out and asked if they were waiting to see him. Shuttlesworth decided he was and asked for an application, remarking, "It certainly is a nice place here." Taken aback, Adams quickly retrieved an application, handed it to Shuttlesworth, and said curtly, "It certainly is. Good day."[14]

Lucy stayed in Dean Adams's office briefly, filling out preliminary papers before heading to the College of Education in adjacent Graves Hall where registration for courses took place. The evening before, M.L. Roberts, the college registrar, huddled for two hours with Dean John R. McClure to decide how to handle Lucy's advisement and which courses she should take. Evaluating her transcripts, they determined she had some undergraduate deficiencies to make up and assigned her introductory classes in political science and geography along with others in children's literature, history of education, and reference service.[15] When Roberts arrived at 7:45 the next morning, students were already lining up inside Graves Hall. He atttributed the unusually long lines to the rain and the historic nature of the day. By 9:00 some 200 students clogged the hallway, and Roberts urged those not in the required alphabetical order to come back at their scheduled times. He welcomed them to wait in line if they wished but thought they would be more comfortable drinking coffee in the "Supe Store." About three-fourths of the students left. By 9:45 another line had developed, and Roberts again urged the students to come back later, this time locking the door to his office to make the point.

Dean Adams meanwhile called Roberts to say that due to the crush of reporters and photographers, he should secure a private place for Lucy's registration. Roberts offered his office. Adams also instructed Roberts not to send Lucy to Foster Auditorium to pick up class cards for her courses. On the short walk from the Administration Building to Graves Hall, a police officer followed Lucy. She spent about fifty minutes in Roberts's office being advised by Mrs. Pauline Foster, head of the Department of School Librarianship. Lucy thought the program recommended by the college entirely satisfactory and asked only that a one-hour class in piano be added. Mrs. Foster persuaded her not to take the additional hour, as sixteen semester-hours already was a strenuous load. Before leaving, Lucy asked Roberts whether she should consider her registration complete before getting a dormitory assignment. He said yes, but Lucy's question showed that she was uneasy about the absence of official word on housing. Leaving Graves Hall, again with police escort, she returned to the Administration Building to see Dean of Women Sarah Healy. There she expected to receive a dormitory assignment and to arrange for meals. Up to that point, all had seemed to go smoothly. She was even beginning to relax. However, the board of trustees had one last surprise.

As Lucy entered Dean Healy's office at 11:50, the university's News Bureau chief distributed a press release to reporters outside. The release was the first public statement made by any university official in the long legal battle. It said that the trustees had instructed university authorities to deny dormitory room and board. When Lucy learned the news from Dean Healy, she immediately asked to call in one of her advisers. Dean Healy gave permission but took the precaution of calling her own assistant, Ola Grace Baker, as a witness. Lucy returned with Emory Jackson. What followed is disputed. White newspapers and Dean Healy reported that Jackson was hostile. When Healy tried to explain that tension was too high as evidenced by four burned crosses on the university campus the evening before, Jackson allegedly shot back, "We don't care how many cross burnings or camp meetings or whatever they have, we demand equal opportunities for her in everything. We do not mean to be belligerent, but we know the law and we are going to get what we want."[16]

Jackson asked to see a copy of the board's statement, whereupon Healy summoned Rufus Bealle, secretary to the trustees. He backed up Dean Healy by emphasizing that she was acting under instructions; whereupon Jackson said that he knew he was dealing with the "lower echelon" and assured Bealle that his party "would be back this afternoon with a court order forcing you to give her a dormitory assignment." "Just give me the paper," he concluded, "and we will leave for Birmingham immediately."[17] Jackson later denied that his behavior was belligerent. He said that he had had occasion to talk with Dean Healy when no one was present. She impressed him with her sincerity, especially her assurance that she would do everything possible to help Miss

Lucy. Sarah Healy had that effect on all who came in contact with her. She had only recently arrived from the University of Michigan where she had frequent contact with black students. In Tuscaloosa she had joined a group at Canterbury Chapel dedicated to peaceful change. Jackson always believed Dean Healy's description of him as "hostile" was the workings of university attorneys, not Healy.

The problems of perception that took place inside Dean Healy's office formed part of a larger pattern of distortion that began the first day and ever after clouded the memory of what actually happened. It is not surprising. On such a day, when every act is gestural, no procedure normal, there are no dispassionate observers. Autherine Lucy arrived in a Cadillac, "they" muttered; she dressed to the nines, "they" nodded; she pushed to the head of the line where she flashed a crisp hundred dollar bill to pay tuition, "they" observed; and, unlike regular students, she did not have to stand in the dreaded lines at Foster Auditorium to pick up class cards, "they" complained. By simply being there, she affronted them, and "they" were offended.

One cannot now imagine circumstances that would not have offended. The committee members who prepared Lucy operated in the dark. They had never known life on a southern campus "for whites only," much less the sensibilities in taste and manners of those who watched their every move. The committee could and did know that society in general valued evidence of material well-being. If a good car was prized, why not a Cadillac; if clothes mattered, why not Sunday best? Moreover, every action had its explanation. They used A.G. Gaston's Cadillac, Lucy later remembered, only after a Chevrolet failed a cold morning start.[18] She overdressed, but would bobby-socks and loafers have appeared too casual? As for the crisp one hundred dollar bill, Jackson and the preparation committee doubted that the university would take a check drawn on their special fund. Avoiding Foster Auditorium and its lines was a decision made by the university, not the NAACP. Whether all the after-the-fact explanations were true, whether there was a Chevrolet or a good reason for the crisp bill, is irrelevant. Every gesture made that day was calculated by the NAACP to avoid giving offense. As Jackson put it, "I have spent most of my career down in this area, and I know better."[19] But the facts also remain. A Cadillac was used, a one hundred dollar bill was tendered, Lucy was spared the drudgery of registration, and white people did not like it.

When Lucy paid the university treasurer, Ernest Williams (himself soon to be a trustee), she became the first African-American to breach legally any Jim Crow barrier erected by the state of Alabama. A reporter noted the precise moment: 12:53 p.m., February 1.

Thursday, February 2

At 9:15 Thursday morning, Judge Grooms called Andrew Thomas, the university's attorney, to say that Shores was in his office with a joint petition in behalf of Lucy and Myers and that Thomas and Frontis Moore should come over at 1:30 to discuss it. Grooms asked Thomas and Moore whether they had read Title 52, Section 492 of the Alabama Code, which said "no person shall be excluded from the full benefit of the university fund, or placed at any disadvantage in the pursuit of his study, who possesses the requisite literary or other qualifications, and is willing to submit to the discipline prescribed for the students." He also noted that the university had a longstanding rule "that undergraduate girls entering the school [which corresponded to Lucy's status because, though a graduate of Miles College, she sought a second undergraduate degree] must live on the campus unless they live near enough to return home nightly." Grooms said that failure to give Lucy dorm and board privileges, while not likely contemptuous of his order, did provide "a basis for contending that the above referred to statute had been violated, insofar as she was concerned." In fact, Grooms said he thought Lucy would win and urged the parties not to set off another round of litigation.[20]

Thomas and Moore asked Grooms whether a room off campus, or at least not in a regular dormitory, would satisfy the law. When Grooms intimated yes, Shores protested vigorously, making it clear that any arrangement outside the dormitory would be contested. When the subject shifted to boarding, Grooms again sought compromise, saying that it was his understanding that "Miss Lucy could obtain facilities for eating without eating in the University cafe." Shores was equally adamant on this point. The nearest cafe for Negroes was at least half a mile off campus. Concerning room assignment, Grooms observed that it might be all right if Lucy were given a corner room on the top floor, not necessarily near other rooms. He understood further that there was no contention that she be permitted to room with a white girl. Grooms, like Martin Luther King in the bus boycott, "was not seeking to end segregation, just modify its terms."[21]

As for Myers, Grooms said that the issue was whether university authorities acted capriciously in denying her admission. He reiterated that he would not and could not sit as a board of admission to pass on qualifications and expressed the hope that Myers's case could be kept out of the press as much as possible. With those observations, Grooms served notice on Thomas and Moore to appear at 9:30, February 9, to answer charges in the Hudson case—Myers's case now being referred to by her married name. Also, Shores expressly agreed to withhold a contempt citation he had drawn up against Dean Healy for denying room and board to Lucy.

While lawyers talked in Birmingham, university officials geared up for the first day of classes, but despite cross-burnings which occurred nightly, they made no elaborate preparations. University police were to handle security. Carmichael himself left on Thursday for a meeting in New Orleans. Before departing he sent a note to his old friend Alvin Eurich at the Ford Fund for the Advancement of Education. "We have finally accepted one of the Negro applicants who enrolled yesterday," he wrote. Then observing better than he knew, he concluded, "The sentiment in the University community is temperate but in the larger community of Tuscaloosa we have had manifestations of intemperateness which has concerned us greatly."[22] That night a cross burned in Dean Adams's yard.

Friday, February 3

In Alabama, when it isn't cold in February, it rains. So with headlines of "Somebody Pulled the Plug, More Rain Due," Lucy set out for her first day of classes on a dreary Friday morning. As if to offset the weather, she wore a two-piece orange dress beneath her raincoat and hat. The Reverend R.I. Alford, pastor of the Sardis Baptist Church, Mrs. Lee, and Reverend Shuttlesworth accompanied her. She arrived a few minutes late (9:05 to be exact) for her first class and entered Eugene A. Smith Hall, the massive yellow brick building that anchored the northeast corner of the quadrangle. In one of those small ironies produced by big occasions, the cornerstone of Smith Hall had been laid forty-nine years before by Hill Ferguson. It was the first building in the "Greater University" campaign that he orchestrated while president of the Alumni Society. Now in room 200, a lecture hall reserved for large introductory classes in geography and geology, Autherine Lucy sat alone, nerves calmed by prayer, preparing to take notes on Professor Don Hays's explanation of the course. She took a middle seat on the front row, hoping other students might join her, though her late arrival (9:08 when she entered the classroom) made that unlikely. When she walked in, a fifteen- to twenty-second hush fell over the students. Then at 9:10, Professor Hays, apparently notified that she was in place, entered to make his remarks.

Hays went through a perfunctory roll call, outlined the units of instruction, explained the examination procedure, and announced the text. He made no special note of Lucy's presence or the significance of the occasion. In the hall a small police guard stood watch while reporters and photographers complained about being refused access to the star attraction. The only incident occurred when a student stalked out with clenched fists and muttered, "For two cents I'd drop the course." This ill-tempered remark, picked up by reporters and published nationally, brought a flood of letters to the university, each with two pennies enclosed. Lucy did not notice the student's rear exit. She could take comfort in the absence of a truly noticeable incident.[24]

As the class broke up, several policemen drove over to the Education Library, an annex of Graves Hall, to take up stations. Photographers caught the loneliness of the day as Lucy walked diagonally across the quadrangle, trailed by a lone officer. Curious students huddled under dripping eaves and watched her solitary progress. Though not yet relaxed, the tension was relieved in her next class, children's literature, when other coeds sat next to her. She even felt comfortable enough to ask the teacher, Jean Lyda, to repeat her last name so that she might get it right. Again, the class was brief, so she went to the bookstore in the Union Building across University Boulevard. There

Lucy received first tokens of goodwill, simple pleasantries, quickly made. A young coed wished her "luck here on campus," and in the checkout line another said, "I hope everything turns out for you here." Lucy later remembered those thoughtful gestures as the highlight of her day. Leaving the bookstore, she went to the home of Frank Thomas, editor of the *Alabama Citizen*. There she and her traveling companions had lunch with Mrs. Thomas and Emily Barrett, the tension of the morning broken by conversation and laughter in racially familiar surroundings.

That afternoon she attended a course in political science taught by Professor Charles Farris, whose sense of outrage at injustice was soon to be tested. This afternoon, however, his performance, unknown to him, was being judged with unusual interest. Accustomed to professors at Miles College conducting themselves in a restrained and formal manner, Lucy watched in amazement as Professor Farris sat on his desk ("instead of just standing in front of the class"), lit a cigarette, and crossed his legs ("not like a lady would ordinarily do, but with his knee up"). From his casual perch, Farris gave an absorbing preview of what the students would be studying. Even though, like Professor Hays, he did nothing to acknowledge her presence, Lucy came away feeling that she would enjoy political science.

The first day had gone as well as could be expected. University officials were elated. They had neither planned for nor expected major disturbances and none had transpired. Nightfall, however, produced ominous signs. Long after Lucy returned to Birmingham, recounted the events of the day for her sister and brother-in-law, and gone to bed, a crowd formed on the university campus. It was hard to know at first just what the crowd had in mind. The time was eleven o'clock, curfew for all coeds living in dorms. Because the women's dorms stood only a short distance from the Union Building, the girls' dates began to congregate in front of the Union with other boys who had come over from fraternity houses. A few were inebriated. Some talked about instigating a panty raid, but there were other, more purposeful elements working the crowd. What may have started as innocent fun for some soon turned ugly.

Leonard Wilson

One of the more determined students that evening was Leonard Wilson, the young sophomore from Selma, Alabama. On hearing talk of a panty raid, Wilson realized that such an adolescent prank "would have been changing the subject [from Autherine Lucy] pretty drastically." He set out to put minds back on business. "I never was in favor of that kind of activity [panty raid] anyway."[25] Indeed, Wilson was a serious, almost rigid young man, whose conservative values included the racial orthodoxy of white supremacy. He expressed his singular determination to preserve those values early and often. His intensity, even as an adolescent, stood him apart from fellow students at

Parrish High School in Selma, students who otherwise agreed unself-consciously with him.

Wilson was an only child reared by his mother. Born March 6, 1936 (the hundredth anniversary of the Alamo, as he likes to note), Wilson grew up on a farm outside Jasper in Walker County. The land had been in the family since 1830. He completed nine years of schooling in Walker County before moving to Selma with his mother in 1951, where he finished his last three years of high school. He took part in student government as treasurer and vice president of the student council and won recognition for oratorical talents. He did not join a debate team but often spoke on what he considered controversial subjects. One such speech concerned whether Hawaii and Alaska should be admitted as states. He was opposed. "I won the debate," he chuckled, "but they both got admitted." As a delegate to the state's Youth Legislature his senior year, Wilson gained notoriety by calling for a measure to send Alabama Negroes back to Africa. He even began compiling scrapbooks on segregation, later saying that he "spent about an hour a day . . . fixing clippings into volumes."[26]

People interested in his activities did not know whether his father had died, deserted the family, or was divorced and living elsewhere.[27] But important men in Selma recognized his potential and took him under their wing. Many of these men also played prominent roles in the formation of Alabama's first Citizens' Council. Between his senior year in high school and his enrollment at the university in the fall of 1954, Wilson was active in the gubernatorial campaign of Bruce Henderson. From Wilcox County, Henderson was an extreme conservative, even for Alabama. He ran a poor third to the victorious and, for conservatives, liberal anathema Jim Folsom. The loss disappointed Wilson but provided him a hero.

Like his hero, Wilson became a failed politician at the University of Alabama. He went out for rush but did not join a fraternity. He ran for student council representative from the School of Commerce but was not elected. He managed the campaign for a would-be student government president but lost. None of this surprised him. He felt himself in a sea of expedient moderation, if not outright liberalism, cut off from the real people of Alabama. Professors prated their liberal propaganda, and students sacrificed principle for grades and future jobs. Though out of his element, Wilson did not lack self-assurance. His element was the state and its people, not an artificial environment of extended adolescence, nourished by impractical and occasionally dangerous professors. They tried to indoctrinate him, but, of course, his views proved "strong enough" to prevent it. Part of his discipline included working at the Warrior Asphalt Company to pay for school (a company known for the number of Citizens' Councilors it employed). In the evening, he returned to his apartment on Abrams Court where he lived with his mother, a woman thought by some to be more radical on race than her son. There, he reportedly would sit in the living room, staring for hours at a Confederate flag on the wall, ruminating about the civilization he believed it represented.[28] He did join the Young Democrats ("too liberal") and the youth group at the University Church of Christ, but his real interest was the work of the Citizens' Councils. At Christmas break, 1954, a month after its formation in Selma, Wilson joined.

In early 1956 Wilson moved away from the centrist faction of the Council, represented by Sam Englehardt of Tuskegee and the Central Alabama Citizens' Council, and was cozying up to Ace Carter and the more extreme elements in the North Alabama Council. He would return to the more moderate fold, but for now, he and Carter made compatible noises. In all likelihood Wilson did not realize the extremity of Carter's views nor his maverick status with the Councils. It was enough that Carter was active in Tuscaloosa and available. How much contact Carter had with Wilson is not known, but Carter and members of the Ku Klux Klan (estimated at about 150 in Tuscaloosa County alone) signed up recruits and planned to make the university a test case in the campaign to maintain white supremacy. If Wilson was unaware of their efforts, he wound up in their traces.

Despite feeling awash in a sea of liberalism, Wilson in fact understood the majority opinion of the student body on the race question. If put to a vote, they would overwhelmingly favor the status quo before *Brown*. So when Wilson turned the crowd's attention back to "the" issue that Friday evening, he merely obeyed the dictates of his conscience. As students gathered to watch a small cross (some sticks wrapped in socks) burn on University Boulevard, they had become a near-mob. With firecrackers and smokebombs exploding, a cry went up to march on the president's mansion. Their number swelled to a thousand. Some shouted for Carmichael. Eventually Mrs. Carmichael appeared on the balcony to say her husband was in New Orleans. Still stirred, but not yet mean, the crowd pushed on to the girls' dorms where it might have devolved into a panty raid but for efforts by some to keep the mob focused. At that point Wilson himself became a focal point. Sarah Healy, spotting his activity, walked up to him, shook her finger in his face, and told him "to get out from there; that [he] was the ring leader in this mess." Healy knew Wilson and believed him to be fomenting the crowd, a circumstance suggesting that Wilson had already been tabbed by the university as a potential troublemaker.

With shouts of "Keep 'Bama white" and "To hell with Autherine," they headed downtown singing "Dixie" and chanting "Hey, hey, ho, ho, Autherine's gotta go." Police made some effort to stop them as they moved westward along University Boulevard into town, but used no force. (The Southern Regional Council, a liberal interracial group headquartered in Atlanta, had evidence that Klansmen had infiltrated the police department.) Still, the half-hearted show of force deterred about half the students. The 500 or so who passed through the police blockade followed a Confederate battle flag to the intersection of Greensboro Avenue and University, a distance of a mile and a quarter from the Union Building. On the way, they rocked a car carrying Negroes and smashed one of its windows. Wilson marched close to the standard bearer, occasionally circling back to encourage the students.

The citizens of Tuscaloosa had erected a flagpole in the middle of the most prominent downtown intersection to honor their war dead. The base of the staff was large and made an ideal platform. Wilson climbed the base and shouted, "The governor will read about this tomorrow. We're in accord with the states of Mississippi and Governor Talmadge [actually Marvin Griffin] of Georgia. We're setting the example for Auburn." Wilson was now bearding his political nemesis, Folsom, to shouts of approval. He later said that had Folsom joined other southern governors in speaking out against integration "perhaps there never would have been a student demonstration at the University." But the governor was silent, or at least equivocal, and the voice of Wilson could be heard clearly—in fact so clearly and so persuasively that the Southern Regional Council detected a new star in the firmament of demagogues. "Wilson is a young man," observed an informant for the SRC, "but his ability should not be underestimated by this fact. . . . all indications are that he has become a professional race-baiter and will be active in the field for some years to come."[29]

After fifteen minutes of shouting and speech-making Wilson called for the crowd to assemble again on Saturday night following a basketball game with Georgia Tech. By the time the students arrived back on campus, it was 2 a.m. and most went to bed. It had been an exciting night, exhilarating for Wilson, and university officials did not know what to make of it. They told the news media that the demonstration resulted from the actions of "a few inebriated fraternity men," but they were clearly concerned. They had enlisted student leaders to calm matters, but it was Wilson, the failed leader, who now stood front and center. The university felt powerless to do anything but hold its breath.

Saturday, February 4

Saturday brought a break in the dreary weather, and Lucy's second day went well. Because her only class was late in the morning, she did not get up early. The husband of Lucinda Brown Robey, the state NAACP's youth coordinator, drove her to campus. Before leaving, advisers told her about the Friday night demonstration, and she agreed that the disturbance probably amounted to little more than blowing off steam. On the way back that afternoon Lucy thought of Shiloh, the community where she had grown up. She planned to attend an aunt's funeral there on Sunday. She thought about seeing her parents and no doubt wondered what to tell them. They had not been involved in her decision to attend the university, and due to the protracted court struggle had had little reason to think about it. Their world and their daughter's were more than miles apart. With her name making headlines, they soon would forfeit the anonymity that made being black in the Black Belt tolerable.

Leonard Wilson also went to the campus on Saturday. He circulated among friends and talked casually about the previous evening's events and the demonstration called for that night. No real planning was involved. None was needed. At the end of his speech on Friday, Wilson made it clear that students who wished to protest should assemble after the basketball game. He hoped the spirit would build and was encouraged by friends to think that it would. He knew campus "politicos" were trying to quiet matters, but he believed that in their heart of hearts they thought just like him, with one exception: they didn't think anything could be done about the situation. In the name of law and order and obedience, these well-groomed leaders wound up conceding hope and playing unwittingly into the hands of those who would destroy the white race.

That night Wilson was gratified beyond expectations.[30] Following Alabama's victory over Southeastern Conference foe, Georgia Tech, many students headed for Denny Chimes, drawn like moths to light as another cross went up in flames near the library. Rebel yells rent the night. Shrill invective and defiance of university officials issued from the lips of students reared in the deferential manners and courtesies of polite southern society. The pitch of the crowd changed too, as some coeds with Saturday night "late permission" privileges joined the renewed chorus of "Keep 'Bama white," and "To hell with Autherine." With increased volume came a change in mood as older men, definitely not students, swelled their ranks and added a meaner spirit. Police estimated the crowd at 2,000 and reported cars stalled for six blocks on University Boulevard. After milling about for a while the crowd suddenly formed for another parade down University Boulevard to the flagpole. Uglier things began to happen. One car, driven by an unfortunate Negro, found its way into the path of the mob and had its window smashed and back door bent so that it would not close. Wilson himself recalled admonishing a student who tried to put his fist through a car window. At another point, demonstrators surrounded a Greyhound bus and rocked it violently until the crowd finally let the frightened driver and passengers through. Further along, as the march neared the bus station, Wilson had to stop another group from making "a run through the colored waiting room."

For the first time demonstrators gave photographers an opportunity to capture images of a southern mob on the loose. Among the pictures flashed by wire across the nation were those of students with flat-top haircuts and ducktails shaking their fists and waving Confederate flags; a grim university president back from New Orleans besieged by a mob; his wife pleading from a balcony while firecrackers exploded on the apron beneath her; and arm-waving, laughing students posing for photographers as they shouted their protest. The most powerful image showed a student stomping on the top of a car in which obviously frightened Negroes sat, but the photograph turned out to be a classic instance of how pictures can lie while telling a larger truth.

The boy whose image would endure was only a momentary participant. Later testimony established that he had had one too many. His coat and tie proved that he had just come from a date. Then caught in the emotional surge of the crowd, he jumped onto the car, at first falling and banging his chin, before successfully struggling to the top. Realizing what he was doing and where he was, the boy jumped down, ran back into his fraternity house and was not heard from again until his father, upset by the event, apologized and paid for the damage done to the car.[31] Still, the particulars do not rob the photograph of its larger truth. The picture gave the nation an accurate portrait of blacks engulfed in a sea of white rage.

At the flagpole, fellow students again hoisted Wilson to the base. He opened with a story about two soldiers who brought a monkey back from overseas and taught it to pick cotton. The monkey became so proficient that one of the soldiers suggested importing lots of monkeys. The other soldier thought it a bad idea because the North would send someone down to set the monkeys free and then Eleanor Roosevelt would get them admitted to public schools. Because the crowd seemed to like the story, Wilson told a few more racist jokes before reiterating themes from his speech of the night before. Characteristically, he cautioned his followers against violence or damage to property, but suggested that students stop cars with Negroes, make them roll down their windows, and ask them if they believe in segregation. "If they say 'no,' " he shouted, "make them believe in it." He urged students to boycott classes on Monday, but when greeted coolly, urged only that they boycott classes in which Lucy was enrolled. He passed the word for demonstrators to assemble at Smith Hall on Monday morning at nine o'clock. On Sunday, Wilson supplied reporters with fatuous assurances that he had intended only that his followers attempt to "persuade" Negroes, not coerce them, and that he had not called for a demonstration at Smith Hall on Monday, only announced the time of Miss Lucy's class.

One other student spoke from the flagpole. His was a voice of moderation. The president of the Student Government Association that year was Walter Flowers, a most-likely-to-succeed young man who later won fame by casting one of two pivotal votes on the House Committee that voted to impeach Richard Nixon. He and other campus leaders put implicit faith in the leadership of President Carmichael and believed the university to be doing only what it was bound by law to do. He courageously faced a hostile crowd that attempted to drown him out with catcalls. Though difficult to be heard above the noise, he shouted his belief that "99 per cent of you are interested in what is best for the University of Alabama." Then, to jeers, he charged that demonstrators were "doing just what a certain group wants done for publicity." As he called for students to go back to their dorms and fraternity houses, the crowd greeted him with renewed shouts to boycott classes on Monday.

The demonstration was not yet over. From the flagpole, the marchers returned to the campus where another speaker, somewhat intoxicated, shouted that the real goal of the "nigger" race was not integration but intermarriage. Declaring that violence was inevitable, he suggested "going down to the nigger quarters and cleaning it out." The students in the crowd, however, pushed on to the president's mansion where they demanded to see Carmichael, who appeared on the balcony. Again pebbles bounced off the portico and firecrackers exploded. He attempted to get them to think about what they were doing and to "uphold the traditions of this great University." As he scanned the crowd below, Carmichael saw unfamiliar faces. Of the twelve to fourteen hundred people spread over the darkened lawn only about 300 appeared vocal, but they made it virtually impossible to be heard. The unruly element did not seem to be students. They looked older, though a few also appeared too young to be college-age. This inner ring of shouting demonstrators forever after allowed the university to blame its violent paroxysms on rubber workers at the Goodrich Plant in Northport, the Holt high school football team, and outsiders in general. It was a true enough excuse—members of these groups were present—but it was an excuse that would not pass without challenge. Carmichael's appearance did have a quieting effect on most of the students, who began to move back to Denny Chimes and to disperse—proving at least that these children of a newer generation could still be moved by the marble-like sadness of Carmichael, a man under siege, whose features, chiseled against spotlights and popping flashbulbs, seemed to belong in a statuary with Lee and those grim-looking lieutenants of an earlier lost cause.

Sunday, February 5

While these night scenes played out in Tuscaloosa, Autherine Lucy slept in Birmingham. The next morning she left early to attend the funeral. She saw her parents, but they talked little about what was going on at the university. For a few more days Milton and Minnie Lucy would not be the targets of threats, and the actions of their youngest daughter still seemed remote. Nonetheless, perhaps out of uneasiness, Milton Lucy decided to go back with Autherine and her sister Ethel when they returned to Birmingham that evening. For the afternoon, however, Autherine took comfort in this trip to a place that had the familiarity of childhood. The three hours from Tuscaloosa to Shiloh measured opposite points in the universe. That evening, as the car hummed along, shadowy trees rushing by, their dark shapes yielding to moonlit pastures of dun-colored grass, Lucy's thoughts turned to the next day.

Leonard Wilson also spent a relatively quiet Sunday. He went to church and spent the rest of the day taking calls from newspapers and supporters. Again, there was not much planning to do. Already the call had gone out to assemble at Smith Hall, and after Saturday night, he had reason to believe a crowd would turn out. Just how much, if any, contact he had that day with the Citizens' Councils or members of the Klan is not known. That he knew members in both groups, and they knew him, is certain. It is also clear that members of both groups were circulating literature on campus and intended to be at Smith Hall on Monday. So Wilson rested from the agitation of Saturday night and prepared for Monday.

By now university officials were alarmed. Carmichael conferred with top officials just before noon. At the end of the conference, he called for a meeting of the faculty at 4:00 on Monday and of the student body at 12:15 on Tuesday—a day originally set aside to herald the beginning of spring football practice. At the same time he threatened to discipline any students involved in the demonstrations. Carmichael's statements added to the growing sense of crisis. Published in the Monday edition of the *Birmingham News*, they accompanied photographs of the young man stomping on the car and of Carmichael standing before a microphone, pleading for calm. One other picture made the front page that Monday—George Corley Wallace. Sounding like the fire next time, Wallace pledged "to jail jury-probing federal officers."

The Saturday night demonstration prompted the university to take additional security measures. To solve a jurisdictional problem and to increase security, the university hired thirty Tuscaloosa policemen to help control crowds on campus. Officials also called the governor's office and got authorization for thirty state troopers, about half to be stationed at the National Guard Armory just east of the campus on University Boulevard and the other half on the campus itself. With the six university security officers, the force available for crowd control came to sixty-six. Rather than keep demonstrators off campus, the university proposed to keep order by dispersing any sizable group before it became a crowd. Meanwhile, in Birmingham, Emory Jackson had received calls threatening Lucy, and, in turn, he called the university to ascertain conditions. Officials in Tuscaloosa satisfied him that adequate precautions were being taken.

The plans that Carmichael and his assistants developed on Sunday were the best they could lay, but they planned without the benefit of precedent. No experience served as a guide, no campus had yet been invaded in a manner that resembled a race riot. The plans laid that Sunday in 1956 were made for an era that would end the next morning.

4　The Mob Is King

When Lucy awoke for breakfast on Monday, nothing indicated that February 6 would be a memorable day. In Birmingham, a series of phone calls allayed the concerns of her sponsors about Saturday night's demonstrations.[1] Henry Nathaniel Guinn, owner of Guinn Finance Company, would take Lucy to Tuscaloosa and return when her classes were over at 3:00. She finished breakfast (a meal she never skipped after reading that it promoted good health) just in time to make the 7:30 departure. They rode alone, for the most part in silence because Lucy was never very talkative. She first sensed something might be wrong when, reaching the outer edge of campus, she saw no students. As they reached the quadrangle and turned toward Smith Hall, she could see a crowd of some 300 in front of the large, yellow-brick building. Guinn drove his Cadillac right up to Smith and let his conspicuous passenger out. She passed within ten feet of several clusters of men and boys, made it all the way to the top of the steps where she paused, turned, then appeared to smile and wave to the crowd before going in—all without being noticed.[2] (What appeared to a reporter as a wave to the crowd was more probably a signal to Guinn that she was all right.)

What distracted the crowd is not known. Perhaps it was the eleven highway patrolmen and three policemen who coincidentally moved in among them at the crucial moment of her passing. Perhaps it was the Dean of Men, Louis Corson, who moments before had urged students to return to classes. The booing that greeted his call had not subsided when Lucy made her unimpeded passage. The fourteen officers were too few to execute what existed of a university plan for crowd control. The idea was to disperse groups before they could form a larger crowd, but by the time the officers waded into the protestors, they were badly outnumbered. Had the entire force of sixty-six troopers and policemen acted prior to Lucy's arrival, they might have broken up the crowd, but most of the thirty-five troopers were still at Fort Brandon Armory, a mile away. Another handful had stopped at a truck stop for breakfast. Soon after Lucy's arrival, it was too late. The crowd swelled to some 500, and the force available, poorly trained for riot control, was insufficient. Moreover, evidence suggests that the troopers were under orders "to make arrests only to protect state property and buildings."[3]

Carmichael, who had walked over to observe the day's progress, took a bullhorn and pleaded with the crowd "to go and disperse immediately. You are disturbing classes." Shouts of "Lynch the nigger," "Keep 'Bama white," and "Where'd the nigger go?" drowned his plea. Inside Smith Hall, Corson, Jeff Bennett, and Dean Healy debated whether to take Lucy off campus for safety or risk taking her to her next class. Realizing that removal from campus would amount to capitulation, they recommended to Carmichael that she proceed with classes. He agreed. Bennett, in overall charge, asked that Dean Healy's car, a late-model Oldsmobile, be brought to the rear of Smith Hall. Meanwhile, Lucy tried to give her attention to Professor Hays's lecture but could not help noticing Healy's movements to and from the door. Lucy decided that Healy might be there to report progress on the request for dorm and boarding privileges—a reassuring thought but wide of the mark.

First escape

When class ended, Healy met Lucy at the door, where she and Bennett convinced her that it would be best to go to her next class by car. Autherine's apprehension was understandable. As they walked out the back of the building with Healy leading, a spotter, standing to the southeast corner of Smith Hall, saw them and yelled, "There they go, there they go." The crowd surged around the building, shouting "nigger lovers" and "hit the nigger whore" as they unleashed a barrage of gravel and eggs. Bennett slid behind the wheel. Healy nudged Lucy into the middle between her and Bennett, who by that time had the engine going. Moving out toward Hackberry Lane, an unidentified woman pulled a car up and stopped, blocking the exit. Bennett blew the horn and thought first of bulldozing her off the road before deciding to put the car astraddle a drainage ditch and race through an adjacent parking lot. As they bounced along, they could hear gravel striking the car and shouts from the mob. Gaining speed, Bennett realized that the more direct route to Graves Hall and the Education Library would be dangerous. He continued south to Tenth Street, circling behind the football stadium, before turning north to the Education Library—a driving distance of about a mile. Advance elements of the mob were already at the library, forcing Bennett to go to the back, a short distance from a rear entrance.

The three rushed for the door through another barrage of gravel, eggs, and rotten produce. They made it in reasonable order, and Healy locked it. Short of breath, they ascended the stairs to the second floor where Jean Lyda taught the class in children's literature. As Mrs. Lyda approached her classroom, Healy and Lucy came around the corner. "They were splattered," Lyda recalled. "Dean Healy asked if there were a restroom there that they could get cleaned up so I took them next door and got them paper towels and things and they wiped the stuff off their clothes and their person. . . ."[4] Meanwhile, Bennett, in his Marine combat-hardened way, decided no crowd would cow him and went back to move Healy's car. He rightly feared physical damage to it. Like other decisions that day, this one was a mistake. He became the mob's target of opportunity. As he put it: "You see, she was gone but I was there. So it was 'nigger-lover, son-of-a-bitch, get the goddamned nigger-lover' and so on."[5] Bennett got in the car and locked the doors, but the mob was on him. It began to rock the automobile. Bennett, temper boiling, threw it into drive, lurched forward, and as soon as the crowd, seeing his intent, parted in front, he sped for daylight. Bricks smashed the rear and side windows as eggs splashed from all directions.

Sitting in Mrs. Lyda's classroom, Lucy gathered her thoughts and feelings. She glanced at her "green, pretty coat" from which she had just cleaned the contents of a rotten egg. She prayed for "courage to accept the fact that I might lose my life there." Outside she could hear the mob's chorus of now familiar chants and obscenities. Mrs. Lyda struggled to control a bad situation. She knew the strain her students were under, "so tense and frightened," and felt obliged to act as if nothing unusual was going on—a façade that proved hard to maintain when the head of the department, Pauline Foster, came from her office, which opened onto the classroom, and locked the classroom door from the inside. Still, Lyda carried on, showing the connection between children's literature and the classics. To illustrate, she read from poems and thought "how ironic it was to be quoting poetry under those conditions." In one corner a girl began to cry, "not aloud or anything . . . just tears." Lyda "could tell they were frightened."

The crowd

Estimates of the crowd's size at its largest ranged to three thousand, changing throughout the day as classes let out and curious students, hands stuffed in jean pockets, joined its fringe. There is little doubt that outsiders were responsible for the bad temper of the crowd. Leonard Wilson had done his job by demonstrating student opposition the previous Friday and Saturday nights. Meaner and more determined elements now did their work. A vicious spirit infested the day. A small, gray-haired lady stood at the edge of the crowd, swung a sweater in circles over her head, and screamed, "Kill her, Kill her, Kill her." An informal tally of license plates showed an unusually high number of cars from Birmingham with a sprinkling from Mississippi, Tennessee, and Georgia. University officials believed the chief troublemakers were Klansmen and Citizens' Councilors from Bessemer and Birmingham, who joined Tuscaloosa brothers from Central Foundry, Goodrich Tire and Rubber Plant, and other operations in the adjacent towns of Holt and Northport.

The Ku Klux Klan fomented trouble. In addition to Asa Carter, Robert

Edward Chambliss worked the crowd. In 1949, police had arrested Chambliss, a resident of Birmingham, on a charge of "flogging," a violent act carried out while masked and robed as a Klansman. Later the city of Bessemer fired him as a garbage collector for "conduct unbecoming a city employee."[7] "Dynamite Bob," a name given him by his Klan buddies, became more than a jocular sobriquet when on a Sunday morning in September 1963 he set the bomb that killed four young girls at the Sixteenth Street Baptist Church in Birmingham—a demented act that took the heart out of massive resistance in Birmingham. But that was in the future. On Autherine Lucy's perilous day, Chambliss provoked the crowd while claiming that he was there only to distribute Citizens' Council literature.

The Klan claimed to be "at wartime strength." Not only did it have informants among the police, it also had a mole in the Post Office. Federal authorities arrested Roy Hobson Hartley, an employee at the Goodrich Rubber Plant, for intercepting mail intended for the Director of the Alabama Council on Human Relations. Hartley was convicted and placed on five years' probation. While on probation, he became temporary head of the Tuscaloosa-based United Klans of America, Inc., when Imperial Wizard Robert Shelton, also a Goodrich employee, relinquished formal control to take a lucrative appointment as "Goodrich's representative to sell tires to the state." The appointment was a payoff for Shelton's support in Attorney General John Patterson's successful 1958 gubernatorial bid—the campaign that gave rise to George Wallace's now famous declaration that he would never be "out-niggered again."[8]

Not surprisingly the connections between labor and the Klan in Tuscaloosa County hurt the cause of labor. In the South management had always used race as a wedge against worker solidarity, and in the supercharged climate of the Lucy episode, race separated workers not only from each other but also from their national affiliations. "Lucy Case Splits Alabama Unions," said one headline in the *New York Times*. Local officials of both the United Rubber Workers and the International Molders and Foundry Workers protested Dean Corson's statement that "the riots were being led by a 'hard core of workers from the rubber plant.'" The unions noted that they employed students on the evening shifts, intimating that if any workers were involved they were also students. The foundry workers in particular observed that absenteeism at the plant during the riot was very low—a claim the rubber workers could not make because they were on strike.[9] The university and the workers were both right. University officials exaggerated outside involvement, and the workers protested too much.

Trapped

M.L. Roberts taught his eight o'clock class in Graves Hall and went over to the Records Office to check some matters pertaining to registration. While there, he heard about the commotion at Smith Hall but paid it little attention. Before leaving he took a call from New York intended for Lucy. The caller, a reporter for the *Times* of London, apparently thought Roberts was trying to deny him access to Lucy and became insistent. Roberts assured him that he would deliver the message to her. By this time it was nearing ten o'clock, and Roberts started for Graves Hall where he encountered two assistants from the Dean of Men's office. Learning of the call from the *Times* of London, they advised Roberts to ask Carmichael about delivering the message. Roberts called Carmichael from a meeting, who told him to let her use the telephone in the Education Library because she likely would be there for some time.[10]

On returning to Graves Hall, Roberts ran into Ernest Williams, the treasurer, who was waiting for some plainclothes policemen. He asked Roberts to show them the way to the library via an underground passage connecting Graves Hall to the library. Before the police arrived, Dean Corson came up and said that the plan now was to keep Lucy in the library and that Dean Healy would remain with her. Corson asked Roberts to stay with them. Roberts and Williams then went inside Graves Hall where the plainclothesmen were waiting. Roberts led them to the second floor of the library, arriving at approximately 10:50. Within five minutes, Mrs. Lyda dismissed her class, and Sarah Healy took Lucy to a small conference room nearby. They would remain there for the next two hours and twenty minutes.

Lucy showed no visible signs of fright, but appeared deeply concerned about the turn of events. When told about the telephone call, she shrugged it off as probably "just another newspaper reporter." She worried for the safety of Mr. Guinn and Ernest Reed, a local black man who volunteered to assist Guinn while he waited for Lucy. They planned to pick her up at 11:45. She feared what might happen if they drove into the mob. Roberts asked if there were any way of getting in touch with them, and Lucy, uncertain, thought they could be reached at the *Alabama Citizen*. A little later, Lucy got up and moved closer to the window, thinking she had seen Guinn's Cadillac. Dean Healy cautioned her not to be seen and closed the blind as a precaution.

The conversations that took place at various times among Lucy, Healy, and Roberts were polite, friendly, and awkward. A bit defensive, Dean Healy wanted to assure the young woman that the university had alerted Emory Jackson to the potential for trouble and to the threats on Lucy's life (though Jackson recalled only talking with President Carmichael's wife). Healy asked if Jackson had told her anything about the danger. Lucy said no, and, according to Roberts, expressed surprise. Healy contradicted herself by saying that university officials had not anticipated an unruly mob. She reassured Lucy that the university was doing everything possible to protect her. Healy added the now familiar disclaimer that it was outsiders and not students. Lucy appeared to agree.[11] Asked why she wanted to go through with it, especially as she was in harm's way, Lucy remembers reiterating the obvious. The university was a tax-supported institution. It was the best in the state for what she wanted. And "there was always that longing to reach the top."[12] Of course, the participants did not say all these things as remembered. It is not surprising that talk between such well-meaning people was awkward. The mob was loose. Trapped on the third floor of the library, they all, for different reasons and for no good reason, felt responsible.

The second escape

After about an hour and a half, Jeff Bennett entered to say that most of the crowd had left the rear of the building. Escape seemed possible; but before they could make a dash, the mob, alerted by a patrol car parked too near the back entrance, reappeared, and Lucy returned to the conference room. University officials desperately needed a diversion; something to move the crowd away from the building. After two hours, it showed no signs of dispersing. Moreover, when attention shifted, ring leaders shouted "there she goes," or words to that effect, to reconcentrate the crowd. Then around one o'clock the university got a piece of luck from an incident that itself portended danger. Officials had not reached Mr. Guinn at the *Citizen*. Already on campus, he had parked his car on University Boulevard, between Graves Hall and Denny Chimes. Lucy might indeed have spotted it when Healy warned her away from the window. Guinn, along with Ernest Reed, now mingled uncomfortably with the crowd. Dressed in a manner befitting his prominence in the black business community, Guinn was dangerously out of place.

Guinn became concerned for the safety of his car and asked his companion to move it for him. With difficulty Reed made it to Guinn's conspicuous Cadillac, but as he tried to pull out, a crowd closed in, rocking it violently. The disturbance brought five police officers, who pushed and pulled people away. One officer was hit in the process, but eventually Reed got the car off campus. Surprisingly, Guinn himself drew little more than catcalls until a one-in-a-million occurrence transformed him from spectator into spectacle. The crowd or the police, it is not clear which, forced a truck on University Boulevard to turn right on Colonial Drive, the street running at a right angle south from Graves Hall and in front of the Union Building. The nervous driver turned sharply, his wheels running up on the curb and toppling some police motorcycles. Guinn stood next to the end motorcycle which, domino-style, knocked him to the ground. At that point, one of the day's heroes entered the picture.

The Reverend Robert Emmet Gribbin, an Episcopal chaplain, circulated through the crowd all morning urging people to go home. Along with the Assistant Dean of Arts and Sciences Hubert Mate, Dean of Men Corson, Father Michael Mulroy, and Circuit Court Judge Reuben Wright, Gribbin engaged groups in debate, hoping to distract them from the larger mob spirit. Some became shouting matches, and in those exchanges, few could equal Gribbin's booming voice. Gribbin's clerical collar helped. As chaplain of Canterbury Chapel, Gribbin was a familiar face, knowing where "all the free coffee and lounges were" and, as a result, was in good position to help.[13] He had noticed the well-dressed black man earlier and felt some concern for his safety. From the steps of the Union Building, he saw the accident that knocked Guinn to the ground. Gribbin got to the side of the older man, who was bruised and brushing himself off. Gribbin talked with him while a crowd encircled them making unattractive remarks about the "nigger" and the "nigger-lover." After a few minutes, the two men separated, the crowd marching with Guinn, hectoring him as he joined a group of maintenance workers. A photographer snapped a picture of Guinn standing behind a worker with his left hand clutching the man's left elbow, face pressed close to his back. By now Guinn needed a shield.

Seeing Guinn's danger, Gribbin knew it would be necessary to get him away. At that same time, Peter Kihss, a rising star at the *New York Times,* saw Guinn's peril and urged him to go to the police standing on the corner. The police thought he should leave and called for a car. When it arrived, Kihss, Gribbin, Guinn, and the police fought their way through the hostile crowd, who kicked and cursed Gribbin. An officer finally got Guinn to the car by holding onto his belt buckle and refusing to let go.[14] Thus occurred the break for which university officials had hoped. The crowd, believing the Guinn disturbance to be the main action, surged away from the Education Library. Jeff Bennett saw all this and realized that Lucy had a chance. For days thereafter newspapers talked about a ruse developed by the university, whereby officials used Gribbin, Kihss, and Guinn to create the diversion. There was no ruse. The accident happened. Gribbin and Kihss went to help, and the police eliminated a potential problem by getting Guinn off campus. Later that afternoon, three men surrounded Gribbin, kneed him in the groin, and splattered him with eggs. These three, along with "Dynamite Bob" Chambliss, were the only men arrested that day for disorderly conduct.

All the while, Lucy waited in the conference room. Mrs. Foster returned from lunch at the "Supe Store" and brought a cheeseburger and candy bars. Lucy thanked her and took the sandwich. After a few bites, she put it down. Soon, Jeff Bennett rushed in to say that the crowd was headed toward Denny Chimes and that they had better go now. Accompanied by Healy and Roberts as far as the stairwell, Lucy went down the steps with Bennett and a policeman and out the back door to a waiting patrol car. Bennett helped her into the back and told her to lie down. Ernest Williams hurried out with her school books. The trooper, tires squalling, sped north toward the Warrior River and Riverside Drive. Another patrol car with six troopers sped behind, bumper to bumper.

Through years of legal battle, through registration, and finally those awkward first days of class, Autherine Lucy had carried herself with calm resolution, poise, and quiet dignity. Only hours before, maybe not understanding, she gracefully turned at the top of the steps to Smith Hall, smiled and waved to her escort and an uncomprehending crowd. Now she lay face down in a car, frightened, moving at high speed to escape a mob.

Back to safety

Emily Barrett sat at her typewriter near the front window of the *Alabama Citizen* going over copy for the next edition. The time was approximately 1:25 p.m. She heard a car pull up and noticed an aerial. That meant police. Soon she saw a patrolman and Autherine Lucy get out. After escorting Lucy to the door, the patrolman left. A flurry of activity followed: people asking what had happened, how she was, whether they could help. All day long people at the *Citizen* had heard reports of trouble but no one knew how serious the situation had become.[15] After supplying details, Lucy was taken next door to Howard and Linton's Barber Shop where a beautician fixed her hair.[16] In the meantime, a carload of white toughs spun into the graveled parking area of the nearby Barbeque King, kicking up rocks as they slid to a stop. They asked where they could find the "black bitch." The white proprieter told them to get away, that the people in the neighborhood were his friends, and that if they came back, they could expect trouble. Nathaniel Howard, owner of the barber shop, had made several calls, and within minutes, black men with shotguns and rifles covered the area. The place teemed with curious onlookers, especially after school when children came to see the woman who had caused such a stir in the white community. Howard also learned that Guinn was at the city jail. Howard, his son, and a group of friends loaded up two cars and drove downtown. Chief Tompkins knew Howard well and was relieved to hand over Guinn, who had been waiting in the chief's office. At the barber shop a caravan formed to escort Lucy to Birmingham. About 4 o'clock, Lucy and Mr. Guinn, followed by a carload of armed black men, started out for Birmingham. As arranged, one of Chief Tompkins's patrol cars took them as far as the city limits.

On campus demonstrators milled about, occasionally arguing with people such as Reverend Gribbin who continued to circulate. By late afternoon, the protesters knew that Lucy had left Tuscaloosa and most drifted away. The events that day raised serious questions about the official response to mob action. Why were firehoses not used to break up the crowd? A fire truck stood by but was not used. Why was the National Guard not requested? It was; twice by NAACP officials in telegrams from regional director Ruby Hurley and national president Roy Wilkins, but their calls probably did not reach the governor's office until late Monday. The university itself inquired about Guard help; if not for that evening, at least for the next day. Judge Wright, after consulting with university and city officials, placed the call from Carmichael's desk sometime between 5 and 6 p.m. Earlier the governor had given assurances of cooperation and asked Major General William D. Partlow to keep his office informed. Folsom told Wright that he would do "whatever circumstances and good judgment suggest," but Wright did not request troops immediately and by Tuesday morning the emergency had passed.[17] The truth is that nobody foresaw the situation that developed on Monday. Over the weekend the governor had gone fishing. As for university officials,

Saturday night's demonstrations alarmed them but not to the point of caus-
ing them to prepare for two to three thousand demonstrators. Nor were
Lucy's advisers fearful of a major calamity. Without sufficient apprehension,
the university failed to take sufficient precaution. Surprised and overwhelmed
by events, they wound up playing the fool to hindsight.

The demonstrations, however, were not over. Monday night following the
Tide's victory over third-ranked Vanderbilt, students and outsiders again
gathered at the flagpole in downtown Tuscaloosa. The police estimated the
crowd at 300, about half of them students. Chief Tompkins let them know
that their presence was not welcomed and took an egg in the face for his
trouble. The demonstrators started back toward campus, picking up the familiar chants, gathering force as they marched. When they reached the Union
Building, turning back cars along the way, they numbered about 600. For a
moment the demonstration seemed to stall, then someone shouted for a
march on the president's mansion. As they converged on the stately home,
they shouted, "Carmichael is a nigger lover" and "We want Ollie."

Mrs. Carmichael appeared on the portico along with university chief
Rayfield. Photographers perched on the arched staircases to get better pictures of the mob and the university's first lady. She tried to explain that Dr.
Carmichael was in a meeting downtown and that only she was at home.
Because of the noise, Rayfield had to repeat her words. As the shouting
increased, she invited a representative to come to where she stood on the
portico. Students guarding the foot of the stairs let a young man pass. After
talking with Mrs. Carmichael and Rayfield, the demonstrator turned and
yelled, "Let's go back to the Union Building." Others shouted, "Don't listen,
don't listen to him." The protestors then lost control, unleashing rocks and
eggs, forcing Mrs. Carmichael and Chief Rayfield inside.

With the targets gone the mob moved back to the Union and began setting
off firecrackers, waving Confederate flags, and burning literature that Bob
Chambliss said was communist-inspired NAACP propaganda. Tuscaloosa
police arrived and joined the university's small force. They waded into the
crowd, trying to break it up into smaller groups. Chief Tompkins decided he
had had enough. "We'll use force," he shouted, and for emphasis, "right
now." With that he fired a tear-gas cannister, which only seemed to spur the
crowd. "Give us some more, give us some more," they shouted, surging back
into the area just vacated. But Tompkins had made his point, and almost
casually, the crowd began to disperse. By 12:15 only a hard core of 50 to 75
demonstrators remained. By 1 o'clock even they had drifted away.

Ironically, at the moment the mob drove Mrs. Carmichael from the portico, her husband emerged from a meeting that by any estimate had done the
mob's bidding. Asserting its police powers to protect students and faculty, the
board of trustees, meeting at the McLester Hotel in downtown Tuscaloosa
since 6:30 that evening, voted unanimously to "exclude Autherine Lucy until
further notice from attending the University of Alabama." No doubt these
grave and, for the most part, gray men had done what was reasonable; but no

reasoning could dispute the student demonstrator who crowed, "Well, we won. It took her four years and the Supreme Court to get her in, and it took us only four days to get rid of her." Carmichael had asked Bennett, "What do you think of the decision?" Bennett replied, "There is no other, but the point is, the mob won."[18]

That night, Autherine Lucy slept. Her brother-in-law and some of his friends stood guard with shotguns.

Reaction

On Tuesday, February 7, Arthur Shores served notice on the university to reinstate Lucy within forty-eight hours or else. The campus itself was outwardly calm. Inside, significant groups among the faculty and students felt betrayed by the university's actions. Some actually supported integration but most did not. What they shared was shame that Leonard Wilson, perennial loser and the darling of radical racists, had won. Carmichael had called a student convocation for Tuesday and a press conference. He canceled both. He did not cancel the faculty meeting scheduled for 4 o'clock. Later he wished he had. A townsman best expressed the feelings of those who were ashamed and embarrassed. Buford Boone, editor and publisher of the Tuscaloosa *News,* wrote an editorial in the early hours on February 7. Titled "What a Price for Peace" and published on page 1 the same day, it would win a Pulitzer Prize. Boone was more than courageous. As editor of the paper and a leading citizen of Tuscaloosa, his relationship with university officials and trustees was close, born of economic, social, and class interests. The people he wrote about that day were members of his clubs and attended the same parties.

He opened with a simple observation, simply put. "When mobs start imposing their frenzied will on universities, we have a bad situation." Such had happened at the University of Alabama. Murder was the near result—the target, Autherine Lucy. "Her 'crimes'? She was born black, and she was moving against Southern custom and tradition—but with the law, right on up to the United States Supreme Court, on her side." Boone queried, "What does it mean today at the University of Alabama, and here in Tuscaloosa, to have the law on your side?" The answer was obvious, "Nothing—that is, if a mob disagrees with you and the courts." And who was to blame? "As matters now stand, the University administration and trustees have knuckled under to the pressures and desires of a mob." They displayed no "firm, decisive action." "Not a single student has been arrested on the campus and that is no indictment against the men in uniform, but against higher levels which failed to give them clear-cut authority to go along with responsibility." Boone's conclusion: "Yes, there's peace on the University campus this morning. But what a price has been paid for it!"

Boone's clear-eyed assessment was echoed on campus that afternoon. At the faculty meeting, Carmichael intended to blame outsiders and to ask for a sympathetic understanding of the decision to exclude Lucy. A similar faculty meeting at 5 o'clock on the afternoon of the riot had gone well. During that Monday meeting, Carmichael, "his mien sorrowful and distressed . . . his voice barely audible," had asked the faculty to give the administration its support and to spend approximately five minutes in each class explaining to students the situation and their responsibilities. He also told them of preparations to increase security. The issue, he said, "has now developed into something more fundamental . . . it involves the question of whether anarchy or law will rule here." He spoke his solemn words to a "hushed" faculty. But that had been on Monday before Lucy's exclusion. Now a number of the faculty were both angry and determined to be heard.

Their anger received a national and international hearing, for by that time, representatives of the world press had descended on Tuscaloosa.[19] The university granted over 260 press passes, but Carmichael had not handled the press well. He called and canceled two press conferences on Tuesday. So when he met with the faculty in Morgan Hall, the press joined in and paid close attention. Carmichael began with a report from Corson, containing the often stated proposition that outsiders had invaded the university. Carmichael seemed especially intent on proving that Autherine Lucy's life had been in jeopardy. He emphasized what became the most repeated statement on endangerment: Bennett's assertion that on two occasions "twenty seconds" separated Lucy from death. Thus, Carmichael insisted, the board's action was necessary to protect life. Other than disbanding the university, suspending Miss Lucy became, "in the circumstances, about the only thing in this crisis that could be done." Carmichael said that the board's order was temporary, so "that we may have an opportunity to work out sensibly and soundly the long-run solution." The exclusion was not a violation of the court order because its "sole object was to secure the safety of students, faculty, and University personnel." Carmichael choked back emotion as he praised the faculty with high words, declaring them the "most loyal, understanding and intelligent" of any with which he had served.

Saying he could not take questions, Carmichael sought to end the meeting. Murray Kempton of the *New York Post* saw the president's eyes as "frightened" and his hands as "pleading"—"everything about him seemed to cry out, please, no questions."[20] The mood begged for a show of trust and support. Instead a revolt broke out. Charles Farris, whose political science class Lucy had looked forward to, protested Carmichael's attempt to avoid discussion, because there "are questions that should be answered now." Farris queried the inadequate police protection and why "only the NAACP" had asked for National Guard protection during the rioting. Carmichael leaned to the microphone and said, "I cannot answer your questions," and attempted once more to end the meeting.

The faculty, a majority of whom supported the president, hesitated. Before they could move, William Lampard, a professor of child development, added his protest of the "no discussion" decision and said that he personally could not acquiesce in the board's action. Scattered applause obscured the rest of his remarks. Dr. Carmichael refused to reply and asked once more for the faculty to leave. Most rose to go but turned to listen as Laurence Calcagno, an art professor, shouted objections to "the University's succumbing to mob rule." Fred Ogden, another dissident from the political science department, insisted that Carmichael at least account for the "inept" police action. Again, Carmichael protested that circumstances necessitated his silence and tried once more to adjourn the meeting, with more than half the faculty now standing by their seats in uncertain compliance.

Determined not to give assent by silence, Farris demanded the floor, moving toward the stage and the microphone as he spoke. Taken aback, Carmichael yielded the microphone and listened stoically as Farris presented four resolutions, the gist of which condemned mob violence and declared that the university should suspend all academic functions if it could not provide for the safety of all its students, including Lucy. By now a majority of the faculty willingly came to the aid of their beleaguered president. Carmichael recognized a motion to table Farris's motion and asked for a vote. The "nays" chorused with surprisingly strong voice but were no match for the "ayes." Carmichael asked those in favor of adjournment to stand. As a majority was already standing (and no doubt would have stood in support of their president), Carmichael declared the meeting adjourned, and the faculty filed out.

That night the students took their turn. They, too, had lost something, especially student leaders and those active in campus life. For the most part, they came from middle-class backgrounds and had every right to assume that obedience to university authority, in and outside the classroom, marked a straight and narrow path to social and economic success. They knew that the reputation of their university affected the value of their degrees. A bespectacled stringbean of a young man, who from appearances seemed an unlikely leader, spoke for these students. But Dennis Holt had more going for him than gangly looks. He was brilliant, eloquent, and popular. The previous spring he and a colleague, Ellis Storey, had won the national debate championship at West Point, defeating Dartmouth. He also won election as president of the Arts and Sciences college council. During the rioting he worked with university officials and personally turned back three demonstrators who tried to mount the staircase to the portico where Mrs. Carmichael stood.

The first opportunity for speeches came at a Tuesday night meeting of the Student Government Association. Holt did not speak first. That privilege went to Leonard Wilson. Wilson distanced himself from the rioting and modestly disclaimed leadership for the demonstrations. Whatever influence he possessed, Wilson argued, had been exerted in behalf of crowd control. He praised the board's decision to exclude Lucy as "wise and considered" and, while deploring any "violent reaction" and any property damage, had good words for those who protested. The demonstrations proved that integration would not work at the University of Alabama. The university must, he said, "remain an all-white school, lest it fall from its present standards."

Most students were not mollified. Like their elders, they wanted to blame trouble on outsiders, but they knew that a few of their own, like Wilson, had allied with the mob. Wilson no doubt enjoyed the confrontation that followed. One student challenged his assertion that admitting Negroes would lower standards and asked, "What about Harvard and Yale? Negroes go to school there." Wilson denied saying anything about standards. Another asked if, during his flagpole speech, he had encouraged the mob to ask Negroes if they believed in segregation and to "convince" them if they said "no." Wilson admitted to similar words. Another asked what Wilson would do if Lucy returned. Wilson vowed to "do all possible by honorable means to discourage her attendance." Whereupon the student asked, "Do you think leading a mob is honorable?" After about an hour of confrontation, Wilson left for a meeting of the Student Religious Association.

It was Dennis Holt's turn. Holt recounted his altercation with three protesters the evening before—"two high school boys and a man so drunk he could barely lurch." When confronted and told, "You're not going anywhere," they fell back, and Holt found a lesson in the incident—"That's all it took—just a little resistance." Unfortunately, Holt lamented, university trustees and state officials had not shown similar firmness. "Our University and its trustees," concluded Holt, "may well be famous for all time for running away from a fight. They have acquiesced to the mob. Let us face it: The mob is king on the campus today. We must all think a little bit about the fact that the mob won. Let us remember that high school boys were able to sway the board of trustees of a great university." Holt introduced a resolution condemning mob violence, pleading that just as America had been called to greatness, "we, too, have been called. Great segments of the world are watching the student government of the University of Alabama on this seventh day of February, nineteen hundred and fifty-six."

Kempton, who soon would become editor of the *New Republic,* observed the proceedings as he earlier had the faculty meeting. He was deeply moved by Holt's performance. When the young man from Birmingham had finished, "they applauded for 35 seconds, which is too long to be quite proper for 60 people. They seemed to keep it going out of some need to affirm what they had waited so long to hear someone say loud and clear." When they had finished, a student asked what the resolution meant. Holt replied, "It means that we are saying to certain responsible officials of this state that when mob violence occurs, it's time for the law to break it up. We have to convince the state officials that they have the responsibility, not just the students."[21] The resolution passed unanimously.

But the student's question was to the point: just what did it mean? Holt himself had narrowed the issue to "law and order," not "integration," and fully conceded that 90 percent of the students opposed the Supreme Court's ruling, though 90 percent of that 90 equally opposed the demonstrations. Most of the petitions and resolutions that swept the campus over the next few days pledged support for Carmichael and the board, while affirming opposition to the mob. Nelson Cole, editor of *The Crimson-White,* spoke the majority sentiment: "We have unquestionable faith in Dr. Carmichael and the Board of Trustees. . . . If ANYTHING can be done, they will do it. If NOTHING can be done, we still believe they will choose—and have chosen—the wisest approach to the problem."[22]

One group of petitioners took the matter a step further. When it became apparent that the board would not readmit Lucy without a court order, two Unitarian students drew up a petition that stated: "We the undersigned, believe in law as well as order. Regardless of our views on integration, we are united in believing that the reputation of the University as a law-abiding institution must be restored." They asked that the board voluntarily lift Lucy's suspension. Only 200 signed this petition, compared with 750 who had signed an earlier one supporting the administration and condemning violence. The chief sponsor of the Unitarian petition was Albert Horn, a law student from Birmingham and member of Phi Beta Kappa. He later payed a high price for his efforts. Despite graduating number one in his class, he received no offer from an established law firm in Alabama.[23] Horn's petition underscored by contrast majority sentiment. Still, to stand for order as most of the students did was no mean feat. To criticize the board, as Holt had done, took courage. Even to support the president, as Nelson Cole had done through his editorial columns and on Dave Garraway's "Today Show," carried risk. Whether standing for order or both law and order, students who became conspicuous in support of either became targets of hate mail and calls.[24]

A few professors eventually left the university. Farris, for example, took an appointment at the University of Florida, though it is not clear that disgruntlement over the Lucy case was the only or even the principal motivating factor.[25] As for the students, some, especially the leaders, learned valuable lessons about speaking in a crisis, and still more learned about values and morality in a racially divided society. Student religious groups and classrooms became forums for talking about "the" issue. Taken-for-granted assumptions suddenly were debatable, and many students who were capable of turning tradition on its head began that youthful process in those climactic days. But the majority did not change overnight. They did not blame themselves. A few blamed those in authority. Most blamed outsiders like the Klan, the Citizens' Councils, and their fellow travelers, and in the days ahead they would follow their elders in tarring the NAACP with the same brush. Because everyone was to blame, no one was to blame; thus most students cast guilt, that principal motivator of moral change, on the discard pile.

Back to court

While the faculty and students talked, the university prepared to go back to court. The February 9 hearing on Pollie Myers's case appeared on the immediate agenda, but trustees seemed eager to fire on all legal fronts. Lawyers for the board received a letter from the Fifth Circuit, dated February 1, denying the university's petition for a rehearing. Andrew Thomas told Bob Steiner that the next step would be the U.S. Supreme Court, but advised, "In our opinion, nothing will be accomplished thereby." Gessner McCorvey insisted on the appeal to the Supreme Court if for no other reason than to show that they had exhausted all legal means. On the same day, Steiner recommended "aggressive legal action against Shores, the NAACP, the girls, et al., to expose publicly the reasons for dismissing Myers and thereby discrediting the whole effort." Steiner acted on advice of White E. Gibson, Jr., a Birmingham attorney.[26]

On the evening of February 8, the board met in the law offices of Burr, McKamy, Moore and Tate. In addition to recommending that the Lucy and Myers case be appealed, board members rehearsed responses the university would give defending its denial of admission to Myers. One question was especially troublesome: why had the university waited until 1956 to study Myers's character? Notes indicate the board prepared to testify that Myers's character "was not called to our attention." In fact, the board had discussed her character thoroughly, including what to do about it, on November 12, 1955, almost three months before taking formal action. Unable or unwilling to admit that it had hired detectives to dig up this information, the board also discussed ways to make their discoveries appear routine. Again, notes reveal a shading of truth: "As to character: Study ea. applicant & all credentials; Hi School principal usually recommends; Also placed rating on most transcripts; Also where there is a lapse of time between periods of school attendance, we check closer."[27]

The next morning, February 9, dawned with a sense of high drama. The riots on Monday and Autherine Lucy's suspension raised public interest in Myers's case. Moreover, the grounds for denying her admission had a salacious appeal. The press appeared in full force, including at least five representatives of the nation's black press. By the time judge Grooms arrived, people already jam-packed the hallway leading to his courtroom. With the press clamoring, Grooms denied a request to adjourn to a larger room, because he did not want to "magnify" a case he already believed out of proportion. He ordered the press to choose twelve representatives and seated them in the jury box. Blacks and whites filled the rest of the room.

It was over in ten minutes. Shores approached the bench and asked for a delay, arguing that it would not adversely affect his client because Lucy's suspension meant that Myers "would not be admitted now anyway." Judge Grooms denied the request. Shores amended to ask for a dismissal of the petition to readmit. Grooms asked if that meant the petition was without merit. Shores said no but conceded that point "for the purpose of dismissal." Whereupon Grooms ordered the contempt charges against Dean Adams vacated. At a press conference, Shores said that the practical difficulties of pressing both the Lucy and Myers cases simultaneously resulted in the decision to drop Myers's petition. When asked whether Pollie's case would be reopened, Shores said the primary concern was having Miss Lucy readmitted.

In later years Shores and others associated with the NAACP claimed to be unaware of Myers's background at the time the board rejected her. Only then did they learn what the board had discovered. Based on that, they decided to withdraw the case.[28] The explanation is convenient, but untrue. They knew from the beginning what the circumstances were. They simply gambled that her early marriage to Edward Hudson would pre-empt inquiry.[29]

For Myers, the events were crushing. The whole business had been her idea. She was the one out in front. Now she was a pariah. Helplessly, she saw the "glamour," in her words, "escheat to Autherine." She tried to participate even after her rejection. She went down for registration. She vowed to stand by Lucy, and of course, she had her own case to prosecute. The signs, however, were telling. When Myers went to Shores's office to discuss her case, he put her in an outer room and told her to come up with a list of character witnesses. It was a meaningless task, but Myers took to it as if her future depended on it. In addition to ministers, she listed "Mrs. Rebecca Ulms, worker at Snack Bar No. 2, 206 Omega St. S." and "Mr. Brown (old respected gentlemen next door to Snack Bar No. 2)." (The university already knew about the Snack Bar through its hired detectives. They reported that Myers's mother worked there during the day and Pollie at night. Informants also advised that the Snack Bar sold both legal and illegal whiskey.) Myers's list even included her husband, whom she was divorcing. Then, poignantly, she wrote, "Mrs. Daniels, sup't. of Even Ridge Nursery to testify that I'm interested in the sound development of son by sending him to said nursery."[30]

Years later she remembered a final symbolic act. Constance Motley had arrived from the New York office to help Shores. Motley called Lucy and suggested she come to the office to work on the case. Lucy, as was her custom, called Pollie, and asked, "When are we going?" Myers did not know what she was talking about. When it became clear that the attorneys had not invited Myers, Lucy insisted that she go anyway, that it was surely nothing more than an oversight. On arriving, they sat for a while in the outer office. When Motley came to the door, she spotted Lucy first and smiled warmly. Only then did she see Myers. Her smile vanished. As the two friends approached their New York lawyer, Lucy in the lead, Motley put her arm around Lucy's shoulder and turned to go into Shores's office. The door closed on Myers without a word. She hesitated a moment and, thinking she might be over-reacting, sat down to wait her turn. Thirty minutes passed, and she knew. When she got home, her mother understood that a time for truth had come. She said simply but firmly, "Get some sense into your head, girl. What did you think was going to happen?"

Only later did Myers come to understand that the NAACP would have to go with the "winner." Shut out and generally cut off from sympathy (except from Emory Jackson's wife), she determined to go to Detroit. "One night, and I didn't know anything," Lucy recalled, "Pollie had left and gone to

Detroit all by herself, driving her own car. See, she had her own car at that time, her own individual car, and she was in Detroit when she got in touch with me."[31] Alice Myers was partly responsible for this result. Myers's father wanted her to stay; but her mother sensed better and told Pollie to go on, that if she got lost, she should stop and ask local police for directions. Her mother's instructions ringing in her ears, Pollie, all alone, got lost only once, around Cincinnati.[32]

Three years to the day that the trustees voted to deny her admission, January 29, 1959, Myers received her first of two master's degrees from Wayne State University.

A charge of conspiracy

Shores, Motley, and Thurgood Marshall were not yet done. After dismissing Myers's case, they filed new charges against the university. They alleged that Lucy's suspension was not necessary for anybody's safety, that no one had injured her nor were other injuries reported, that sufficient police force had been available. The only ground for her suspension, therefore, was race. Dean Adams, with other university officials and members of the board, had "conspired to defy the injunction order of this court." Moreover, the university entered into conspiracy with unidentified persons in the crowd and specifically with R.E. Chambliss, Earl Watts, Ed Watts, and Kenneth Thompson, the four arrested in Monday's demonstration. These co-conspirators acted "in concert" with the university "to assimilate the air of riot and disorder and rebellion on the campus against the Plaintiff in order that same may be used as a subterfuge for refusing to permit her to pursue courses of study at the University." Lucy's attorneys asked two thousand dollars in damages on this charge.[33] They also cited Dean Healy for contempt in denying dormitory and dining privileges and asked one thousand dollars for the expense of traveling to and from Birmingham and the cost of eating out.

Shores called a press conference to explain the charges. Originally scheduled for his office at 2 o'clock that same afternoon, it did not get under way until 3:29 in the regional offices of the NAACP, housed in the same building. By then, ill-tempered reporters noted the precise time. Approximately thirty-five members of the press crowded the room, including camera crews from CBS and NBC. Both Shores and Motley read and discussed the charges. One reporter recalled that Motley did most of the talking.[34] Indeed, most observers pointed to Motley as the one who framed the conspiracy charge. Those associated with the case also pointed to her, perhaps scapegoating her because the charge proved so counterproductive.[35] But, both Shores and Marshall signed the petition. Complaints, they knew, were made to be amended. A charge such as the one they constructed would get the board of trustees back into court along with Dean Adams; and if by court time the evidence did not support the charge, they could drop it. To them, there was nothing personal in the charge. It simply represented the way they chose to work the case.

However, the idea that the plaintiff could level such wanton, baseless, and vile charges against their character shocked, even mortified, university officials and members of the board. They were among the state's "best men" and held positions of great trust. Many were lawyers sworn to uphold the law and the courts. Now they stood accused of conspiring with Klansmen and rioters to defy a court order. Forgotten were the embarrassing discussions of November 12 about inviting criminal contempt charges against Dean Adams in order to get a white jury to acquit under circumstances that would necessarily involve perjury. Forgotten were the prevarications, cooked up only the evening before, designed to make the board's denial of Myers's admission appear routine—fabrications intended for a federal court. Forgotten were all the efforts, legal and extra-legal, to sabotage the efforts of Myers and Lucy, whose only crime, in Buford Boone's words, was to be "born black" and to move "against Southern custom and tradition—but with the law, right on up to the United States Supreme Court, on [their] side."36

Though "shocked," the university was prepared. The day following the conspiracy charge, Carmichael held a press conference to express his outrage. Answering questions submitted in advance, Carmichael denied "the Negro woman's allegation that University authorities allowed 'an air of riot and disorder and rebellion on the campus.'" He characterized the conspiracy charge as "untrue, unwarranted and outrageous."37 Carmichael's indignation matched that of his board, but at the same press conference, he began to drift perilously close to an independent course. His resolve to do what the law required seemed to stiffen. He told a meeting of the Faculty Council on the 10th that "he knew what course was right . . . and planned to put his ideas before the board." He told Bob Bird of the *New York Herald-Tribune* "that the return of the Negro co-ed was requisite to his staying on as president." (Under pressure, Carmichael later disavowed the report.) He told a reporter for the *Tuscaloosa News* that he planned to be more insistent on his own ideas.38

John Caddell, whose full appointment to replace Gordon Palmer the board would confirm within days, described the overall effect of the *Herald-Tribune* article as "considerate and favorable even with the false impression left in." Caddell added, "I wish more than anything that it would be possible for us to bring the woman back without bringing in troops." He even suggested deputizing students as a way of preserving order.39 Caddell expressed the sentiment of most associated with the university, perhaps even a majority of the board, though not its most influential members. As moderates, they would have welcomed a peaceful settlement, but like moderates in all seasons, they did little to accomplish it. Though many shared the sentiment to have Lucy return, no one risked his fortune or standing to make it happen. Now that the case was back in court, resistance reasserted itself. The university would take no action unless a court ordered it.

A question of character

Although the university's official family was quick to guard its own character, it was not reluctant to attack the character of the two women. The university had hoped to use the February 9 hearing on Myers's case to make public its view that hypocrisy, duplicity, and a want of moral fitness characterized the two women. They prepared Dean Adams to testify that Lucy and Myers had no legitimate purpose in seeking admission, that the NAACP used them to advance the organization's ends. When Shores dismissed Myers's case, the scheme went awry, but not for long. Adams proved perfectly willing to expand out of court what he had prepared to say in court.

Adams's willingness, even eagerness, may have stemmed as much from personal pique as from institutional loyalty. From the beginning segregationists ridiculed him for inadvertently sending letters of acceptance to the two women. Malicious phone callers asked if he were too ignorant to know that Miles was a Negro college. He now defended himself to a New York *Times* reporter. He said that the university had not officially accepted the women and that the letters of welcome they received resulted from a simple office mistake. Adams attacked Lucy. She was not interested in getting an education at the university. The fact that both she and Myers had recorded the other's field of interest on the application form proved their lack of sincerity and "that someone was engineering the applications and made an error." Lucy never arrived on campus "as an individual." She always appeared "at the head of a delegation of lawyers, ministers and wealthy friends from Birmingham." She was not, concluded Adams, "a poor little Negro girl who is just trying to get an education to better herself."[40]

Nor did the university have to worry about making Myers's private life public. *South: The News Magazine of Dixie,* an organ of Alabama's business community, published Edward Hudson's police line-up pictures, his and Myers's marriage certificate, and their son's birth certificate. Hudson had two convictions, one in 1947 and another in 1948 for daytime burglary of an uninhabited dwelling. Myers's father had a conviction in 1938 for moonshining. The magazine found delicious irony in the university's denial of room and board privileges to Lucy, noting smugly that the university had "hurled back at the Supreme Court some of its own language: applicants were studied . . . on 'sociological' grounds."[41] While not so blatant as *South,* the state's dailies made clear the grounds for Myers's dismissal and the denial of privileges to Lucy.

Whatever damage the NAACP inflicted with its conspiracy charge, the university and its supporters trumped with assertions of hypocrisy, duplicity, and immorality.

Week's end

The week of February 6 to 10 began in riot and ended in uncertainty. In the days and weeks that followed, the university would get rid of Autherine Lucy, though not the principle she and Myers won. The university had not conspired with the mob but had followed a policy that paralleled its will. Placed in context, the actions of the university prior to February 6, wittingly for some and unwittingly for others, created the outcome of suspension. Events after February 10 proved that the university, willingly for some and unwillingly for others, continued to do the mob's bidding, up to and including the permanent expulsion of Lucy.

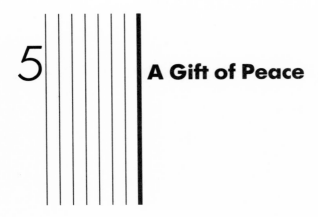

5 | **A Gift of Peace**

When Dean Adams went public with the university's view of Lucy and Myers, he turned to Wayne Phillips of the *New York Times*. Of all the reporters in Tuscaloosa, none took a more sympathetic approach to the university's problems. Phillips was a maturing reporter who, along with Tad Szulc and Bernard Kalb, occupied the second row on the rewrite bank at the *Times*. These three were gifted, "very fast and lively," and the city editor did not object when they hung a sign over their desks that proclaimed: "Greatest Bank in the World—Human Interest Compounded Nightly."[1] Phillips displayed his talent in a feature for the *New York Times Magazine* titled "Tuscaloosa: A Tense Drama Unfolds."[2] "Tuscaloosa," he wrote, "could be a middle-sized city in any part of the United States. Along its broad, straight, main streets one sees the same familiar store fronts—Woolworth's, Sears, Penney's, Grant's—and the theatre marquees advertise the same movies. Walk its quiet, tree-shaded residential streets and one might be in Westchester or Bergen County." Part of the halcyon picture included race relations. Negroes voted and owned middle-class homes, and where they lived in impoverished conditions, it was no worse than the unhappy poverty of poor whites. An Interracial Committee of the city's Religious Council, made up of ministers, school principals, some businessmen, and a handful of university professors, kept lines between the races open.

Through the efforts of this Interracial Committee, "a quiet revolution was taking place." White professors at the university served as part-time faculty at Stillman, the Negro college in the city's west end. Negroes served on governing boards of the United Community Fund and the Red Cross. A recreation center functioned under the aegis of the United Fund, and the city gave the center money proportionate to the Negro population. The Chamber of Commerce sponsored an annual dinner for the best Negro farmer, and the school board built a completely modern $2,000,000 high school for Negroes. Similar good things were happening in the unions. In the CIO unions, Negroes and whites "belonged together, met together, sat down together at the bargaining table." The AFL unions segregated Negro locals from white locals, "but the representatives of each sat together . . . and were equally vocal." "In church, school, store, on the street," Phillips concluded, "the segregation the Northerner sees is only more rigid, but no different than the purely social, unenforced segregation he knows in the North."

In Phillips's estimation, Tuscaloosa was a best-case example of "separate but [truly] equal." Normally, this city of 56,000 was tranquil and friendly. But the town was no longer normal. "Beneath the same peaceful surface that has remained unchanged, there now runs a pitch-black current of hate and fear." Phillips saw race relations in Tuscaloosa as its white inhabitants would have desired. What he saw issuing from the Autherine Lucy episode, the deep current of hate and fear, was a truer depiction.

Radicals

In the wake of Lucy's personal ordeal, the battle for the state's hearts and minds began in earnest. On Friday night, February 10, the Citizens' Councils staged the largest rally they would ever hold. Meeting in Montgomery, crowd estimates ranged up to 15,000. Senator James O. Eastland of Mississippi, a University of Alabama alumnus, gave a two-fisted speech. "I am sure that you are not going to permit the NAACP to control your state, and you are not going to permit that organization to use your children as pawns in a game of racial politics." Sam Engelhardt, president of the Central Alabama Council, sounded the keynote: "It is surprising and sickening to see many white people, so-called, even including some in high office, refusing to be listed on either side of the color line." Governor Folsom became the principal target. State Senator Walter Givhan from Dallas County pointed the finger: "When we soon reach our goal of 200,000 Alabama members, I promise you that our next governor will not have [Negro] Congressman Adam Clayton Powell [of New York] riding in a state car or drinking at the governor's mansion."[3]

The demonstration released accumulated frustration and hostility. Victory in Tuscaloosa four nights before emboldened the organizers, and over the next months the Citizens' Councils in Alabama experienced phenomenal growth. One week following the Montgomery rally, Tuscaloosa got its own Council. A crowd estimated by police at 1,800 packed the largest courtroom in the county courthouse. "They were all there," observed the ubiquitous Phillips, "over-clad farmers, business men in conservative suits, burly industrial workers, and white-shirted university students—jammed together in a sweltering 85-degree room."[4] Senator Givhan gave a twenty-seven-minute speech. "The best friends our darkies have," he declared, "are in the South, and most of the good darkies don't want to change their way." Putting the issue on a global scale, he blamed the Communists for attempting to "amalgamate the races" and to corrupt "pure Anglo-Saxon blood."[5] At Givhan's right hand sat an earnest and solicitous Leonard Wilson. Already elected temporary chairman of the newly formed West Alabama Council, Wilson wanted to make a good show. As he stood, Wilson sniffed the air in search of government officials. "The tribe is assembled," he proclaimed, "where are the chiefs?" A voice from the crowd shouted: "Home under their beds." Wilson wanted to know: "Do I look like a mobster to you folks?" to which someone said, "No, a real Southern Gentleman." They laughed. Wilson had found his element.

There were actually two meetings in Tuscaloosa that night. The Tuscaloosa Committee for Political Education was celebrating the merger of the AFL and CIO, and labor leaders throughout the state assembled. Among the speakers was the peripatetic George Wallace. The soon-to-be "fighting judge" from Barbour County still showed some of his Folsomite progressivism by declaring that "the new South is going to be built on the shoulders of labor." Wallace's effort to curry favor with labor, however, proved to be another sign of the times. The next day's *Tuscaloosa News* gave top billing to a picture of Leonard Wilson conferring with Walter Givhan. The bottom of the page showed Wallace with state labor leaders. Wallace still had not maneuvered himself into the right place at the right time. The right place, of course, was over at the Citizens' Council rally where "Big Ed" Robertson, a union member at Goodrich, held forth against Walter Reuther and George Meany, saying they did not speak for labor. Big Ed would later become a powerful legislator and a Wallace floor leader. But on that Friday night in 1956, Big Ed, not Wallace, knew where to be. One dejected union official put it best: "Today, if we were to endorse a candidate and the Citizens' Councils were to endorse another, our candidate wouldn't have a chance."[6]

Moderates

The Alabama Human Relations Council, the state's only organization for rallying moderate thought, met in Montgomery on Saturday, the day after the pro-segregation extravaganza. Two hundred brave souls gathered on the campus of Alabama State College to discuss ways of promoting racial harmony. They had a standard around which to rally, but no standard-bearer. Governor Folsom's populism, though attractive, reduced him to humorous prattle in the face of massive resistance. When asked what he thought of the White Citizens' Councils, Folsom replied, "I'm for white councils and black councils and red councils and yellow councils and brown councils if there are any. Nothing built on hate can exist for any length of time in a Christian democracy. I hope there's none of the councils who have based their beliefs on hate."[7] Folsom had attempted to build his first administration, 1947–51, on grand ideas for a resurgent democracy in Alabama but faltered in inept execution. His second, 1955–59, began with the promise of better organization and leadership. Before a year had passed, however, the bus boycott and the Lucy episode killed his hopes for a reapportioned legislature and for the extension of basic human services.[8] Sitting in the ruins, drinking more bourbon than branch water, Folsom's gift for humor devolved into buffoonery.

If the handful of interracial stalwarts could not look within the state for leadership, the nation offered little better. Eisenhower washed his hands, and Adlai Stevenson quickly seconded the President's inaction. At a February 11 press conference, an NAACP representative asked, "What do Mr. Stevenson and Mr. Eisenhower mean by moderation? As a Southern Negro I have been taught moderation from my mother's knee up. I know it is a terrible thing to say but she was wrong, my own mother." "Any moderation," he concluded, "means delay, delay which will let the White Citizens' Council get stronger."[9] The only national leader to emerge with some standing in the black community was Richard Nixon. On February 14, at a GOP dinner in New York, the vice president proclaimed that "a great Republican chief justice, Earl Warren, has ordered an end to racial segregation in the nation's schools." Despite later developments in the Kennedy years, Martin Luther King, Jr., continued to think well of Nixon for what King believed to be a credible commitment to civil rights.[10]

In the absence of clear signals from any corner, moderates put their trust in Carmichael, whose statements on Friday gave indication of vigorous leadership. Not only did he condemn the mob and guarantee law and order upon Lucy's return, he vowed to punish students who acted as ring leaders. On Saturday, the 11th, he solicited the faculty through their deans to name students known to have rioted. To investigate student participation, he established a committee headed by Hubert Mate, an assistant dean and former Navy intelligence officer. Moderates and liberals rallied to Carmichael's side. Church leaders met and voted confidence in his leadership. A young reporter for the *Birmingham Post-Herald* used company letterhead to commend the president. "There are many things to be ashamed of in Alabama today, and very few to be proud of," wrote Clarke Stallworth. "I would like to say that I am very proud of you and the way you have handled the problem." Virginia Durr wrote her girlhood idol to say that "Cliff and I have been so distressed at the terrible situation at the University, and we feel that you will bring order out of chaos if anyone can."[11] There was even good news for restless alumni. The NCAA had put the Auburn football team on probation for attempting to buy recruits.

An unfortunate incident

The mood of resolution was not to live long. On Tuesday evening, the 14th, an incident occurred just off campus that recharged Tuscaloosa with fear. Two black men, Arthur and John Washington, ages twenty-four and nineteen, accosted Samuel Perrin Taylor, a nineteen-year-old student, as he walked down the middle of Third Avenue, heading toward campus. The Washingtons lived nearby. The brothers grabbed Taylor and took him between two houses where they told him they intended to get even for what had happened to Lucy. Arthur, a Korean war veteran, held Taylor, which was not difficult as the 5'9" lad weighed only 135 pounds, while his brother kicked Taylor on the leg and hit him once on the forehead before he broke free.[12]

The brothers intended only to frighten Taylor. They did not pursue him and inflicted no serious damage, only a small bump above the eye. They were so little concerned that they strolled over to campus the same evening where police picked them up as they walked by the Union Building. Taylor identified them as his assailants, and the police hauled the surprised young men to city jail for booking. The initial charge was assault and battery, but when police found a partially opened pocket knife, they charged assault with intent to murder. Authorities later moved the brothers to the Greene County jail, about thirty miles southwest of Tuscaloosa. Tension mounted. The Tuscaloosa Council on Human Relations published a letter declaring Tuscaloosa to be an "emotional powder keg" and urging blacks and whites to do nothing that "might imperil the safety of both races." The Council noted that whites frequently perpetrated similar offenses against blacks, most of which went unreported. Buford Boone thought the incident worthy of another page-one editorial. Besides urging calm, he published a picture of Taylor with a caption noting that the "only visible injury was a small knot over one eye."[13]

Indicted on March 3, the accused pleaded innocent and innocent by reason of insanity. Despite an effort by officials to reduce the charge to relieve tensions, the Taylor family refused all entreaties and demanded a jury trial. Circuit Judge Wright, the same judge who had placed the call for National Guard protection from Carmichael's office, heard the case. Arthur Washington appeared in uniform. Except for discussion about the seriousness of the injury, the facts were not in dispute. The defense attorney made his case by begging the jury not to make martyrs of the Washington brothers, because to do so would only create another fund-raising opportunity for the NAACP. The jury deliberated twenty minutes and found the young men guilty of the lesser charge of assault and battery. Judge Wright imposed a sentence of six months in jail and fines of $500 each (or an additional 140 days in jail). It was the stiffest sentence permissible by law.[14]

Convocations and meetings

Two days after the Washington brothers' run-in with Taylor, the university staged a large convocation to calm its nerves and renew its commitment to law and order. Because the faculty and student body would not fit in Foster Auditorium at one session, faculty and non-military students convened at 11:00, while the the corps of cadets assembled on the quandrangle at 11:20 and marched to the convocation at 11:45. They filed in to stirring strains from the university band and filled with pride as their own 1955 Miss Alabama, Pat Huddleston, sang the national anthem. The mood was perfect for what Carmichael had to say, and he was perfect saying it. His gray mien, which looked sad and beleaguered in the face of the howling night demonstration, now appeared wise and sage, still sad but resolute. He gave a "factual account" of the long litigation that preceded the recent happenings on campus. He justified the board's decision to suspend Miss Lucy and denied that it was abdication before the mob. He praised the students, who "though opposed to integration, would not like to have our university choose the side of lawlessness." He believed that the "active minority" who had participated in the demonstrations, no more than "three or four percent" of the student body, "were not aware of the issues involved." And "the issue," he declared, was "not segregation vs. integration but law and order vs. anarchy." Carmichael vowed punishment for the ringleaders of the demonstration but only after careful investigation. He closed with a plea for mutual effort and invoked God's favor. A standing ovation brought Carmichael back a second time for sustained applause.[15]

The arrest of two photographers marred the occasion. George Tames of the *New York Times* and a student attempted to take pictures in violation of Carmichael's "no pictures" order. Carmichael allowed the print press in. The exclusion of photographers was the only sour note, but it represented a continuation of bad press relations.[16] Carmichael did not lose much thereby. The press was going to support his brand of moderation under almost any circumstance. But poor press relations, expecially in the week following the riot when Carmichael tried to assert himself, diminished the appearance of effective leadership. That week, from February 9 to February 16, represented a fleeting moment of opportunity for Carmichael. He stepped out front because somebody had to represent the university, and the board was happy to let him have the stage as it pondered its next move. For the remainder of February, Carmichael gradually receded from view in deference to his trustees.

Expulsion

A decision to expel Lucy permanently for the allegation of conspiracy formed gradually over a nine-day period beginning February 10. An original impulse had been to counter-sue for malicious prosecution.[17] However, by the 13th, Andrew Thomas advised that the best they could hope for was disciplinary action by Judge Grooms "when the conspiracy charge is proved false on the 29th." Thomas argued that should Judge Grooms fail to discipline the NAACP lawyers, the state bar might act against Shores as John Caddell had suggested in an earlier letter. On February 17, a Friday, Gessner McCorvey wrote the lawyers, with copy to the board, congratulating Thomas and Moore on "the swell job" they were doing and recommending permanent expulsion for the conspiracy charge. It was the first written suggestion that expulsion be used and came a day after the board's executive committee had met at Bob Steiner's office in Montgomery to lay plans for the February 29 hearing. The final decision to expel came two day later, again at a meeting of the executive committee in Montgomery. Among other things, they agreed on a full board meeting in Birmingham at 7:30 p.m., February 28, the eve of the trial. They instructed Rufus Bealle to make overnight reservations for board members at the Tutwiler Hotel and voted Caddell onto the executive committee in place of the ailing Gordon Palmer.[18]

So it is clear: even as Carmichael spoke to the students and faculty on February 16, pledging the university's resolute determination to obey the law and to provide for the safety of Autherine Lucy, members of the board moved to expel her. Only the question of timing remained. The board did not keep Carmichael in the dark about the plan; it simply did not invite him to participate in its development. As a consequence, he never seemed certain that the board would carry the expulsion through. Carmichael believed that Lucy's safe return was essential to the university's reputation and retained some hope that the board would bow to the inevitable court order directing that Lucy resume her studies. At Carmichael's direction Jeff Bennett led a team of city, state, and university officials who came up with a plan for securing the campus. It was an elaborate plan that involved closing the campus to all but essential traffic, using state troopers and national guardsmen with fixed bayonets but no ammunition, radio checkpoints at stratgic locations, and clear central command control of the security force.[19] Bennett coordinated his efforts with the governor's office. Folsom flew in to Tuscaloosa on Sunday morning, February 26, for an 8 o'clock breakfast with Carmichael. They spent an hour and a half discussing Lucy's reinstatement. Folsom told reporters that he would do "all in [his] power" to maintain order; he would not submit to the mob. In fact, Folsom had endorsed the Bennett plan.[20]

Having met with Carmichael and talked to reporters, Folsom flew to Birmingham where he conferred with some members of the board. Folsom and Carmichael still thought, perhaps even hoped, that the prospects for Lucy's safe return were favorable. Whatever the chances, decisions reached at the pre-trial hearing on Tuesday the 28th doomed them. In an ex parte meeting with Judge Grooms early that afternoon, the university's lawyers outlined their defense, which included asking for separate jury trials for each defendant (to which they were entitled because "conspiracy" involved criminal contempt). Counsel for the university further asserted the right to expel Lucy for her "scurrilous and scandalous charges" against the board. On hearing this, Judge Grooms was emphatic. He "did not want such issue raised at the hearing on Feb. 29." The purpose of the hearing, he insisted, arose from the events of February 6 which led to Lucy's suspension, not the conspiracy allegation that arose subsequently. Grooms asked if he could "advise Thurgood Marshall of the attitude of the Board of Trustees as just expressed to him." Thomas and Moore agreed.

Later that afternoon, in a meeting with all attorneys present, Grooms told Marshall, Motley, and Shores of the university's intentions. Marshall "acquiesced in the opinion of Judge Grooms that the right of the Board of Trustees [to expel] should not be made an issue in the hearing on February 29," and Thomas and Moore agreed to redraft their filings "so as to delete therefrom any reference to the conduct of the plaintiff in making her charges [of conspiracy]." The NAACP lawyers knew they were in a bind. They could not prove conspiracy, and Judge Grooms had decided to leave the question of expulsion to another round of litigation. The case no longer involved criminal conspiracy. There would be no jury trials as the university desired, but at the same time Grooms had flung the door wide for expulsion.[21]

February 28

On leaving chambers, the university's lawyers knew they had won the battle of Autherine Lucy, if not the war over desegregation. Barring some hitch, within twenty-four hours they would have what they wanted: Judge Grooms would not hold the university in contempt; he would order Lucy reinstated; whereupon the board would expel her. When Hill Ferguson called the trustees to order that evening, he was, no doubt, the most satisfied. Thomas explained the agreement reached in the afternoon and spoke of it as a "possible settlement." The university would get what it wanted, a clear name and the right to go forward with expulsion. The NAACP would get what it wanted, the principle that Autherine Lucy was legally entitled to reinstatement. After general discussion, McCorvey introduced the resolution of expulsion, and Frontis Moore discussed its timing.[22] Still, the drama of the 29th remained to be enacted, and no board member would rest easy until the final curtain.

The meeting of the 28th is central to understanding the siege mentality that had developed among members of the board. For the second time in four months, the board kept a meeting secret, the other being the meeting of November 12. The move for secrecy seems to have come from university officials. On March 7, John Caddell agreed with Rufus Bealle "that no minute record should be made of the meeting of the Board held on the evening of February 28th." Another trustee, Brewer Dixon, agreed to secrecy provided the minutes of the 29th be amended to reflect receipt of the student/faculty petition circulated by Albert Horn. Dixon wanted it on record because "it is entirely possible that the Board may desire to take some disciplinary action against said members of the faculty who undertook to embarrass the Board. . . ." "In fact," Dixon concluded, "I might request that at your convenience check through the list and send me the list of the faculty members who signed such petition, including the names of any associate professors or other persons employed on the payroll who signed. . . ."[23]

There was one other occurrence on the 28th. Leonard Wilson spent two and a half hours under interrogation by the committee investigating student disorders. He wore his ROTC uniform. He assured reporters that he was in no danger of disciplinary action because he had not been guilty of violence.

February 29

Since her exclusion, Lucy had spent most of February on the campus of Talladega College about forty miles east of Birmingham. The Congregational Church founded the college and opened it to all races, though few whites ever attended. The president, Arthur Gray, and many of the faculty actively participated in the Alabama Council on Human Relations. They happily provided Lucy a haven from the numerous threats on her life. Lucy enjoyed the peaceful setting. She watched professors stay up all hours playing bridge and developed a keen, yet never fulfilled desire to learn the game. One other thing came of her stay. A visiting lecturer from Africa, named Angela, inspired the name of Lucy's first daughter. The Klan and other resisters discovered her location but limited their response to occasional rides through the campus. Young college men stood guard during these affairs and welcomed the opportunity to return taunt for taunt.

The threats did not stop when Autherine returned to Birmingham for the February 29 trial. On the short drive from Shores's office to the Federal Building, armed blacks positioned themselves in stores and atop buildings along the route.[24] Arriving at the courthouse, Lucy and her escorts found the corridors jammed and the courtroom packed. One estimate had blacks outnumbering whites three to one. One black man shied from a camera, saying, "Don't take me standing out here. My boss might see it in the paper. I'm supposed to be working." There was a large press contingent. "Southern accents and English accents mixed together in the corridor." Off to one side "a nattily dressed photographer . . . leaned against a post and ate a pink, candy Easter egg." Emily Barrett, the secretary/reporter for the *Alabama Citizen,* arrived late from Tuscaloosa but got in on a pass given her by a reporter for the *Chicago Defender.* She sat with reporters, pretending to scribble on her pad, "really faking because I wasn't writing anything."[25]

Judge Grooms gaveled the hearing to order at 9 o'clock. As the university feared, Thurgood Marshall opened with a surprise. He amended the plaintiff's petition to drop all conspiracy charges. He said that "after careful investigation we are unable to produce any evidence to support those allegations." For emphasis, he said, "This amendment takes out every single allegation of conspiracy." Having anticipated this possibility, Andrew Thomas moved quickly to reset, or at least to make clear, the grounds for expulsion. Objecting to the deletion, he said the charges had been "relayed throughout the world," and he demanded the right "to prove the falsity of each and every one of these scurrilous charges." Following this curious demonstration, Grooms accepted Marshall's deletion and consequently denied the defense motion for separate jury trials—there no longer being any criminal charge pending. Of course, Thomas's display was to set up the trustees' resolution of expulsion.

Thomas spent the remainder of the morning interrogating Lucy. Wearing a pink knit suit, she responded to each question in her slow, methodical manner, a manner that seemed almost detached. Thomas wanted to establish that she feared for her safety during the Monday demonstrations. He asked about an article attributed to her entitled "My Day of Terror." He wanted to know if "terror" were her term. She replied it was not. When Thomas quoted from the *Birmingham News,* Judge Grooms stopped him, saying, "This case has been tried in the newspapers already, I don't feel we should try it from the papers." Thomas turned back to Lucy and asked if she had feared death that day. She answered yes.

"Did you pray?"
"Yes."
"Did you pray, 'Oh, Lord, don't let me be lynched?' "

Marshall's patience snapped. He angrily interjected: "Certainly, your honor, we are perfectly willing to stipulate that she—as anybody else in those circumstances—was frightened."

Establishing Lucy's fear served the legitimate purpose of avoiding a contempt citation. The university needed to show that circumstances justified suspension. However, Thomas used most of Lucy's time holding her up for public ridicule. He asked what had prompted her to apply in 1952. Marshall objected to the questions's relevance, and Judge Grooms sustained. Rather than concede, Thomas expanded on the purpose of his question for the benefit of the media. The whole business, he explained, had been for publicity, both for Lucy and the NAACP. Still, Judge Grooms was not moved; so Thomas turned to Lucy's $2,000 request for damages to discredit her motives. Because part of the $2,000 covered attorney's fees, Thomas asked who paid for her litigation. Lucy became evasive, declaring at one point that she was paying, at another that she did not know how Shores was being paid, later that a committee was paying, and finally only that someone must be paying. Her answers did not sound good, but neither did Thomas get anywhere with his effort to show that the NAACP was financing it all.

Judge Grooms recessed for lunch at 12:53. In the afternoon, the defense called Jeff Bennett and had him repeat his opinion that on two occasions Lucy was seconds from death. In cross-examination, Marshall attempted to show that the university was negligent in providing security. He observed that early in the day on Monday there was ample force to disperse the mob. Bennett replied that by the time they realized it was a mob, the crowd had swelled to 500. Marshall wanted to know why the university had not arrested the ringleaders. Bennett said four men had been arrested, but, more to point, it was difficult to tell who the ringleaders were. Marshall asked about the fire truck, and Bennett explained the decision not to use it. Bennett parried convincingly each thrust from Marshall. In addition to being Carmichael's chief assistant, he was a professor of law.

When Marshall asked about Lucy's future security, Bennett did not tell the whole truth. He told the truth when he said that the campus itself posed no threat and that state officials had promised to provide assistance. He further spoke the truth when he said that outsiders posed the gravest danger, that it was difficult to secure a campus crossed by a federal highway. But in telling Marshall that he did not know whether security forces could isolate the campus, he neglected to mention his own elaborate and sensible plan to do just that. The reason for such incomplete disclosure was simple. Attorneys for the university had decided that in addition to blocking the contempt citations, they could make a good case for continuing the suspension on grounds that Lucy's safety remained in doubt. So Bennett fell in line.

John Caddell advanced the security argument by reporting that police had found a rifle inside a car on campus. Nothing could have prevented the man who drove that car from killing her. "Steps to protect her in the event of her return," he said, "have not been taken because we think nothing can be done." Marshall had had enough. To Caddell he said, "We have one person acting in a lawful manner and the other side in an unlawful manner. The law-abiding person was ordered from campus. That being true," he asked, "what did the board do to the unlawful group?" Caddell dropped his usual reserve and shoved back. "I'm not at all sure the actions of Autherine Lucy and persons accompanying her were not the very cause of the demonstrations. Autherine Lucy came in a Cadillac automobile, she had a chauffeur, and walked in such a way as to be obnoxious and objectionable and disagreeable." Taken back by this attempt to blame the victim, Marshall lamely inquired whether anyone else on campus had a Cadillac. Caddell said he did not know.

Judge Wright followed Caddell. Judge Wright talked of an uneasy calm in Tuscaloosa. He mentioned the impending trial of the Washington brothers and noted that they had to be removed to Greene County for their protection. He, too, said Lucy was in danger of being killed. "Maybe not on the first day, but they are determined she will not return." Wright reiterated Caddell's testimony about Lucy's obnoxious and objectionable behavior and concluded that it was "the general sentiment that she should not have been there in the first place." Wright might have added that Tuscaloosa's black community was as determined as its white citizens, only for different ends. As he spoke, city workers had scheduled a one-hour work stoppage in protest of Lucy's treatment.[26]

In all, the defense subpoenaed some twenty witnesses. The Reverend Gribbin, who appeared for the university but did not testify, described the boredom. "We have sent out for cokes, read the newspapers several times through, and the floor is covered with cigarette butts. . . . The whole business is very sad."[27] Despite Judge Groom's best offers to stop conjecture about prospects for violence, Bennett, Caddell, and Wright got their say. Testimony continued until 4:25 whereupon Grooms recessed the hearings and went to his chambers. In little more than an hour he returned to announce his decision. About 100 blacks had waited. He did not disappoint them. Grooms refused to hold the university in contempt, saying that while officials were guilty of "underestimating the fury of the mob," they were not derelict in so doing. The violence was without precedent. No other attempts at integration had produced similar results, and thus university officials had done the best they could, which as it turned out was inadequate, but nonetheless the best they could. Lucy's suspension was justified for her safety.

On the request for readmission, Grooms dismissed testimony that the university could not guarantee her safety or "that law enforcement agencies of this state are unwilling or inadequate to maintain law and order at the University." To accept that view would render any court powerless before a mob. "There are some people," he observed, "who feel that this court should carve out a boundary here in Northern Alabama, mount the battlements and from the ramparts defy the Supreme Court of the United States. That this court will never do." Accordingly, he ordered Lucy readmitted by 9 a.m., Monday, March 5. He took the petition for dorm and board privileges under advisement and said he would announce his decision the next day. It was over just that quickly. Lucy emerged from the courtroom at 6 p.m. One black man exclaimed loudly: "He [the judge] really laid it on the line." By 6:30, empty corridors, strewn with paper coffee cups, chewing-gum wrappers, and cigarette butts, echoed a hollow victory.

Frontis Moore, Andrew Thomas, and others in the university's delegation walked one block to the Tutwiler Hotel where board members were gathering. All were present except Folsom, Austin Meadows (the state Superintendent of Education), and the ailing Gordon Palmer. Also attending were Carmichael, Dean Adams, and Rufus Bealle. Within minutes they approved the resolution of expulsion. In part it read: "No educational institution could possibly maintain any semblance of discipline if any students, whether they be black or white, guilty of the conduct of Autherine J. Lucy, be permitted to remain a member of the student body after making such baseless, outrageous and unfounded charges of misconduct on the part of the university officials." The board named Caddell its spokesman.

Reaction

The expulsion of Autherine Lucy climaxed a crescendo of opposition to desegregation in Alabama. After the board's decision but before a public announcement the next morning, Ace Carter called for a Friday night rally of "more than 15,000" persons in the municipal auditorium. He promised "a solution to Judge Grooms' decision" and an added attraction, Leonard Wilson. The lawmakers in Montgomery matched Carter's defiance. The house ordered Lucy to appear before them on Monday (the day readmission was to take effect) to determine whether the NAACP was "directed or controlled by Communists." The senate adopted unanimously a resolution calling on Congress to set aside funds to move southern Negroes to the North and Midwest—"areas where they are wanted and needed and can be assimilated." Another house resolution demanded that Carmichael publish the names of students and faculty who had supported the Horn petition for Lucy's re-

instatement. The house also considered a bill to cut off the state's $350,000 appropriation for Tuskegee and the $82,500 allotted for out-of-state scholarships. Another measure would punish with fines and jail terms those who filed contempt of court proceedings "without good and sufficient cause." Finally, a bill was introduced to require affidavits of "fitness and character" from three alumni of the college or university to which any student applied.[28]

The hysteria that crested with Grooms's order to readmit, combined with testimony about the dangers that awaited her return, convinced Lucy's lawyers that she would never be a student at the university. For public consumption, they expressed shock at the expulsion and their determination to pursue re-admission. Privately, they knew it was over. Circumstances strongly suggest that Lucy had not planned to return to the university whatever Judge Grooms ruled on February 29, that at some point before her expulsion the NAACP lawyers made plans to take her out of Alabama. When on Thursday reporters asked Shores about the expulsion, he first replied with "no comment" because he was off to Montgomery to file new proceedings in the bus boycott. Later in the day he said the board's action did not surprise him because "it had been intimated." Though not suprised by the action itself, NAACP attorneys expressed dismay at the haste with which the board acted. Andrew Thomas and Frontis Moore arrived early Thursday morning to tell Judge Grooms about the trustees' decision. Judge Grooms cooperated by saying that he did not consider the expulsion a defiance of his order of readmittance. He further noted that the board's action now rendered moot any consideration of room and board privileges. Grooms's quick acquiescence appeared motivated, at least in part, by concern for Lucy's safety.[29]

Marshall tried one last maneuver during the hearing to effect Lucy's reinstatement by withdrawing the charge of conspiracy, but even there he simply laid groundwork for future litigation. In Alabama the law of libel said that the aggrieved party must first demand a retraction and that if the retraction is made then no punitive damages would attach—though real or actual damages could be assessed. Marshall no doubt hoped that a public retraction would stop the expulsion. Under normal circumstances it might have, because the reason for expulsion had been publicly retracted. However, these were not normal circumstances. It was too late, and Marshall knew it. Judge Grooms had given the Board of Trustees a green light. A year later he would declare the expulsion perfectly permissible and not taken in defiance of his order.[30]

There was more to it for the NAACP than bowing to the action of the trustees. A cloud had descended over the whole question of whether Lucy's return was wise. The NAACP appeared shaken by what it had seen and heard over the past month, a fear confirmed by the parade of university witnesses who painted a grim picture of danger. It was as if the NAACP's lawyers had walked to the edge, peered into the abyss of violence that would mark the next ten years, and backed away. Before leaving for New York, Marshall said, "We leave the final decision to the good people of the nation." On learning of the expulsion, Lucy declared her "shock," and at a later press conference in New York, said, "I cannot see any reason to abandon my sole purpose of obtaining an education within the meaning of the decisions of the Supreme Court of the United States."[31]

But this was little more than posturing. The prospect of Lucy becoming anything more than a test case died with the Monday riots and her suspension. Moreover, she showed signs of physical deterioration. Arriving in New York, Marshall told reporters that Miss Lucy "appears to be very calm but her insides are a mess. She is here to see doctors. She needs rest, peace, and quiet." As newsmen and welcomers pushed in around her, they got a look at the circles under her eyes, and she seemed to tremble. She said only that she would be in New York "for a few days." Then she whispered to Motley: "Please get me out of here." Marshall told the crowd, "She is at her rope's end. Her nerves are shot and she needs a rest."[32]

Years later Motley summed it up: "She wouldn't have gone in anyway, even without that statement [about conspiracy]. Who was going to guarantee that she would go in? . . . by that time, the South, and the rest of the country realized that we were facing far more than the admission of one single black student to the University of Alabama—we were facing a *social revolution* of major importance and consequence."[33] Lucy's expulsion had become a safety valve for all parties.

A gift of peace

Wayne Phillips, still in Tuscaloosa, said that the town's people greeted the news of Lucy's expulsion with "the greatest sense of relief the city had known since Miss Lucy first appeared here." The mood prevailed especially among those who favored gradual desegregation. The Tuscaloosa Council on Human Relations agreed that the university had done all it could and that expulsion now seemed the wisest measure. From the other side of town, the *Alabama Citizen* echoed the sentiment, welcoming the opportunity to take stock in an atmosphere of calm. As usual, Buford Boone took the best measure. In an editorial entitled "Could We Have a Gift of Peace," he argued that Tuscaloosa now faced a more certain prospect of violence than when Lucy first enrolled. He said that Lucy had it within her power to give her native state and its people the greatest possible gift, to "stay away from the University." He acknowledged that it would be a "gift" because she would be giving up "her hard-won right to attend."[34]

As moderates leaned first left, then right to maintain a steady boat, they again found themselves in the company of massive resisters, who also praised the trustees—but for their "cleverness." Ferguson, McCorvey, and company confirmed their belief that good white men, if truly worthy of the trust they held, would find ways to frustrate desegregation. Massive resisters now had the best evidence they would ever get that their tactics worked. In the moderate hierarchy of values, order occupied the top rung—order with or without segregation, but nonetheless order. As if to reaffirm their own decency, moderates in Tuscaloosa followed up the Lucy debacle by extending a hand of goodwill across Greensboro Avenue to the black community. The New York *Times*'s Phillips saw it as "a swing back toward the good race relations that characterized Tuscaloosa before Miss Lucy arrived." A downtown department store added "the prefix Miss or Mrs. to the names of Negro women customers." The faculty at the University of Alabama gave 400 books to the Stillman College library. Some white business men joined Career Day proceedings at Stillman. Despite these happy signs, Phillips believed "the veneer of calm" to be "so thin no one can be sure of what could happen."[35] It was precisely that kind of vague unease that justified moderate control.

Not done yet

Moderates need not have feared Lucy's return. Neither she nor the NAACP had plans for that soon. Arthur Shores did file a petition on March 9 to have her enrolled for the fall semester, but more to keep the principle alive. Any quick return to Alabama faced obstacles. The legislature had summoned her to testify about Communist connections with the NAACP. Moreover, the brothers Ed and Earl Watts, Kenneth L. Thompson, and Robert E. Chambliss, four defendants in the NAACP's conspiracy charge, were now suing the NAACP for four million dollars. The presence of these men at the February 29 hearing had embarrassed the university. "The [conspiracy] petititon put them with us," Jeff Bennett remembered. "We fought the bastards three days and here they were, sitting with us as co-defendants. They even tried to reach out and shake hands, and said, 'How Do You Do?' "[36] Though an embarrassment for the university, their lawyers in the four-million-dollar suit now included some of Birmingham's more respected members of the bar—Frank Bainbridge, George Rogers, and T. Eric Embry—men whose conservative and segregationist credentials were not in question, but who were unlikely to take on Klansmen for clients. It is improbable that this kind of suit would have occurred to Chambliss and his friends. However, it was precisely the kind of suit Steiner recommended to the university's lawyers in early February—a suit designed to harass the NAACP and defame Lucy and Myers by accusing them of malicious prosecution and libel.[37]

Among other allegations, Chambliss and friends charged that the NAACP paid both women $300 a month. Neither was "fit" to attend the University. At the time they sought admission they "were intimate friends, associating constantly with each other." Pollie "was a woman of loose morals and unfit to associate with the students." She married on November 14, 1952, "a felon then on parole from conviction of burglary." The fact that she bore her child "in wedlock, six months and four days subsequently, did not ameliorate her lack of fitness." The suit further alleged that Pollie's conduct "rendered said defendant Lucy by association morally unfit."[38]

Attorneys for the four litigants got some of their information through Myers's husband, Edward Hudson, and Morel Montgomery, an attorney hired by Hudson to represent him in the divorce proceedings Pollie had brought. On March 1, Hudson told reporters that she parted with him "so she could become a 'guinea pig' for the NAACP." His attorney further claimed that Hudson did not abandon her "but left because she told him to do so." They charged that the NAACP was paying Myers $300 a month and that she was "not a fit and suitable person to have care, custody and control of the child." Three days later, Hudson, in an affidavit, denied having made the allegations reported on March 1. He said that he simply had hired counsel to represent him in the divorce proceeding and that he did not say she was a "guinea pig" or anything else. About a month later, Hudson agreed not to contest the divorce and to give Myers custody of the child.[39]

University trustees and prominent members of the Birmingham bar made a sport of publicly smearing the reputations of Lucy and the NAACP by digging into Myers's personal relations. They did not care much about the sources of their information. Edward Hudson, a "convicted felon" contesting a divorce suit, became a reliable source against his estranged wife. Nor did the lawyers seem to mind the backgrounds of their clients. Chambliss was notorious for violence. On March 13, the *Birmingham News* reported that Tuscaloosa authorities held another plaintiff, Phillip Earl Watts, in the county jail on charges of burglary and grand larceny for stealing plow tools from a nearby Negro farmer. Bond was $2,000. The suit dragged on for months with discovery being used to harass the NAACP.

Two for the show

Ironically, while some of Birmingham's best lawyers were busy suing the NAACP in Chambliss's behalf, "Dynamite Bob" was threatening Carmichael. In a "Message to Dr. Carmichael," Chambliss gave his telephone number and address and proudly proclaimed that he was "the first one in Jefferson County to join the White Citizens' Council." Calling Carmichael "a self-made s.o.b.," he said he had heard of threats to punish students "down there who are taking up for their rights" and warned that he would "take it on [him]self to do something about it."[40]

On Tuesday night, March 6, Leonard Wilson held enthralled a crowd of 2,500 members of Birmingham's White Citizens' Council with an all-out attack on university officials. It was an excellent way to celebrate his twentieth birthday and the 120th anniversary of the Alamo. To Carmichael's claim that the issue was no longer segregation but "law and order," Wilson cried, "Hogwash." The crowd roared back: "Impeach him," "Throw Carmichael out." Wilson smiled triumphantly. He told them that the latest rumor indicated he would be expelled within five days. "I hope you'll remember it's your university," he shouted. "I think this inquisition committee might be scared. They gonna wake up and realize who their employers are." Wilson's audience was a middle-of-the-road gathering of the Associated Citizens' Councils of Alabama. Outside Ace Carter set up pickets with his radical followers from the North Alabama Citizens' Council. "This is the real stuff," said one of the pickets, "those fellows inside are just politicians."[41]

Wilson had freed himself from the Jew-baiting, religious bigotry of Carter's group, but he did not stop "nigger-baiting." He leaned over and whispered in mocking tones that Carmichael told northern reporters that the university would eventually admit a qualified Negro; to which he got the anticipated reponse: "There ain't no qualified nigger." Everybody laughed. They laughed louder when Judge Hugh A. Locke said that Thurgood Marshall "reminded him of a monkey trained to roller-skate." And so it went. The liberal columnist Murray Kempton wrote, "Wilson was the warmest biscuit served up by the Citizens' Council, whose faithful gave him a 30-second ovation and lingered to shake his hand in preference to the adult orators. From Roy Cohn to Leonard Wilson," Kempton concluded, "troubled, middle-aged people in all latitudes in this decade are pushovers for crusading children."[42]

Wilson correctly predicted his own expulsion. During the second week in March, the university disciplinary committee reported its findings. The trustees discussed the report on Sunday, March 11, but because state law forbade formal action on Sundays, waited until just after midnight to ratify their conclusions. As in the Lucy expulsion, Caddell was spokesman. They gave nine students letters of reprimand, placed eight on probation, indefinitely suspended four, and expelled Wilson. Wilson had gone to work at 6 o'clock that morning and arrived back on campus for his 9 o'clock class. He was in Mrs. J.H. Menning's class on business correspondence, when assistant dean of men Henry J. Sikir called him to the door. When Wilson went back to get his books, he told a friend, Jack Winfield, that they wanted to see him at the administration building. (Winfield, who received an indefinite suspension for his part in the demonstrations, also was discharged from the National Guard three months later because of a racial incident at Fort McClellan). At the dean's office, Louis Corson told Wilson about the board's action. Reporters shortly found Wilson in the university barber shop, but he refused comment, saying that he had scheduled a news conference for 4 o'clock that afternoon. His mother learned of the expulsion from newsmen. When asked whether she planned to stay in Tuscaloosa or move back to Selma, she turned away without answering and burst into tears.[43]

Despite warnings not to use university property for his press conference, Wilson spoke from the steps of the Union Building. A number of students jeered his arrival, and when he spoke, hecklers tossed pennies at his feet. He thanked them. A few students cheered his remarks, but most came to razz him. When he said he had "just begun to fight," one heckler cried out, "Don't give up the ship." Wilson might have expected derision from his fellow students. The *Crimson-White* had run a cartoon that featured the body of a young man, standing at the base of a flagpole, with a toilet bowl for a head. The caption read: "Now there's a fellow with a head on his shoulders."[44] What Wilson did not expect was to be criticized by former supporters. The Dallas County Alumni Chapter (his home county) wrote that while they liked him personally, his intemperate remarks about the university constrained them to withdraw their support. They warned that he was hurting the cause of segregation. Alston Keith, an organizer of Alabama's first Citizens' Council in Selma, "commended the trustees for expelling Wilson. He told the 800-odd persons meeting at Demopolis that the 'trustees couldn't do anything else.' " And when Marvin Griffin, Governor of Georgia, offered a free education for Wilson at the University of Georgia, the Alabama legislature introduced a derisive resolution suggesting that Griffin take over the education of Autherine Lucy in "a package deal."[45] Despite Wilson's loyalty to middle-of-the-road massive resisters, his only support came from the intemperate Ace Carter, who condemned Carmichael and the board. The politicians of massive resistance knew that Wilson had to go. Their reasoning was simple. If Lucy had been the only person expelled since the nineteenth century and if the grounds were slanderous charges against university officials, then to leave Wilson untouched would have been politically unacceptable to a national audience, an audience that had to be appeased. Wilson came to adopt this view himself. He gladly accepted his expulsion as a small price to pay to get rid of Lucy.[46]

Some, perhaps most, in the university family believed the expulsions were a procedurally correct way to protect the institution's authority from insolent attacks. No matter the motives, the public perceived the expulsion of Lucy as "one for the money"; of Wilson, "two for the show." In a letter to Carmichael, Lillian Smith, the Georgia novelist and quiet advocate of human rights, expressed the public view of moderates, but a view that with slight alterations fit massive resisters. She felt bad about Miss Lucy's expulsion, but concluded, "I am glad, very glad, that the trustees tried to balance her dismissal with that of the ring-leader."[47] All, save the most relentless segregationist, could agree. It was, in fact, a balancing act.

Wilson continued to play his role as the darling of the Citizens' Councils. He spoke whenever called and warmed up his audiences with continued attacks on Carmichael and the university. Prospects for a major impact on state politics must have seemed bright on his twentieth birthday, but he was too young to know how young he was. Thrust to the front by the Lucy episode, he receded with the ebbing of that issue and played minor parts thereafter. He eventually became executive secretary of the Alabama Citizens' Councils, a post he held until the dissolution of the state organization in 1969. He sat for a time on the executive committee for the state Democratic party, but in the late sixties and early seventies he switched to the Republicans. He ran unsuccessfully for public office several times. The Lucy affair terminated his education at the university, and by uncorking his political career prematurely, turned his prospects flat.

Autherine Lucy

Reporters now spoke of Lucy's calm and measured tones. A more strident voice would have made it easier for moderate whites to replace shame and guilt with self-justification, but her natural reserve guaranteed that she would become the victim, not the perpetrator. The young coed, who agreed for reasons simple and unfathomable to go with her friend Pollie, became a national symbol of virtue defiled. She embodied in her private nature the public philosophy that Martin Luther King came to espouse. In so doing she helped nudge forward the guilt-actuated gradualism that changed the South and the nation over the next generation.

Her contribution came at a steep price, part of it being the toll it took on her parents. When a reporter for the *Selma Times-Journal* found Milton and Minnie Lucy late in February, the aging parents had no recourse but to disavow their daughter's fight. "Why," Mr. Lucy said, "I keep asking myself, out of all the colored folks in Alabama, did this have to fall to my baby daughter's lot." He said he had tried to dissuade her and added, "Before all this came up, I had plenty of white friends that I could go to their house and knock on their door any hour of the night I needed help, but I wouldn't do it now." Mrs. Lucy said, "I can't sleep and I can't half eat. I'm so worried. Every time a car drives up and a stranger gets out, I'm afraid he's come to tell us something bad has happened." "I wish," she continued, "Autherine would come on home to live with us or else go to some school that's all right for colored folks." When reporters tried to get Lucy to comment on the interview, president Gray of Talladega College referred them to Arthur Shores, who in turn begged the newsmen to have the decency to let her "rest and study."[48]

Lucy's mother showed the effects of Parkinson's disease, and ever after Autherine traced its worsening to the frightful experience her mother endured. Lucy never saw the newspaper interview as a true depiction of her parents' feelings. She always believed her father was proud of her. She knew that she was "pretty to her Daddy" and that "he was always quite proud of me." To the end, wherever she was, he would send her a box of peanuts because "he just got the biggest kick out of watching me eat them, because he knew I loved them, and I still do." The end came for Milton Lucy on December 28, 1960. Autherine's mother died in September 1962. Fifteen years later, Milton and Minnie Lucy's baby girl said, "I'm very close to them."[49]

The harrowing experience of her parents was but one of the burdens Lucy carried as she flew to New York. It was her first plane ride, and as Marshall said, her outward calm was deceiving. On landing, she went to a doctor who gave her a checkup and a mild sedative. She spent the next two weeks happily in the homes of Marshall and a friendly physician. She liked Marshall's personable manner. At one point he told her that he and his wife, who was expecting a baby, would name their child Autherine if it was a girl. She did the normal things any tourist would do. She went to Radio City Music Hall, had her picture taken in front of the United Nations, and took in the sights. The physician gave a tea in her honor and introduced her to a number of famous black leaders. She also endured several press conferences. At the first, she vowed to fight her explusion, but her demeanor undercut her declaration. Marshall took a sheet of paper from his pocket and handed it to her. A reporter described her as "quivering" and "blinking." "She lowered her head, appeared to be on the verge of tears and said: 'I can't read this.'" Finally, after gentle urging, she read the statement "at a rapid pace." NAACP officials called another conference to have her disavow any connections with Communism. From East Germany to Radio Hanoi the communist press played her treatment in Alabama to the propaganda hilt. Flanked by Ralph Bunche, Channing Tobias, Arthur Spingarn, and other notables in the NAACP, Lucy said that she was a Christian and for democracy and that because communists were apparently the opposite, she would do nothing to support them. Voice of America recorded the statement and broadcast it to Southeast Asia.[50]

Lucy had other, more important, things on her mind. She announced her plans to marry Hugh Lawrence Foster, a divinity student at Bishop College in Tyler, Texas, whom she had known since college days at Miles. Her special circumstances prolonged their courtship for what seemed an eternity. They kept up with each other through correspondence and several times set aside plans to marry, but with the expulsion order, he urged that they go ahead. She announced her intentions while in New York. She arrived in Texas about three weeks before the wedding and stayed with a young couple who had befriended her husband-to-be. The Reverend Foster and Autherine Lucy married on April 22 at the St. John Baptist Church in Dallas. The wedding drew a large number of reporters from the press, radio, and television.

Newspaper accounts indicated that she would undertake a speaking tour of Texas for the NAACP. They also noted her comment that the University of Texas might be "an appropriate place" to continue her graduate studies, a notion that drew a chilly response from Texas officials. Roy Wilkins doubted the wisdom of her "making speeches for us in Texas" and sought Marshall's opinion. Marshall apparently did not object, and the new Mrs. Foster was busy over the next months making speeches before local branches of the NAACP in Texas and at least one in Oklahoma. She also spoke at a mass rally at Madison Square Garden where she met Eleanor Roosevelt and again saw Martin Luther King. The New York rally raised money for the bus boycotters in Montgomery and for victims of racial violence in Mississippi. In August, the national youth director for the NAACP urged her to speak in Chicago at a meeting of the National Student Association. On these occasions she received expenses and usually a small honorarium, but the money never was much.

After the Chicago meeting, she had to prompt reimbursement, saying, "if it weren't that this money is needed right away it wouldn't be necessary to bother you."[51]

The speaking engagements and the notoriety faded as news cameras passed to other events. Autherine Lucy Foster settled into being the wife of a minister in east Texas and lived in a number of locations, including Shreveport, Louisiana, when the Reverend Foster took a pastorate in nearby Wackson, Texas. She and her husband made a speaking trip to Philadelphia two years later, but for the most part she drifted out of sight. Lucy once assessed her role in the struggle. "Progress," she noted, "was the product of the hands and the hearts of many persons, not just those who might, by accident, become a symbol."[52]

Aftermath

On April 16, Wayne Phillips wrote Jeff Bennett to thank him for his personal assistance while Phillips worked in Tuscaloosa, Birmingham, and Montgomery. He commended Bennett for his "personal courage, both in the help you gave Miss Lucy in the midst of the riots, and in the way you maintained your personal convictions while faced with the necessity of placating not only a predatory press but a riotous legislature." Phillips also wanted to say thanks for the special attention given both Bob Bird of the *Herald-Tribune* and him, "and I'd like to think that because of that help we were able to give the university a better break than some did." Phillips promised to remember "Tuscaloosa as a lovely university town with some fine and gracious people. . . . Compared to what I found in Birmingham and Montgomery it was certainly an oasis of enlightenment."[53]

Bennett thanked Phillips for understanding that most of our people are "good folk," then mused, "Spring has really come to Tuscaloosa. Our campus is perfectly beautiful in its new green 'clothes.' It is difficult to look across the quadrangle and clearly remember the violence of the past or be conscious of the tension at the present."[54] In fact, from a public standpoint, the university not only found it difficult to remember, officials consciously turned it into a non-event: euphemized in Carmichael's annual report and vanished from the campus yearbook. The rest of the state refused to forget. The legislature continued its paroxysm of resistance. Sam Engelhardt, who said he owed "98 per cent of my living to them [Negroes]," stood on the steps of the capitol and pulled out a sheaf of bills designed to preserve segregation, one of which, authored by Charles McKay of Talledega, made it a high state crime for any white and any Negro "to play or be seated together in any game of cards, dice, dominoes, checkers, baseball, softball, basketball, football, track or in swimming pools." When asked whether there was some room for negotiation, Engelhardt responded: "Negotiate? There's nothing to negotiate. The niggers don't know how far back they're going to go. . . ."[55]

Engelhardt stood for those who desperately wanted to turn the calendar back to before December 7, 1941. For them, the infamy was what had happened to blacks since. Not only did the legislature want to conduct a communist witch-hunt in the NAACP, the attorney general, John Patterson, advanced his own fortunes by subpoenaing its membership rolls to check for communists. When state leaders refused to surrender the rosters, Patterson got an injunction from Circuit Judge Walter B. Jones, an arch-conservative and segregationist from Montgomery. When the NAACP still refused, Judge Jones enjoined them from further action in the state and fined them $100,000. The NAACP went out of existence in Alabama on June 1, 1956, and did not re-emerge until 1964. Alabama's black community remained in considerable disarray until Martin Luther King, Jr., came to Birmingham in 1963.

While politicians postured over the attempt to desegregate the University of Alabama, individuals suffered. Six white thugs assaulted Nat "King" Cole on a Birmingham stage. Though he returned to a ten-minute ovation from an audience of 3,500 whites, he did not finish the performance. That night he gave a full concert for a black audience that packed the house in admiration and support. Cole's assailants faced prosecution only for assault and conspiracy to assault. However, they received the maximum fine and 180 days in jail. Still another casualty of the times occurred in Tuscaloosa. A liberal white judge, running against a hard-core segregationist, yielded to pressure and gave a pro-segregation speech. He went home, confided to his maid that he was being given "a devil of a time over segregation," and put a .22 caliber bullet into his brain. On a train from Birmingham to Mobile, John L. LeFlore, a thirty-five-year-old Negro mail carrier and executive secretary of the Mobile branch of the NAACP, sat up in the smoker and forced himself to carry on a conversation with whites just "to show them we're not afraid."[56] All over Alabama, the lights flickered out.

In September, Bennett sent Carmichael a note and enclosed a clipping that reported Lucy as uncertain whether she would return to the university or perhaps go to the University of Texas. Carmichael wrote at the bottom of the note: "Thanks for letting me see this. It's so hard for her to make up her mind about a matter already settled for her. It is an odd performance."[57] Carmichael was right, the case was settled, though not legally resolved until January 18, 1957. On that day Judge Grooms ruled that the university's trustees, because it had the right to discipline a student, did not hold his court in contempt when its members expelled her. but Carmichael was also wrong. Her performance was never odd.

A year later, the NAACP attempted to sum up its experience with Autherine Lucy. John A. Morsell, an assistant to Roy Wilkins, drafted the NAACP's official version of events in response to an inquiry from an Episcopal minister in Texas. First, Morsell noted, the NAACP needed to make it clear that it had not "recruited" Autherine Lucy. Her [actually Myers's] activities with the NAACP youth chapter at Miles College no doubt "intensified" her interest "in securing her rights," but the NAACP offered no legal assistance until she manifested that interest. Moreover, though Lucy "seemed to us in every way a desirable student prospect, it was also probably that we would not have selected her had we been in the market for plaintiffs as has been charged." Morsell observed that there would have been "tactical advantages in our looking for a younger woman from the top standings of the local Negro high school in Tuscaloosa. That this did not occur should be added confirmation of our bona fides in the matter." Morsell also emphasized that the NAACP never financed Lucy. The NAACP had no involvement in her employment as a teacher and temporarily as a secretary. A private foundation (the Jessie Smith Noyes Foundation) paid for her tuition and fees. Morsell's anger flared when he declared it "almost embarrassing to make note of such minor items as the way Autherine was dressed when she went to register; the make of the automoblile in which she was driven there; and her special treatment instead of being required to stand in line." Morsell offered the standard explanations—she dressed for the occasion, she rode to Tuscaloosa in cars of a variety of makes, and her expeditious registration resulted from a decision by university authorities.

Morsell did admit a serious mistake. "It was an error to charge school authorities with any degree of complicity in the actions of the mob, although even this was not done capriciously. It was an error of judgment based on observations which indicated that the allegation might have substance. Its possible consequences should, however, have been more thoroughly considered, in which case it would not have been included in the brief."[58] In a letter to Lucy, Thurgood Marshall admitted the same. "The last ruling of Judge Grooms [January 18, 1957]," he wrote, "has placed the expulsion order on such technical grounds as to cast considerable doubt on the possibility of a successful appeal." "Whatever happens in the future," Marshall closed, "remember from all concerned, that your contribution has been made toward equal justice for all Americans and that you have done everything in your power to bring this about. Bearing this in mind, we continue to look to the future."[59]

Marshall never put eloquence to more truthful ends, but it was left to Emory Jackson to make the best summation, perhaps because Jackson remembered Myers's as well as Lucy's contribution and remembered both women from the beginning. "Those two friends," he said, "opened doors that will never be closed. In more sober days when sanity has been restored, the University of Alabama's board of trustees is likely to try to make peace with history and square itself with the conscience of humanity." He urged Autherine and Pollie to take comfort in the fact that "they contributed coincidentally to a reluctant, hesitating and resisting University of Alabama to move unwillingly toward full standing in the family of universities of unquestioned status."[60] Jackson's error was in the thought that it would come soon.

6 And Four to Go

Friday, February 17, 1956, was bitterly cold in New York. Icicles hung like daggers from grilles of parked cars, the sky was bleak, and a freezing rain turned to ice as it hit the ground. Thurgood Marshall sat slouched, his tall frame awkwardly assembled in an airplane seat as workmen prepared the Electra for a 4:35 flight to Atlanta. After greeting a reporter who joined him for the trip, Marshall exchanged pleasantries while the ground crew de-iced the plane. Three and a half hours later, Marshall and the reporter landed in comparatively balmy Atlanta. A gathering of regional NAACP officials occasioned the trip. Conversations centered on recent events in Alabama. Old hands praised the pacifism of Montgomery's blacks but wondered whether Martin King's Gandhi-type tactics would work in the face of violent resistance such as occurred in Tuscaloosa. The question of the weekend was: "What would the N.A.A.C.P. do if it found in the future that it won lawsuit after lawsuit and court decision after court decision but couldn't get the decisions put into effect?" Calling it the "sixty-four-thousand-dollar question," Marshall said, "I don't know what we'd do. That's something I can't even contemplate. It would be anarchy. It would be the end of the country. I can't imagine it coming to that."[1]

At the moment Marshall propounded his answer, university trustees devised their plan of expulsion. Marshall would experience shortly what he could not imagine. Not only did the university find a way around Judge Grooms's order, it did so with his acquiescence. Moreover, the university found other ways to discourage and frustrate applicants. They discovered that without the active backing of the NAACP, no applicant could succeed against informal pressure. For almost seven years, those wishing to take advantage of Lucy's and Myers's hard-earned rights would do so without resources sufficient to counter the university's tactics, tactics honed in the six months following Lucy's expulsion.

Ruby Steadman Peters

From late February to early September, the university seriously considered three applicants. The first, Ruby Steadman Peters, wanted to take courses at the university's extension center in Birmingham. No firebrand, she asked to take classes by correspondence or even by television in order to allay the fear whites had of interracial contact. Her sole purpose was to bank credits at the extension center and transfer them to Alabama State College in Montgomery, where she was pursuing a degree in social work. Her plan was simple and, as events demonstrated, naïve. Peters was in her forties. She had separated from her husband some five years before after a domestic quarrel ended in a shooting that injured both. Later, Henry McCain, a prominent Birmingham realtor, befriended her, and they began a courtship that lasted until her death in 1981. While working for McCain's company and catering parties for extra money, she took extension courses offered by Alabama State at the Industrial School in Birmingham. It was a happy time. McCain and Peters enjoyed entertaining in Peters's home at 900 Central Street, and distinguished members of the community, such as Emory Jackson, frequented their dinners.

Jackson and McCain had been friendly rivals in high school. As president of the senior class, McCain won the school's scholarship to Knoxville College in Tennessee—a success for him that determined Jackson to go back to his native Georgia and the more prestigious Morehouse. When Jackson, McCain, and Peters got together, the talk turned to issues of the day but never to Peters's own ambitions. Peters did not plan to integrate the University of Alabama; she wished only to earn college credits in the most convenient way.[2] Though mindful of the turmoil in Tuscaloosa, she believed that personal goodwill would be enough to turn aside hostility. Living in a black world that accepted her as friendly and agreeable, it was impossible for her to fathom a white mind that could not see her as her own race did. She would win whites over as easily as she won the friendship of white bankers and mortgage-holders who did business with McCain Realty. She never sought NAACP support, and Emory Jackson, who made it his business to know everything, never learned of Peters's singular determination.

The university knew and seemed willing to reciprocate her cooperative spirit. During the same February 29 meeting at which the board permanently expelled Lucy, it ordered the extension center in Birmingham to let Peters pursue courses "if she be found qualified in all respects." The only dissenting vote came from the redoubtable Ferguson, and Rufus Bealle communicated the board's wishes to Richard T. Eastwood, the extension center's director. The university probably told Judge Grooms as evidence of its good faith but gave no public notice of the action. Ferguson was right in thinking that an announcement would provoke Klansmen and Citizens' Councilors who had been appeased by the Lucy expulsion and who would renew their threats of violence.[3] The decision to admit Peters signaled a board wrestling with its duty, if not conscience, but the nod to compliance was short-lived. Peter's ambitions would be set aside by the machinations of Ferguson and by the university's lack of planning.

At 5:30 p.m. on Tuesday, March 6, the day after Lucy's readmission was supposed to have taken effect, Eastwood called Peters to his office. In a fifteen-minute interview, he explained the procedures for completing new forms for application. He said that if previous courses were credited from Daniel Payne College and Miles, she would have almost sixty-four hours, the maximum anyone could take from an extension center. Peters still wished to take as many as she could to facilitate her graduation from Alabama State. The next morning, Eastwood was in President Carmichael's office begging for a plan of action. He argued that the Court's ruling made clear the future admissibility of blacks, and he urged "a definite plan to comply." Eastwood believed the university might work out a formula for gradual compliance acceptable to the courts.[4] Three days later, John R. Morton, dean of the university's Extension Division, became even more insistent with Carmichael. Morton feared that Tuscaloosa was dumping its problem on Birmingham. The extension center was more dependent than Tuscaloosa on the goodwill of the community and therefore more vulnerable in the integration crisis. Morton did not wish to thwart Peters's effort and proposed a formula for gradual desegregation similar to her proposal. Morton believed, not without some authority, that the courts might go along.[5]

The pleas of Morton and Eastwood were trees falling soundlessly in a forest. Not only was the university without a plan, its officials were being dragged back into the Ferguson scheme for resistance. Background investigations on Myers and Lucy convinced Ferguson that routine use of private detectives should be a first line of defense.[6] Information so gained could be used in a campaign of dissuasion. The scheme worked quickly in the Peters case. On March 19, Eastwood informed Ferguson that Peters had withdrawn her application. Elated, Ferguson wrote Judge Clarence W. Allgood and James A. Green, president of the Avondale Savings and Loan Association, thanking them "for the excellent manner in which you have handled the negotiations leading up to this happy conclusion. Please convey our thanks also to our other friends, who assisted you in this undertaking."[7]

Green held the mortgage on Peters's house and did business with McCain Realty. The deal smacked of coercion, but the facts suggest something less. Peters was on friendly terms with Green, a product, no doubt, of her easy personality and the fact that McCain Realty did business with him. She never suspected duplicity and took Green's counsel as well-meaning.[8] Green acted at the insistence of Ferguson, but his counsel against her application would have needed no prompting. Green knew the dangers of following so closely on the heels of the Lucy fiasco and so advised Peters. Only Ferguson played the devil's hand. He brought the application to Green's attention and designed the campaign of dissuasion. What role Judge Allgood played is not known, but given Ferguson's propensity to use information from detectives and informants, Allgood and Green likely brought up Peters's domestic past.

117

Whatever was said and however it was received, Peters knew who was talking and what failure to comply could mean for her and for McCain.

Ferguson's methods soon became *de rigueur* for the board and university officials. No one talked openly about how to accomplish the inevitable desegregation. Even those who knew that threats of violence should not abrogate the rule of law got caught up in extraordinary efforts to please the chief resister. One ugly act made the next easier, until good men found themselves committed to unconscionable acts they would sooner forget, and in fact have. Though racked with private doubt, they could not imagine a receptive audience for planned desegregation. Taking counsel of their fears, they placated radical racists. The applications of Billy Joe Nabors and Joseph Louis Epps were to show that complicity with Ferguson made evil-doers out of otherwise good men.

Allan Knight Chalmers

The story of Nabors begins with Allan Knight Chalmers. Chalmers, former pastor of New York City's Broadway Tabernacle and a member of the faculty of Boston University's School of Theology, had long been a diplomat of the civil rights struggle. He was no stranger to Alabama. Appointed to chair the Scottsboro Defense Committee in 1935, Chalmers spent the better part of the next decade working with Grover Hall, Sr., of the *Montgomery Advertiser* and other Alabamians to secure the release of the Scottsboro boys. Chalmers maintained a persistent, almost perfervid, optimism.[9] At the time he entered the University of Alabama case he chaired the fundraising Committee of 100 for the NAACP Legal Defense Fund. His other connection to Alabama was as a trustee of Talladega College, whose president, Arthur Gray, actively promoted the Alabama Council on Human Relations.

In early March 1956, Chalmers wrote Gray about a recent weekend in Alabama which he spent "occupying [him]self fully in the Montgomery and then eventually the Tuscaloosa end of things." He was especially pleased to meet several white ministers who seemed ready to help. "As an indirect result of my having seen some of them," he continued, "one of the trustees of the University of Alabama, who became aware of my presence in the state, wishes to talk with me because he likes what he knows about my attitude and methods of working. . . ." Chalmers did not name the trustee, but he believed a meeting could be arranged. A week later, in another letter to Gray, he confirmed that the ministers would pick him up in Talladega at noon on Saturday for a conference with "top people in administration and in University circles" and would take him to Tuskegee in time for him to speak there on Sunday. He planned to spend Sunday afternoon and Monday morning in further negotiations. Talladega, he concluded, by serving as a "cover" for conferences with university officials, "is thus able to make a real contribution . . . in this quite unexpected and informal way."[10]

Talladega already had contested segregation at the Capstone more directly. Gray, the first black president of the college, proved his courage by allowing Lucy to rest on campus in the wake of the Tuscaloosa riots. Among those who kept vigil over her then was Billy Joe Nabors, a senior, who had given serious thought to attending law school at Tuscaloosa. He had requested a bulletin in the fall of 1955 and on March 12, 1956, wrote for an application. At the same time, he kept his options open by applying to law school at Howard University. He was just the kind of young man, or so it seemed, to succeed where Lucy had failed. Nabors's ambition and Chalmers's plan for desegregation now intersected. A young black man and a veteran of the human rights struggle would take the walls of segregation by burrowing underneath.

A visit to Tuscaloosa

Chalmers got things rolling by trying to convince the university's administration to trust him to find the "right" applicant. He rounded up three white ministers in Alabama—Dan Whitsett, John Rutland, and Charles Prestwood—and went straight to Carmichael's office. The three were Methodists who had already demonstrated courage by speaking out for interracial cooperation in achieving desegregation. Whitsett, who pastored a church in Sylacauga, had been president of the Alabama Council on Human Relations, and Rutland, whose Woodlawn congregation included Bull Connor, had run afoul of that arch-segregationist more than once. Prestwood was assigned a church in Eutaw and already had a reputation as a "nigger-lover" for sponsoring biracial meetings and encouraging a black to run for office.

Surrounding himself with such people typified the way Chalmers operated. He inspired hope in dark hours. One night, after Klansmen burned a second cross in John Rutland's yard and after the Bishop called to say he had gone too far, the young preacher went to bed thoroughly discouraged. Early the next morning he woke to a ringing phone. The caller said, "John Rutland, this is Allan Knight Chalmers. Would it be helpful if Dan Whitsett and Charlie Prestwood and I came over to see you?" Rutland almost cried with relief. His delegation complete, Chalmers arrived at the university Saturday morning, March 3, three days after the expulsion. Carmichael and Jeff Bennett welcomed them. Prestwood, who would leave for Albright College in Reading, Pensylvania, when things finally got too hot, proved the most insistent. He pressed Carmichael for action, saying at one point, "If you had enough courage, you would do it." Chalmers, ever the diplomat, stepped in to calm Prestwood. Speaking softly, he cautioned that "maybe we need the courage to back off and look at their point of view." Chalmers then spoke of Carmichael's accomplishments. ("Chalmers was prepared," Rutland thought, "I wasn't.") Chalmers gently urged Carmichael to work with the ministers and with the people at Talladega College to find the right candidate for admission. He used all his "tremendous personality; tall, wavy hair, eyes that would just bore through you, a twinkle in his eye; and the courage of a lion; but soft spoken. . . ." Moved by Chalmers's performance, Carmichael asked the minister to pray before they left.[11]

Chalmers thought it possible to avoid more litigation. Thurgood Marshall and Arthur Shores, he told Carmichael, could sit this one out. Carmichael promised Chalmers not to impede the progress of the selected candidate.[12] When Chalmers arrived back in Boston a letter awaited him from one of his young lions in Alabama. Dan Whitsett spoke for the rest when he said, "One telephone call, one visit, one letter can mean a great deal to those who are facing difficulties." To the man who had left his prosperous New York congregation for Boston University so "he could combine a radical conception of Christian social ethics with instruction in preaching and in the tactics and strategy of the minister's life in the community," such words amounted to confirmation. No doubt Chalmers had his brave young ministers in mind when at the end of the year he wrote, "Always in any bad situation you will find good people who are on the edge of doing necessary things, even if dangerous, to correct injustice."[13]

Billy Joe Nabors

The central figure in the drama, Billy Nabors, was born June 28, 1933. He graduated from Talladega's West Side High School in 1952. The fourth of ten children, his father, a pipe shop worker, was frequently unemployed, and his mother worked as a maid. President Gray said that Nabors had been "personally and almost solely responsible for meeting his educational expenses at Talladega. Each year he has left school with heavy arrears but has invariably made payment to the College at the end of the summer." Despite this burden, Nabors's activities on campus were the envy of all. He was president of the NAACP chapter for two years, represented the college at a national NAACP conference, was active in the YMCA and Little Theatre, served as president of his fraternity and vice-president of the senior class, and chaired the College Council Community Cooperation Committee (a group seeking to improve town and gown relations). He led a membership drive on campus for the Alabama Council on Human Relations and joined the debating team. Given all of this, there was no surprise when Dr. Gray observed that Nabors's grades were only "good." "Only his intimate acquaintances can know how he has managed against such odds."[14]

Nabors was yet to encounter his fellow resident of Talladega, Brewer Dixon. On May 15, Dixon picked up a copy of the *Birmingham News* and learned for the first time of Nabors's intentions. He chided Dean Adams for having had to learn about Nabors from the papers and asked for a copy of the application. When Dixon asked whether Nabors had received the necessary endorsements from practicing attorneys in the county (one of the impediments designed to eliminate Negro applicants), Dean Adams replied that

Nabors had acquired elsewhere affidavits as to his moral character. Where-upon Dixon began his own background investigation. Dixon heard precisely what he wanted to hear. One informer said that Nabors had "a bad reputation as a malcontent while a student in the City High School." More alarming, and therefore more to Dixon's liking, was a story told by John Norman of the Quality Laundry. While Nabors was the laundry pick-up and delivery boy on campus, Norman accused him of embezzling some $75 paid to him by students but not remitted to the laundry. Norman said that President Gray offered to intercede rather than have a student at the college involved in a criminal prosecution, whereupon Nabors began remitting $10 a month but failed to follow the schedule until just before graduation when he paid the balance in full. Dixon believed "the fact that he paid the balance in no wise mitigated the crime." He was convinced that Nabors paid only when made aware of the "local investigation" into his background and when threatened with "probable indictment by the grand jury."[15]

Dixon's discoveries threw a wrench into the informal machinery set up by Chalmers and university officials. On July 27, Chalmers, along with Whitsett and Rutland, was back in Carmichael's office. Carmichael assured Chalmers that he still wanted "to have a qualified Negro student accepted by the Law School." However, Carmichael insisted that the allegations uncovered by Dixon had to be cleared up. Chalmers agreed to set the facts straight. He talked with President Gray and another trustee of the college, both of whom "had felt so strongly about the qualification of Nabors that they had already granted him a fellowship for study at the university of his choice." A week later Chalmers assured Carmichael that he had also talked directly with Nabors and received the following explanation about the laundry charge: "Near examination time [December 1955] I found that I could not continue the job. At this time I had let out about $50.00 worth of clothes for which I had to account to the laundry man. I told him the story and assured him that I would have him paid before school closed. I have a receipt showing that the debt was *paid in full* on June 2, 1956."[16]

Nabors contended that the $50 was not for personal benefit but resulted from extending credit to fellow students. Mason Davis, who would figure in Chalmers's plan a little later, also picked up and delivered for a laundry and worked out similar lines of credit.[17] The facts, however, did not matter because neither Nabors nor his supporters played to an audience that wanted explanations. On August 7, Carmichael wrote Chalmers a "cordial letter" in which he stated "that Mr. Dixon's 'conviction' about Mr. Nabors' unsuitablity had to be cleared up directly with Mr. Dixon, not only on the laundry situation but on certain other rumors going back to conduct in High School." Chalmers did not understand that Carmichael was washing his hands of the case. Perhaps he was an aging diplomat, the victim of too many skirmishes, compelled by years of frustration to take gestures for substance. So Chalmers urged President Gray to set up a meeting with Dixon. It is not known whether the meeting took place, but it is clear that Dixon decided Nabors's fate. In addition to the laundry charge and the allegations about conduct in high school, Dixon also learned about a shouting match just before graduation between Nabors and the college's acting comptroller (a white man). During this confrontation Nabors allegedly assaulted the comptroller.[18]

Apparently Nabors did have a quick temper, and on that score began to lose the support of President Gray. During August, Chalmers became increasingly aware that there might be a problem. On August 22, he wrote Nabors assuring him that everything was moving as expected and compared the young man's situation to that of Jackie Robinson. "You have something of his fire and I hope you also have some of his intelligence. He knew he had to keep himself under tight rein for all personal things. . . ." Chalmers suggested that Nabors talk it over with President Gray and Arthur Shores "but do remember that our job this time is one of persuasion and we must not mix up that method with legal pressures, particularly of the 'chip on the shoulder' variety." Chalmers remained optimistic about Nabors's chances. On a blind copy of his letter to Nabors, he scrawled a note to Gray. "Billy's answer to Dixon's objection was *full* and reasonable. Hope it turns out that way in your mind too."[19]

Nabors clearly did not push his own case at the expense of the overall cause. He assured Chalmers as late as August that he was "willing to yield wherever and whenever necessary. I have no intention of impeding progress." He had already been accepted at Howard University. Moreover, there was another candidate at Talladega whose record, according to Gray, was "less vulnerable." His name was Mason Davis. He was an attractive young man with excellent grades and wide-ranging extracurricular interests. The only problem with Davis was that he seemed less interested in pressing the cause than in furthering his education. He had already been accepted by the University of Buffalo Law School, where he would establish a fine record.[20] (Davis, partner in a distinguished Birmingham law firm, is today an adjunct professor in the University of Alabama Law School and member of the President's Cabinet).

By the end of August, Chalmers realized the situation was hopeless. President Gray had withdrawn his support of Nabors, and Davis was not interested. Gray's turnabout vexed Chalmers. "I appreciate your reluctance to share it with me," he wrote, "but wish you had felt able to tell me in July when I pressed you for reasons. You were so indefinite then that I wondered if it would turn out to be a 'rumor' as in the case of the laundry affair." The "it" referred to Nabors's temper. "We can not afford," sighed Chalmers, "to have a person in that situation who is not in full control of his emotions & actions under pressure."[21] A Jackie Robinson was needed; mortals would not do.

Thirty years later, one of the hopeful young ministers could see the futility of the last meeting between Carmichael and Chalmers. Reminiscing about that hot day in late July, when he and Dan Whitsett again accompanied Chalmers to Carmichael's office, John Rutland remembered that "it was about the same as the first meeting; except . . . even then the Board of Trustees was putting heavy pressure on Carmichael. . . . I know he [Carmichael] was more cautious, and he was real careful, and he had somebody to take down what we were saying. And he was very, very cautious about it. I left the first meeting admiring him greatly. I left the second meeting with a question mark; wondering, really, if he were doing what his conscience told him to do." In this last meeting, Carmichael turned to Bennett with embarrassing frequency to say, "You answer that."[22]

As the ambassadors of goodwill met for the last time with Carmichael, Brewer Dixon enjoyed a much needed vacation at the Cloister on the Georgia coast. The next day, Saturday the 28th, Chalmers was in New York seeing the Broadway production of *No Time for Sergeants,* while Carmichael made his report to the board, minus Dixon.

Good people

Such a bewilderingly sad end to diplomacy was predictable. It had never been more than a phony exchange, but too many people had invested too much to think otherwise. At stake was personal esteem itself and nowhere more evident than in the character and action of those closest to Carmichael. They were men and women of unusual ability. They not only knew the law, they had a sense for what was right. Three key figures in the administration, Jeff Bennett, Sarah Healy, and Louis Corson, were also members of the Canterbury Episcopal Chapel, and as such engaged each other and kindred spirits in discussions about how to make desegregation work. Their chaplain, Emmet Gribbin, already had proved his courage before the mob. Together they formed a proto-human relations council in Tuscaloosa. During the February crisis, they met often and talked long about what they could do to restore the good name of their university and to serve the interests of justice.

So taken was the Reverend Gribbin with their conviction, their courage, and their goodwill that he came up with the idea of offering himself as an intermediary between them and legal counsel for the NAACP. He believed that if only the NAACP could know the good intentions of these university officials, a relationship of mutual trust could be built. Like Chalmers, Gribbin took heart in the notion that through Christ all things are possible. In that spirit he wrote an acquaintance, John Burgess, the Episcopal Chaplain at Howard University. In his letter he spoke of the good people at the university and of their desire to see the problem through to an effective and lawful resolution. Gribbin offered himself as a person who might bring the contending parties together and suggested that Burgess use his connections with Thurgood Marshall to get things moving.

Gribbin spoke the best intentions of those who surrounded Carmichael. "We are discussing these things night and day here," he confided, "and I feel I am giving you an accurate picture of the attitude and convictions of those here who sincerely want Miss Lucy to come back."[23] Burgess's reaction penetrated the dark glass of good intentions. "I can understand," he wrote, "that because of your association with several members of [the] Administration, you know their sincerity and integrity. However, you must try to realize that such an acquaintance with white people is impossible for Negroes. The N.A.A.C.P. has no reason to suppose that the officials in Alabama are any different than officials in Mississippi, Georgia or South Carolina. And you certainly know that things have been less than honorable in those states."

Without flinching Burgess spoke his own "truth in love." "It is one thing to call on Christian motivation in this situation," he charged. "It is another to realize that within the area of race relations, Negroes are not accustomed to find white people willing to bring the terms of the Gospel into the consideration of the subject." Saying that Gribbin was "a *white, Southern minister,* and each word is a strike against you," Burgess cushioned the blow only with an acknowledgment of Gribbin's "courageous witness." "But I imagine most Negroes, while admitting the strength of your witness now, would say that they doubt if you would carry on when the pressures get too great and the chips are down. And this cynicism is born of bitter experience."

Burgess had seen enough in neighboring Virginia to convince him "of the clay feet of 'men of integrity.' " He did feel, however, that in the "series of tragedies" that lay ahead "Christian forces have one of the greatest opportunities in history to demonstrate the power of love in softening the blows that will be struck and in healing the wounds that will be incurred."[24] Burgess was right, of course, but it would be as wrong to applaud him for perspicacity as it would be to condemn Gribbin for hoping against hope. Each played a role in a drama that had many roles ranging from the beatific to the diabolic. Burgess was also right in his belief that tragedies lay ahead. One of the more painful would be played out by a man of integrity, one of Gribbin's own parishioners, Jefferson Bennett.

Joseph Louis Epps

As miserable as was the university's conduct toward Peters and Nabors, it did not compare with the treatment given Joseph Louis Epps. Epps grew up in Dolomite, a mining community on the northwestern edge of Birmingham. He established an enviable record at Westfield High School and completed two years of pre-med at Morehouse in Atlanta. He knew Emory Jackson, and Jackson encouraged his application to the university. Epps first applied in September 1955, but failed to send his transcript. On July 31, 1956, his mother wrote President Carmichael in behalf of her son. Carmichael responded on August 2, saying that he should write to the Dean of Admissions, which he did. On September 4, Epps wrote Carmichael asking about the delay. University officials knew they had a serious application. Epps's record at Morehouse was beyond reproach, and privately Carmichael acknowledged him as the most qualified applicant to date.[25]

Epps's application coincided with mounting pressure on Carmichael about his handling of the Nabors case. In a letter to Hill Ferguson, September 5, Dixon criticized Carmichael for calling a board meeting to decide the Nabors case and for having expected him (Dixon) to conduct the investigation. He believed that the university had the resources to handle investigations without involving members of the board. Ferguson endorsed Dixon's view and in a letter to Carmichael said that Nabors's "application should be denied by Dean Adams as a routine matter and not dignified to the extent of making a Board case of it." In a testy reply, Carmichael said that summary rejection of Nabors would result in a suit and therefore should come before the board. Moreover, Dixon's letter gave "the impression that we had asked for a meeting . . . to consider only the Nabors case. As you will, I am sure, recall the Epps case bothers us considerably more. . . . Mr. Bennett has made a preliminary investigation of applicant Epps without being able to find any evidence that would warrant rejection of his application." Carmichael felt that if the board were meeting to discuss Epps, it might as well decide Nabors, especially since the board had requested that all cases involving Negroes come before them because of the possibility of lawsuits. "Indeed," he observed, "our attorneys in the Lucy, Hudson [Myers], and Peters cases dictated the wording of the reply to be made by the Dean of Admissions."[26]

Carmichael's irritation did not mean that his administration would be less diligent in discouraging Negro applicants. If anything, it was prepared to dig deeper than Dixon had dug in the Nabors case. On the same day that Carmichael sent his hot note to Ferguson, Bennett began keeping a log of his efforts in the Epps case. Under the heading 9/6/56, Bennett recorded that he had made first contact with Sam Christian, a detective hired to investigate Epps, and with Dr. James Sussex, chief of psychiatry at University Hospital in Birmingham. Sussex was treating Ruby Anna Epps, Joseph's mother. The next day Sussex called Bennett to say that he had talked with the mother who had just been visited by Emory Jackson. She said she did not want her boy to go to the university, that she expected him home tomorrow, and that she would bring him to Dr. Sussex "if necessary to keep him in Atlanta." Bennett also noted conversations with Christian and Ferguson.

On the 9th the university got a letter from Epps *"demanding* action," whereupon Bennett went to Birmingham to see Sussex. Sussex got in touch with the mother, who agreed to have her son call when he arrived. Meanwhile, Sussex requested a delay or postponement of the application. The next day Mrs. Epps reported that her son had arrived. She told Sussex that her son had agreed to go back to Atlanta and made an appointment to see the doctor with her son the next day at 1:00 p.m. But at 2:30 the boy came in alone. Bennett's notes create a graphic sketch of what happened: "[Epps] had over one hour w/Dr. Boy uncertain what to do. Word is out among NAACP.

[Emory] Jackson and others putting heavy pressure. Tried to keep him from seeing Dr. Mother shifts from one side to other—typical of paranoid. Dr. impressed on boy that his admission might put mother in asylum. Boy agrees to give decision to Sussex at 8 tonight."

At 10 o'clock that evening, Sussex had not heard from the boy. He called Epps's mother "on pretext that he had been out," and she may have tried to reach him. Sussex learned that Epps had been with Emory Jackson since leaving the doctor's office. The next day, Sussex reported that "he had called the boy at home" and that Epps had pledged to go back to Morehouse but said he would not withdraw his application. Sussex speculated "that Jackson has gotten the boy to leave door open, hoping to change his mind over weekend." Bennett then observed, "We can hope he doesn't, but should be ready with decision if he does."

Epps pushed no further. If he had, it would not have mattered. The board was not about to admit a Negro, and Bennett had raked through enough muck to satisfy a Brewer Dixon, even a Hill Ferguson, that university officials had risen to their task. Other notes gleaned by Bennett revealed that the mother was "obese, 43–44 yrs of age & looks like the 'Aunt Jemima type.' " She was proudest of "Joe L.," who was well behaved and did not drink or run around. The father, Willie James, was a coal miner, a "good man" but "runs around." From the private investigator, Bennett learned that a white man described the Epps family as an "average negro family, the mother is crazy & police quiet her from time to time." The investigator had interviewed two Negroes: "one of these was chased by mother w/knife—1 negro lives close by—mother's crazy—drinks liquor in front yard & curses—doesn't care about what children do."[27]

This mother, who reportedly did not care about her children, had written Governor Folsom on August 5 asking him to intervene. She told of her letter to Carmichael and of his "most promising note" in reply. She also told of a letter to Superintendent Austin Meadows who had suggested "a college in the East" with state aid. She already had asked for that and been refused. She told Folsom that she had lived a Christian life and raised her children to do the same. Mrs. Epps assured the governor that she opposed "mixed marriage," but supported "democracy." "I believe in what is right," she explained, "but to deny my son the right to a good education is hard to bear." "Stick to your guns," she urged. "God can't use no cowards in his kingdom."[28]

Reflection

Bennett's notes record the actions of a thirty-five-year-old man at work for his university and his career. He was a moderate man doing the bidding of those in power; not his boss, Carmichael, who by that time had all but abdicated, but Ferguson and Dixon and McCorvey and all the alumni who equated civilization with white supremacy and believed their university should be its conservator. Bennett received a letter from F. Edward Lund, president of Alabama College at Montevallo, Carmichael's old post. Bennett considered Lund a mentor. Lund wrote that when they last met, Bennett looked "if not depressed, at least harassed." He predicted bloodshed ahead and thought the time at hand "when your counsel of moderation can be heard." "Meanwhile, you should not lacerate your own anxieties or guilt too much. It will only play hell with your judgment."[29]

Bennett was grateful for the words and supposed that "the reason for my appearing harassed is the knowledge that the motives of the President and myself are questioned by people to whom we should be able to look for moral support and sympathy." At moments like this Bennett may have wished for a cleric's collar so that his heart and his mouth might say the same thing. But he was ambitious and savvy enough to know what could and could not be done if one wished to get to the top or even hold one's grip. Moreover, if Epps had to be sacrificed, Bennett could still press for some future candidate. Like the good Reverend Chalmers, Bennett could hope for next year, and if not next year, then soon. In fact, Bennett would be there in the end, seven years later. In October 1959, he wrote to Lund, then president of Kenyon College. Bennett was happier, certainly confident in his expanded role within the university. "Our only unsolved problem," he concluded, "is 'the problem' which for the time being is lying fairly dormant."[30]

Carmichael would not be there in the end. His days were numbered. His principles, too, had taken a back seat to exigency, but unlike Bennett he had no future, no time to justify the sacrifice of conscience or to rectify the moral balance sheet. He was a man drifting in time, cut loose from all the certainties of earlier ambition, and now able to perceive only dimly the consequences of his actions. Carmichael had lost his grip, and he would go.

Carmichael

When Carmichael decided to leave is not clear. His brother said that from the beginning he planned to take a Ford Fund offer to study higher education in British Commonwealth and English-speaking countries. In December 1955, exhausted by the desegregation turmoil, he considered a leave of absence, but Steiner talked him out of it.[31] Nine months later, correspondence revealed how hopeless his presidency was. A letter to Steiner acknowledged receipt of a bill for $418.90 from the Bodeker Agency for investigating Epps. "I did not know that the agency had been employed," Carmichael protested lamely. "I agree, however, that since the chairman of our Executive Committee [Ferguson] arranged for the employment of the agency we should make prompt payment." Steiner, too, claimed no knowledge, but objected only to the amount, thinking it too high. Carmichael's professed ignorance is doubtful. The president's chief assistant knew. Bennett received reports and conversed regularly with Ferguson. On the same day Bennett made contact with the private investigator, Carmichael referred to Bennett's preliminary investigation of Epps, in part to disprove Dixon's allegation that the university was lax in ferreting out information about Negro applicants. If Carmichael had trouble seeing or acknowledging his complicity, Ferguson suffered no ambiguity. When told that the bill was high, he replied, "I am satisfied the Bodeker Agency did good work in helping us achieve the desired end, and from that point of view, the expense is very nominal."[32]

Uncertain of his authority, Carmichael seemed to drift. He failed to devise any plan. His brand of moderation never had a chance because he staked out no center to which people could hold. The best that could be said came in his own words. "We have sought honestly and earnestly to find the least undesirable of solutions for each separate problem that has arisen." When Harvie Branscomb, his old wartime friend who succeeded him as chancellor at Vanderbilt, offered the services of the Southern University Conference, Carmichael turned the offer aside. "I have no idea what the Southern University Conference could do that would help in this situation. Indeed I am quite confused as to what anybody could do except to start a long process of education."[33]

Lonely

Carmichael came to see his travail as intensely personal. Emotionally, he was always a private man. To one acquaintance, he wrote, "I have, of course, been greatly humiliated by the occurrences here." To another, he confessed, "You can scarcely imagine my chagrin and humiliation over it."[34] Carmichael's burden distressed his secretary to the point of tears. Looking out her window from the Administration Building, she watched him walk to and from the president's mansion in growing isolation and despair. His only friends were those who worked for him and who saw on a day-to-day basis his compassionate, quiet dignity. Years of success outside Alabama left him with confidants around the country but no one close by to share the almost daily disasters. He loved golf, but since coming to Tuscaloosa there had been no companions, and now the loneliness seemed oppressive.[35] His personal code left him no option but to bear the slings and arrows in silence.

Over the next months, faculty, distinguished associates from all over the country, and members of the board all voiced their lack of confidence in him. Faculty members held faster than most. They had a natural sympathy for what they came to see as their president's martyred administration and were more prone to read his behavior as he read it. Still, they urged leadership. At a University Council meeting on February 8, faculty members began to speak up. Carmichael listened quietly, penciling occasional notes on a copy of the agenda. York Willbern, a highly regarded political scientist soon to depart for Indiana University, spoke of the faculty's frustration. John Pancake, a young historian who divided his teaching time between the university and Stillman, was thought by Carmichael to be one of the "leaders of the disturbing group" on campus. Pancake insisted on clear signals from the top. Even Sarah Healy talked about the insufficient police action on the day of the riot.[36]

Many on the faculty found themselves caught in a professional whipsaw between stands taken by their academic disciplines and university policy. In May 1956, the department of psychology addressed a letter to the president of the American Psychological Association endorsing the association's statement of ethics which condemned segregated education. They hoped a letter in the *American Psychologist* would give sufficient evidence of compliance with the association's code, while avoiding the personal and professional dangers of wide publicity. In July 1955, Leigh Harrison, Dean of the Law School, made it clear that "in view of the Supreme Court decisions which directly affect state university law schools, and the necessity of maintaining our relationship with the Association of American Law Schools, . . . an early reconsideration of the University's policy in regard to the admission of Negroes to the Law School should be undertaken." Harrison also worried about the university's loose construction of non-discrimination agreements in federal contracts.[37]

Faculty protest took many forms, from the open protest mounted by Charles Farris in the faculty meeting of February 7 to secret cabals offering aid to the NAACP lawyers.[38] Yet opposition to Carmichael was more the exception than the rule. Many of the faculty shared historian Frank Owsley's view that dissidents, especially members of the political science and psychology departments, were "screw-balls." But even those who supported Carmichael confessed bewilderment. Perhaps the best summary of feeling came from John Pendleton, the man Carmichael brought from Virginia to conduct the institutional self-study. "The faculty . . . feel somewhat frustrated because they would like to be of help to you but do not know quite how. I think it is not only important to keep them informed of what is going on but more particularly to tell them what you would like them to do. I believe they would welcome such guidance."[39]

Forsaken

For a man whose personal code could make a virtue of being misunderstood, declining faculty support may have confirmed Carmichael's sense of rectitude. Less confirmatory was his loss of reputation among the nation's elite. Having developed a currency among the best and the brightest upon which he traded from time to time, Carmichael, more than ever, needed assurance, but it was not to be.

The university had planned originally to kick off its 125th anniversary celebration on either February 7 or 9, 1956. After Arthur H. Sulzberger, publisher of the *New York Times,* and Grayson L. Kirk, president of Columbia University, declined Carmichael's invitations to address the opening convocation, the university changed the date to April 14—a fortunate change as the events of early February were to prove. In the meantime, Carmichael, working through Robert B. Scott, Jr., chairman of the chemistry department, had secured Linus Pauling to open the ceremonies, but that, too, was before Black Monday. Now, on February 11, Pauling wrote Scott saying that he had planned to talk about "the alpha helix as the principal way of folding polypeptide chains in proteins, and also would discuss abnormal human hemoglobin in relation to diseases such as sickle cell anemia." Moreover, he would have mentioned Professor Herman Branson (his collaborator on alpha helix) and Dr. Harvey Itano, who had done most of the work on hemoglobins and sickle cell anemia. However, Pauling said, ". . . my reference to these non-Caucasion coworkers of mine, . . . , might possibly be considered provocative, and I would not want to be in the position of being provocative in this matter." Pauling said that coming to Tuscaloosa "might indicate that I approve the actions that have occurred, whereas in fact I do not approve."[40]

Scott responded immediately, expressing both regret and shame and asking the distinguished scientist to hold his decision until he heard from Carmichael. This exchange occurred at the precise moment Carmichael gave evidence of assuming control in Tuscaloosa, and his letter urged Pauling to reconsider, assuring him that "though we are caught in a very bad spot we are determined that right shall prevail and are making progress." Carmichael promised a resolution of matters in the next two weeks and asked Pauling not "to add to our burden." Pauling waited two weeks. His response was terse: "The *Pasadena Star-News* of 1 March 1956 carried the statement that Miss Lucy had been expelled by the University of Alabama, after Federal Judge Hobart Grooms ordered her readmitted to the University. This action by the University of Alabama clearly makes it impossible for me to take part in your celebration."[41] A more understanding Robert Calkins, president of the Brookings Institution, finally agreed to speak, but Carmichael had unmistakable evidence that his national reputation was damaged.

Abandoned

The crippling blow was delivered by trustees. Not that the relationship had ever been good, but Carmichael understood these men and knew their institutional value. A lack of trust had been apparent for some time, even to those board members who defended the president, but now the breach widened. Whatever else might be said of Ferguson, he had little patience with those who did not prosecute their causes. He clearly believed Carmichael had no stomach for a real fight. His frequent letters to Carmichael cunningly talked of their mutual agreement concerning segregation, but the show was designed more to box Carmichael in than to stress agreed-upon principles. Following the February crisis, Ferguson dropped pretense and displayed his lack of respect without caution.

When the Citizens' Councils renewed old attacks on Carmichael for his 1954 Miami remarks and for his participation on a 1947 commission on "Higher Education for American Democracy," Ferguson wrote Carmichael, duly noting Carmichael's previous evasions and rebuttals to the charges but conveying at the same time as much regard for these public denials as Marc Antony for Brutus's honor. Ferguson concluded, "I think it unnecessary for you to answer this sheet, but I do think our board might give consideration to making a statement, that the board itself assumes full responsibility for the admission or non-admission of negroes to the University, and that this is not a responsibility of Dr. Carmichael or his staff."[42] Ferguson restated established policy, but the reiteration, coupled with the litany of what he considered the president's missteps, made unmistakable Ferguson's belief that Carmichael was a liability.

Disaffection soon erupted among other members of the board. The question of academic freedom caused the stir, an issue for which Carmichael, to his credit, showed more enthusiasm and leadership. Not surprisingly, the faculty came under fire from massive resisters who wanted the university to get rid of known integrationist sympathizers. Carmichael met with other university and college presidents in the state to develop an acceptable statement defending the freedom of faculty and staff to discuss, reasonably, issues arising in the classroom.[43] Timidity finally killed this effort, but the University of Alabama faculty had been working with Carmichael and his staff to draft its own forward-thinking proposal about academic freedom. It came as part of a statement of "Basic Aims and Objectives of the University," which Carmichael forwarded to the trustees in May 1956.

The "Aims and Objectives" document itself was part of a program of institutional self-study and evaluation, required for re-accreditation. It represented the kind of planning for which Carmichael showed the most talent and gave him a chance to play his long suit. He did not expect the storm about to break, nor did the blast originate from Ferguson. On June 2, Steiner expressed shock that the board had unwittingly approved the "Aims and Objectives" by accepting the report which contained the document. Carmichael responded that the board had merely received the report, not ratified it, and that the document simply encouraged the faculty to think about broader aims and objectives—a healthy and desirable process.[44]

Not mollified, Steiner acquired an ally in fellow Montgomerian Judge Lawson. Lawson objected to any statement on "academic freedom" that did not involve board representation in the process. He feared "a political crisis the like of which the University has not encountered in more than fifty years"; after all, the university belonged to neither board nor faculty, but to the people whose "prerogatives" the document usurped. The following day Steiner fired off a three-page letter with copies to all board members. He charged that some faculty were "more conscious of and ready to assert their freedom to think, teach and say what they please than ever before and [now wanted] to effect a tenure contract or agreement by which they cannot be censured or their utterances or teachings censored, or the members disciplined for such utterances." Steiner called it a "spirt of rebellion" touched off by the "recent Lucy trouble."[45]

Carmichael came as close to exploding as was possible for him. It was one thing to bow to the trustees on racial policy; quite another to accept their preachments on affairs for which they demonstrated profound ignorance. He scolded Steiner for alleging a " 'spirit of rebellion' in the faculty." The actions of two or three faculty members did not constitute "a rebellion, particularly since their colleagues promptly voted them down." Carmichael deliberately understated faculty dissension, but he was after Steiner now and, through him, the board. Carmichael defended the statements on academic freedom as consistent with basic trends in higher education and sent copies of his rejoinder to board members.[46] Before Brewer Dixon received Carmichael's reply and three days before he launched his investigation of Nabors, the Talladega attorney had written to support Steiner. He felt that "no one would wish to unduly limit the rights of an individual as to his private ideas and beliefs, but in view of the fact that the State of Alabama has a direct interest in the University and its teachers, I feel that those members of the faculty who have been guilty of teaching and acts which are opposed to the policies of the University Board on this negro question should go. And there is a pretty definite feeling throughout the state," he added, "that we have several such members of our faculty."[47]

Dixon's attitude, like that of Steiner and Ferguson and McCorvey, was expressed best in Judge Lawson's reply to Carmichael. Lawson had read Carmichael's rejoinder but was unwilling "to abrogate to the American Association of University Professors or any other similar group the responsibilities and duties which I understand have been placed by the Constitution of this state on myself and other members of the Board of Trustees of the University."[48] Lawson confused his duty with his personal opinion, a confusion typical of the conservative mindset on the board. For these men, yielding to the judgment of outsiders (opinions with which they did not agree) amounted to abrogating the constitutional procedures by which they exercised authority. Notwithstanding Judge Lawson's professed belief that the university belonged to the people, it had become an institution limited to and by its board.

Denouement and déjà vu

There was nothing left for Carmichael but to play out his time. As early as April 1956, he doubted that he would be around another year.[49] Many believed his troubles owed as much to losing football games as racial turmoil.[50] But to influential alumni the central theme was race. In the wake of the Lucy episode, Wallace Malone, then chairman of the board of the First National Bank of Dothan and later one of the state's most powerful bankers, wrote a letter to the *Tuscaloosa News*. He said that integration was "a first step toward the amalgamation, miscegenation and mongrelization of the two races in our state," and he laid blame at the feet of institutions like "the Carnegie Foundation, which, according to Senator Eastland's undisputed published statements, paid for the publication of the Gunnar Myrdal book, upon which the Supreme Court admittedly relied in its recent unconstitutional decision about our schools." A month later, Malone wrote Carmichael asking for assurances that Carmichael did not believe in integration.[51]

Carmichael knew that men like Malone, members of the board of trustees, and officers of the alumni association had become the University of Alabama. When Carmichael assured Malone that he "did not believe in the wisdom of desegregation in the South,"[52] he may have spoken his conviction but not his better judgment. Ferguson, Steiner, Lawson, McCorvey, and Dixon were forceful men. They could sense weakness as a dog smells fear. They needed a new president.

No one could know, least of all Carmichael, that this was a scene played before. Despite his credentials, Carmichael's talents were those of a board chairman, not a chief executive officer. He saw himself as a corporate servant, not a leader. This Achilles heel first manifested itself at Vanderbilt. In 1943, a handful of liberals on the faculty and staff helped shape a Religious Emphasis Week program. None of the general convocations touched a controversial subject, but a few concurrent sessions addressed the issues of labor and race and attracted progressive speakers. When something of a tempest blew up, led by an especially reactionary member of Vanderbilt's Board of Trust, Carmichael "as always bent with the wind." Summing up his administrative style at Vanderbilt, the university's historian wrote:[53]

> He maintained a very harmonious relationship with the board, although he never gained from it the respect, even the devotion, granted [former Chancellor] Kirkland. Board members sensed that they worked with a clever politician. Individual board members seemed a bit wary or suspicious of Carmichael. Kirkland guided the board, even educated it; Carmichael consulted it, even negotiated with it. When necessary he altered his policies or his language to keep harmony. The results did not seem so different, but in subtle ways the policy-making process was very different—less predictable, more open, tentative, and precarious.

Whatever Carmichael's talents, and they were considerable, he was an unfortunate choice for the crisis at Tuscaloosa. His personal tragedy matched that of the alma mater he hoped to serve. In the fall of 1956 he resigned, and left office December 31.

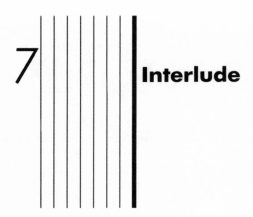

7 Interlude

Four hours north of Tuscaloosa, through Alabama's coal country and into the rolling Tennessee Valley, lies the battlefield of Shiloh, a place of tranquil beauty and profound introspection. That bloodiest two-day fight of the Civil War, which raged on the Tennessee–Mississippi border, forms a metaphorical backdrop for the confrontation in Tuscaloosa. The clash between good and evil in 1956 was indeed like the two 1862 armies stumbling upon each other before either was prepared to fight. The resulting battle staggered both sides. In 1956, the forces of desegregation, soon to be irresistible in their national armor, were met by an angry foe, determined to maintain by massive resistance its regional prerogatives. As the combatants crawled from the field and prayed for dark, the Resisters, like Beauregard on that first day at Shiloh, seemed to have the better of it. The NAACP was in full retreat, soon to be routed. Moderates had thrown down their guns and run for cover, cowering like Grant's raw recruits beneath the bluffs that towered above Pittsburg Landing. Reinforcements in the form of national sentiment thrashed about in the woods, just as General Lew Wallace did on that near fatal day in April 1862, unable to make contact with embattled virtue. A long night descended on the field, as a small band of white southerners, whose voices so recently rang out in unison for interracial cooperation, sent up agonizing cries in desperate isolation, desembodied by an impenetrable darkness.

A campus minister, writing two days after the Monday riot, said," . . . the important thing is that most of us who had not been through an experience of this kind have been pretty thoroughly shaken, if not taken off our hinges. We did not know quite what to do, felt we had to do something and yet did not trust ourselves to move out to do it, being aware of the possiblity of a 'backfire.' "[1] Like raw recruits, whites who supported change hunkered down and waited for someone to take the lead, and if no one did, so much the better. Nor did resisters show much more enthusiasm for doing battle. Writing a year later, the head of the Alabama Council on Human Relations found that "Negro emphasis on voter registration has served to calm immediate white fears. Thus the state is still in a state of vague anticipation of a 'show down' that is regarded as still distant with segregationists not showing any particular desire or ability to precipitate a round of activity yet. In all likelihood they are of a mind to wait upon Negro initiative which may increase at any time."[2]

In fact, black initiative in Alabama ebbed in 1957 to rush out the remainder of the decade. The Montgomery Improvement Association spun its web of limited demands in the bus boycott and withdrew cocoonlike while the Southern Christian Leadership Conference struggled to be born. The NAACP, paralyzed by Judge Jones's injunction, went out of existence. Fred Shuttlesworth formed the Alabama Christian Movement for Human Rights to replace the defunct NAACP, but the old guard did not like him. W.C. Patton, leader of the old guard, formed the Alabama State Co-ordinating Association for Registration and Voting, but linkage to NAACP placed it under the same cloud. If moderate whites and radical segregationists awaited the Negro's next move, the wait would be longer than expected. Little Rock caused whites to tense, but briefly. As blacks busied themselves with voter registration, whites bided their time, unthreatened by the glacial progress of black suffrage.

The interim presidency of James H. "Foots" Newman contributed to the lack of direction on campus. Newman, class of '29, served as Dean of Administration from 1950 to 1956. The more influential trustees liked him. Those in the university community who believed trustee leadership had failed the university did not like him. Newman made a fetish of efficiency and especially irritated those given freer rein during Carmichael's tenure. Jeff Bennett, who often worked long hours for extended periods, especially when the legislature was in session, took particular umbrage at Newman's request to account for time away from the office. That kind of leadership rankled.

Malaise in administration combined with legislative harrassment to demoralize faculty. The *Montgomery Advertiser* attributed "the loss of some thirty or more top-notch educators" at the university to the Lucy incident and the threats of reprisal that followed.[3] The number may have been exaggerated, and low pay accounted for some departures; but whatever the reasons, better members of the faculty continued to leave. Among them was York Willbern in political science. In September 1957, he asked Bennett about legislative efforts to curtail academic freedom. Bennett said the university had been "extremely lucky." "A number of dangerous pieces of legislation were quietly killed in committee, with the full encouragement of the Board of Trustees, except for one member [Ferguson] who vehemently urged that we 'do something' [to curb academic freedom]."[4] In these legislative fights, Bennett worried about Newman's lack of direction, and Newman fretted about what the trustees might think. Bennett told Newman that "it would be extremely embarrassing to the institution and to each of us should I find that I am pursuing a course which our board does not support, or that the institutional position is one which I could not support as vigorously as might some other individual."[5] In the end Bennett had his way with Newman and the legislature, and the university escaped unscathed.

Rose and the "Bear"

As the nation's attention switched to Little Rock, peace returned to Tuscaloosa. A year and a half had passed since the Lucy turmoil, and events seemed to confirm that desegregation would be a long time in coming. The mood brightened when the university named Frank Anthony Rose its twentieth president. Rose was not that well known, but he made an instant impression on all who met him. Tall, handsome, a magnificent speaker, he could charm a crowd or win the confidence of the most skeptical dinner companion. A speech before a national meeting of the Kappa Alpha social fraternity in the summer of 1957 brought Rose to the attention of trustee Ernest Williams, who along with Caddell and Ferguson formed the search committee. At age thirty, Rose had been the youngest college president in the nation. Appointed to head his alma mater, Transylvania University in Lexington, Kentucky, he served that institution capably from 1951 to 1957. The National Junior Chamber of Commerce named him one of ten outstanding young men in America in 1954. Another honoree that year was Robert F. Kennedy, an acquaintance who would serve Rose well during the 1963 crisis.

Others began to learn about Rose. Caddell received a call from his brother-in-law in Louisville who spoke in glowing terms about Transylvania's young president. This was good news indeed, for the search committee had developed no viable candidate since Carmichael's resignation, and "Foots" Newman was anxious to be considered despite an earlier pledge not to become a candidate. Caddell called Paul W. "Bear" Bryant for a reference. Bryant had established a winning football team at the University of Kentucky before doing the same at Texas A&M. A groundswell was already developing to bring Bryant back to Alabama, and the "Bear" was rumored to be interested, especially since his wife, Mary Harmon, hated living in College Station. When Bryant returned Caddell's call, he learned that the subject was not football but Frank Rose. Bryant remembered Rose from his Lexington days, but felt he should call a former administrative friend at Kentucky before commenting. Bryant soon got back to Caddell with high praise.[6] The trustees moved quickly and at a special meeting of the board in Montgomery on September 5 named Rose the new president.

The first weekend in October found Rose in Alabama accepting accolades as the university's redeemer. The press compared him to George Denny, who was credited by many with having rescued the university from an academic and athletic wilderness in the twenties. On Thursday Rose spoke to the Elks in Montgomery before journeying to Nashville where he addressed the Tide faithful on the eve of the Vanderbilt game. "Not only did the personable Dr. Rose look the part of a fine educator," wrote an admiring sportswriter, "but a striking athlete with many of Johnny Mack Brown's facial features." (Brown had been a star on 'Bama's first Rose Bowl team in 1926 before becoming a Hollywood cowboy star of the thirties and forties). "Foots" Newman introduced Rose to the alumni as an "18-hour-a-day worker" who would usher in an era of greatness, and Rose returned the compliment, assuaging Newman's disappointment at being passed over by declaring him a first-rate administrator around whom he could build a top-flight administration. The football team itself seemed to rise to the occasion by fighting to a 6-6 deadlock with heavily favored Vandy. The next morning, however, world events took unusual precedence over sports and cheated Rose of further publicity. Headlines blared the unthinkable: "Red Space Victory is 'Stunning Blow.' " Across from the news of Sputnik another headline queried, "Is Race Strife a Commie Plan?"[7] With federal troops encamped for a second week in Little Rock and all eyes scanning the evening sky for a glimpse of the made-in-Russia steel ball, Rose slipped back into Kentucky to mull his sudden good fortune and plan his return.

The new president's highest earned degree was a Bachelor of Divinity from the Lexington Theological Seminary. After a four-year pastorate at the First Christian Church in nearby Danville, Kentucky, he had accepted the appointment at Transylvania. He erased the school's debt, built a library, doubled faculty salaries, and imbued the campus with optimism. His abilities as a fundraiser and administrator were precisely the qualities the board sought. If nothing else, his fresh youthfulness would be a change from the tired, gray image Carmichael left behind. So on September 5, 1957, the board announced the appointment, effective January 1, 1958, and rectified the absence of an earned doctorate by awarding an honorary Litt.D. the following spring. (Rose also had an honorary LL.D., 1951, from Lynchburg College in Virginia.) Those who wrote to endorse Rose questioned his academic credentials but thought his administrative style more important for the job at hand. But even his style created some pause. Trustees wondered about his ministerial background but were relieved to find that he was neither a fundamentalist nor a teetotaler. They also found him a bit "sugary and polite" and worried whether he was "tough enough," but in the end the trustees felt good about their choice.

They paid him $20,000 plus use of the mansion. His salary at Transylvania had been $19,600, but that included housing. In addition Rose got a $3000 unaccountable discretionary fund for entertainment outside normal university functions (a year later the fund rose to $5000 or roughly the amount of an assistant professor's salary), a car with travel expenses reimbursed, and the right to keep speaking fees for engagements not paid for by the university (a privilege he had enjoyed at Transylvania). The mansion was redecorated by a wealthy Tuscaloosa family, and the kitchen enlarged for the new president's family of six. The departure from Transylvania apparently created some awkwardness in Rose's personal finances, for on December 6, he wrote to request that both he and his secretary, Marian Coates, be placed on the payroll as of

December 1. By then the board was happy to oblige, and lest anyone forget the reason, Rose added a gentle reminder. "I sincerely hope," he concluded, "that everyone is delighted with our hiring Paul Bryant as our new coach and with his contract." (By comparison, Bryant's contract called for a salary of $17,500, a house, a $4,000 discretionary fund for athletics, and eventually a $2,500 personal fund; plus, of course, complete control over the athletic budget.)[8]

Indeed, everyone was delighted. John Caddell wrote back to say that the hiring of Bryant "was the most popular thing that could have happened from your standpoint and makes people believe you can accomplish anything. I have told them that that is true and that you are going to make them, 'the people of Alabama,' put up the money for adequate salaries for our faculty." (During his interviews, Rose told the trustees that he was going to hire his "own" coach but, of course, acknowledged their interest in Bryant and let them know that the "Bear" would be at the top of his list.)[9] Rose came in on a wave of high expectations, and he did not disappoint his many admirers. Between January 1 and March 7, he made fifty-four speeches, talking earnestly about the need to make faculty salaries competitive.

Rose's message fell on receptive ears. The state legislature never wanted to punish the university and did not blame the university for desegregation, though some faculty were always suspect. In fact, Rose and the university benefited from a growing desire to help the institution through these years of turmoil and transition. With a national finger pointed at racial strife and more generally at the sorry state of affairs in Alabama higher education, Rose deflected embarrassment by talking with all the confidence of a national leader. By the spring of 1959 he launched a $5 million Greater University Development Program to boost private giving. With increased emphasis on federal matching funds (a reaction to Sputnik), dollars seemed to grow exponentially. The university's assets tripled over the next decade. Faculty salaries improved as did the number of faculty holding the Ph.D. degree. Enrollments in the three-campus university system grew from 12,250 to 20,000. A major building campaign transformed the Tuscaloosa campus, including a new administration building named for Rose and a high-rise dormitory named for Rose's equally charming wife, Tommye. The old administration building where Autherine Lucy had registered was eventually named for Carmichael.

In the early years Rose did not have to talk about race and desegregation; but when it came up in conversation before and after speeches, he had the remarkable ability to convince everyone and anyone, from the staunchest segregationist to the most determined liberal, that he was on their side. He was a charismatic leader, not a hands-on administrator. He established policy, others executed. The "18-hour-a-day worker" took naps religiously every afternoon at 2 o'clock, while others worked through. He made thorough use of the perks of office, the driver, club memberships, airplanes, and associations with the rich and powerful. From the one-horse operation of John Moran

Gallallee five years before, Rose ushered in the administrative style of the modern corporate university. Board members, especially the older ones who had lived in more frugal times, had difficulty figuring the ethics, if not legality, of discretionary funds for which there was no accounting, and eyebrows shot up when he asked for $100,000 to launch his campaign to raise $5 million.[10] Still, they anted up, convinced as they were that Rose had his finger on the future. Eventually his presidency would come to be viewed as imperial, and he would lose favor. But in the late fifties and early sixties he was precisely what the university needed, a charismatic figure with a take-charge image.

A lore has developed that the trustees brought Rose in to take charge of desegregation and to manage the transition. In this telling, trustees viewed Rose as a native southerner, born in Meridian, Mississippi, in 1920, but one whose professional life had been outside the deep South. His border state experience, the board is said to have reasoned, equipped him to deal with the problems of desegregation.[11] This legend is doubtful. For one thing, Hill Ferguson never gave up his quest to "keep 'Bama white" and would not likely have been party to a hidden agreement to facilitate desegregation. It is probable that conversations between Rose and the board centered realistically on the Lucy experience and the future of desegregation, but the board was steadfast in its determination not to relent without another court order and none seemed imminent. In the September 5 meeting where the decision to hire Rose was made, no trustee mentioned his qualifications for dealing with the momentous issue, though it is probable that discussion about his "toughness" had some bearing on the question. Whatever the private understandings, it is clear that Rose took no steps in his first four years to move the process along and, as will be seen, continued an active resistance until late 1962. Rose no doubt believed peaceful desegregation in the best interest of the university, but like his board and his predecessor, he was not willing to act until compelled.

By the time Rose and Bryant arrived in the winter of 1958, it was evident that the South was in a hiatal period for desegregation. The shock of Little Rock made people wonder what price the nation would have to pay to desegregate the South, district by district, fight by fight, and *a fortiori,* if in Little Rock, then how much worse in Birmingham or Jackson. A Gallup poll tested sentiment in the South twenty-two months after a similar poll in February 1956, and found that while segregation was still uppermost in people's minds, "there is a greater reluctance to discuss it openly."[12] The award-winning *Alumni Bulletin* never mentioned desegregation; the campus literary and humor magazine, *The Mahout,* never discussed it; and the *Crimson White* never reported it. During this time of indecision and silence, the president and the football coach healed the wounds of the Carmichael period. Winning football returned in the fall and desegregation became the stuff of fantasy as students talked knowingly among themselves about how if integration were to come, it surely would begin with the first grade until gradually, by increments, the school system would be integrated and then only by the handful of

Negroes who wanted it anyway. In such an environment no one was about to step out of line and rock the boat, not Rose, not the students, not the faculty.

Writing in the *Yale Review,* the university's most distinguished philosopher, Iredell Jenkins, elaborated an apology for inaction. He acknowledged the moral dilemma of serving an institution that violated personal principle. Such a condition, he reasoned, opened three courses of action to a faculty member: resignation, acquiescence, or intervention. Jenkins said that ultimately institutions must act "in a coldly practical manner," and they must do this because "institutions are infinitely less expendable than individuals; they are in fact, literally indispensable. When an institution is seriously weakened, it takes a long time to recover; and in the meantime it leaves a vacuum in which its functions go unperformed and its values unserved." A university, he concluded, "is justified, and even obligated, to bow to the expedient and to cultivate prudence in a manner that would be clearly immoral in an individual." Jenkins did note that "as individual moral agents" professors "feel obliged to serve the right as they see it," but it was no more than a felt obligation, for membership in the institution created an equally compelling obligation for a faculty member "to consider its [the institution's] judgment of what is expedient and practicable in the light of its vital interests."[13]

Reorganization and resistance redux

In the coldly practical world of institutional survival, the university and Rose needed Jeff Bennett. He was energetic, smart, tough, and ambitious. He made it his business to know anyone and everyone who could help or hurt the university. In the Lucy crisis, he was a rock. When Carmichael lost his effectiveness, Bennett kept things going. When the university faced a reactionary legislature and a funding crisis in 1957, Bennett worked night and day to salvage the university's position in Montgomery and succeeded. Moving in a world of hardball politics, Bennett dealt daily with the seamier side of institutional life. He adopted, because he had to, a pragmatic approach to institutional survival. He knew what the Klan was doing, what the White Citizens' Councilors planned, and how these and other extremist groups influenced politics in Montgomery. A dedicated churchman, he was as committed to human rights as his colleagues who joined the Human Relations Council and the Open Forum, a group of seekers who met at the Canterbury Chapel, but he tempered his commitment by a clear-eyed assessment of reality.

On learning of Rose's appointment, Bennett hoped for an expanded role. He wrote a letter detailing his present duties and saying that his effectiveness in Montgomery could be enhanced if more directly involved in the decision-making process, a role he had enjoyed under Carmichael. If such were not in Rose's plans, Bennett was inclined to accept a more lucrative position with a law firm in Birmingham.[14] Rose needed Bennett for the transition and gave the necessary assurances. Bennett continued to play a key role but not the

141

central part he perhaps deserved. Rose's style would not allow it. "Foots" Newman stayed on as Vice President for Administration and Alex Pow handled contracts and grants and eventually became Vice President for Development. With these and other vice presidencies Rose created, the new president guarded against being upstaged or outflanked; not that Bennett, or anyone else for that matter, had such ambitions.

The arrival of Bryant and Rose augured well for university morale but its importance, at least in the near-term, was no greater than two departures from the board of trustees. In 1959 Ferguson stepped down after forty years of service, and Steiner died. They would be replaced in turn by Ehney Camp, a Birmingham insurance executive, and Winton "Red" Blount, an emerging mogul in the international construction business. Both would play constructive roles in the final drama. Ferguson stayed on in an emeritus capacity; a nuisance, but relatively harmless.

The players were changing but not the game. As before, black applicants paid a price. Fewer applied in Rose's early years, and they were backed by less organizational support. Whatever strange courage drove them, they applied nonetheless. University officials worried most about their center at Huntsville. Overnight, federal money had transformed that sleepy little village in northeast Alabama into a burgeoning center for space-related research. With these dollars came scientists, engineers, and desk-jockeying soldiers. The University Center itself was a response to this influx of talent, but of course, some of this talent was black. In the fall of 1958 two Negroes applied for admission, only to be chewed up in the same machinery that had spit out Myers and Lucy and Peters and Nabors and Epps. And as before, the levers were pulled by an elite corps of university and community leaders, joined this time by top brass at Redstone Arsenal.[15] A studied conclusion had developed among the region's conservative leadership that the less said, the more effective the resistance. Of course, that was precisely the way they always dealt. As businessmen, lawyers, bankers, industrialists, and generals, they never conducted business in public. A university president was expected to act in the same manner, and Rose did not disappoint his conservative friends.

By the end of 1958 the Supreme Court had confirmed the wisdom of silence. In the Birmingham school-desegregation case, a three-judge panel ruled Alabama's pupil placement law constitutional "on its face." When the Supreme Court upheld the lower court, it set an extraordinary burden of proof for showing that blacks had been excluded from public schools on account of race and allowed officials to use elaborate criteria for placing students. The net result, observed Arthur Krock of the *New York Times,* was to accept "by plain inference the prospect that, in many Southern areas, generations, not years will elapse before Negro pupils in general can no longer be constitutionally excluded from white schools on at least one of these criteria."[16] Five years later, the Supreme Court reversed itself. However, Christmas 1958 found conservatives cheerfully inscribing, carte blanche, the terms of desegregation.

University officials seemed content to drift with the current, like a leaf on the bosom of a broad river, unable to direct or influence its flow. If they objected to the course of events, it was never in writing; not even among themselves in envelopes marked personal and confidential. Those envelopes were reserved for detailing their complicity with massive resistance. In 1959, Rose suggested naming individuals in each county to keep tabs on black applicants. Former police commissioner Clyde C. Sellers, whose handling of the Montgomery bus boycott bordered on ruthless, was mentioned as a contact. Rose also considered adapting an exam given to "incorrigible" students (an exam that had "flunked out 106 students") as an entrance test for black applicants. By 1960 university complicity with resisters ran so deep that applications from black students were routinely turned over to the State Department of Public Safety where clumsy, untutored investigators ran amok in the private lives of young men and women. In fiscal year 1960–61, the Department of Public Safety established a Security Section in response to "the racial push and communist activity in the South and in this State." In the first year, investigators set up files on 130 individuals and doubled the number the second. In addition they kept files on 47 organizations and attended over 25 public meetings "which were suspected of teaching or advocating policies which were detrimental to the interests of the State of Alabama."[17] The Rose administration submitted lists of its black applicants to the file-happy Security Section.

As testimony to conservative success, students entering the university in 1957 graduated in 1961 believing that very little had changed from their parents' day. For all they knew, the Autherine Lucy incident was but a blip on the screen, over before it had begun, and, if anything, Lucy's unhappy experience proved the impotence of the NAACP when faced with determined resistance. Campus publications never mentioned the Lucy episode, much less the current state of race relations. By the fall of 1957, even the Open Forum had closed, no longer providing a vent for the handful of frustrated moderates and liberals who thought change inevitable. A great silence fell across the university as conservatives throughout the state concluded that the best way to preserve the status quo was to say nothing. So the happy freshmen who enrolled in 1957 planned their weekends around resurgent football, perhaps even took a spring picnic basket to Shiloh where minie balls could still be gleaned, souvenirs of a heroic past. They would graduate four years later expecting even more heroic things to come, like a national championship— which Paul Bryant delivered in the fall of their first homecoming.

Autherine Lucy. (Courtesy *Birmingham News*)

(*above*) Jefferson Bennett. (*Courtesy Corolla*)

(*opposite, above*) Board of Trustees, 1963. Hill Ferguson,
Brewer Dixon, Winton M. Blount, Thomas D. Russell, William H.
Key, Frank Rose, W. A. Lecroy, John Patterson, Ehny A. Camp,
Eris Paul, Ernest G. Williams, John A. Caddell, and Gessner T.
McCorvey. (*Courtesy Corolla*)

(*opposite, below*) Emory Jackson and Pollie Myers with Lucy
during registration. (*Courtesy Birmingham News*)

(*above*) Excitement and terror on University Boulevard, 4 February 1956. (Courtesy *Birmingham News*)

(*opposite, above*) Students burn NAACP literature on steps of the Union Building, 4 February 1956. (Courtesy Jim Oakley)

(*opposite, below*) Leonard Wilson at the flagpole, 4 February 1956. (Courtesy *Birmingham News*)

Oliver Carmichael addresses the faculty.
(Courtesy Robert Kelley, *Life* Magazine, 1956, Times Warner Inc.)

Students ponder the issue. Leonard Wilson at left.
(Courtesy Robert Kelley, *Life* Magazine, 1956, Time Warner Inc.)

(*above*) Vivian Malone. (*Courtesy Birmingham News*)

(*opposite, above*) Autherine Lucy, Constance Motley, Thurgood Marshall, Arthur Shores, Birmingham, 29 February 1956.
(*Courtesy Birmingham News*)

(*opposite, below*) Frank Rose takes questions from reporters.
(*Courtesy John Burton*)

(*above*) A rally south of Tuscaloosa, 8 June 1963.
(Courtesy John Burton)

(*opposite, above*) The schoolhouse door ringed by troopers,
Foster Auditorium, 11 June 1963. (Courtesy John Burton)

(*opposite, below*) National Guard, 11 June 1963.
(Courtesy John Burton)

(*above*) General Graham's "sad duty."
(Courtesy *Birmingham News*)

(*opposite, above*) Katzenbach confronts Wallace.
(Courtesy *Birmingham News*)

(*opposite, below*) Katzenbach prepares to escort Vivian Malone
to her dorm. (Courtesy John Burton)

James Hood and Vivian Malone, New York, June 1963.
(Courtesy Tom Dent)

Vivian Malone escorted by Joseph Dolan to the schoolhouse door. (Courtesy John Burton)

(*above*) James Hood signs up for classes, 11 June 1963.
(Courtesy John Burton)

(*below*) Vivian Malone in class. (Courtesy *Birmingham News*)

8 Class of '65

The class of '65 experienced campus life in a radically different way. Transforming events took place during their school days. There was still time to join fraternities, to "twist and shout," to wear madras shirts, Villager dresses, and Bass Weejun shoes; still time to sweat through hot September football games and take your girl to the drive-in theater, where an unlucky fellow might see *My Fair Lady* in its entirety without distraction. But slow-dancing to "Moon River" or whiling away time at Pug's were happy hours enacted against a backdrop of profound change; change with which each student in the class of '65 would have to come to terms.

For would-be campus leaders, the times were littered with signs. Many came to college as Young Democrats, loyal to the party of their fathers, inspired by John Kennedy's promise of a New Frontier, but also uneasy about the new President because their daddies had cursed him for cozying up to Martin Luther King in 1960. Their unease, however, would yield to the enlightenment of the classroom and to bull sessions where clinging to the past seemed old-fashioned, even if embracing the new meant abandoning segregation. For some this nudge beyond parochialism transformed the religious zeal of Sunday school and church into a secular crusade for doing good. For others it meant resignation, a conscious effort to accept change without prejudice, thereby not foreclosing future options for advancement or leadership. Still others would rally to Governor Wallace in a vain attempt to preserve the laws and customs to which they felt entitled by birth. No one would remain unaffected.

Most had heard of the new civil rights groups, but had difficulty separating them. The NAACP remained the catchall term for organized agitation, though most students had heard of the SCLC. Everyone knew Martin Luther King's name but generally associated him with the NAACP, or simply with civil rights trouble. Newer terms vied for attention. No one knew what to make of SNCC (Student Non-violent Coordinating Committee), but it sounded sinister. Most thought of CORE (Congress on Racial Equality) as

145

new despite its comparatively long history. If the acronyms were confusing, the actions sponsored by these groups were not. The sit-in movement, begun in February 1960 in Greensboro, North Carolina, stayed out of Alabama until 1962 but everyone talked about the lunch-counter demonstrations and what they might portend.

Mother's Day, May 14, 1961, signaled an end to segregation's Indian summer in Alabama. Freedom Riders rolled into the state. Not even high-school prom nights or graduation exercises could blind the university's next freshman class to what followed. Just outside Anniston on Highway 78 a Greyhound bus was firebombed. Hours later, in Birmingham, a mob of Klansmen surrounded a second bus and beat passengers savagely while Bull Connor's police, by agreement with the Klan, waited nearby to give the mob time to do its worst. Enraged whites repeated the beatings a week later in Montgomery. The sheer savagery of these attacks encouraged parents of 'Bama's rising freshman class to blame the victims. Governor Patterson beat the drum for this transference, while the Kennedy administration, showing little patience for Freedom Riders, made a deal with Patterson and Mississippi's Ross Barnett to end the confrontation. In return for safe passage out of Alabama, Kennedy officials acquiesced in the Riders' subsequent arrest and imprisonment in Jackson, Mississippi.

The Kennedys were just what the class of '65 needed—young, charismatic, clarion in their call to serve the future; they also were pragmatic, tough, and Janus-faced when it came to race relations. By their reluctance to antagonize the southern wing of the Democratic party, the Kennedys made it possible for young Alabamians to embrace the party of their fathers while planning their future. The Kennedys might be cursed at home, but how much worse the Republicans, still the party of Reconstruction and Depression? President Kennedy sealed his relationship with 'Bama's class of '65, when on December 5, he met with Bear Bryant and members of the 1961 national championship team at the Waldorf-Astoria in New York for the presentation of the MacArthur Bowl. The President also made a telephone address to a Sugar Bowl pep rally in Foster Auditorium the same evening. To cries of "We're Number One," the champions later defeated Arkansas, 10-3, for an umblemished '61 season.

Of all the forces shaping the class of '65 none was more powerful than television. Southern whites had long benefited from the movie industry's portrayal of the region's racial dilemma. From *Birth of a Nation* to *Gone with the Wind,* Hollywood justified the "superior" race's dominion. But television, like the eye of Big Brother, would see everything all the time. Most families had acquired sets when members of the class of '65 were turning ten or eleven, before the new medium hit its full entertainment stride, and teenagers could take it or leave it depending on what was going on at the high school or church or what was playing at the drive-in. Then in 1960 *Bonanza* blazed across the screen in full color to replace Dinah Shore with high production-value drama, and *Gunsmoke* became a one-hour, psycho-western thriller. Soon news itself went to thirty minutes as the networks learned the entertainment potential in national self-absorption. The biggest beneficiary of instant visualization would be the civil rights movement, as a national audience saw for the first time a moving pictorial dramatization of segregation.

The class of '65 took all this in and were shamed or angered by what they saw, but whatever their feeling, they knew their regional soul was being laid bare before a national audience. Exposure made it impossible not to feel something; not to act or react in some manner to the South's latest permutation of its "peculiar institution." Fortunately the image industries served up role models that were not entirely negative. Hollywood helped when it produced Alabama alumna Harper Lee's 1961 Pulitzer Prize-winning *To Kill a Mockingbird*. The book's hero, Atticus Finch, represented the region's patrician ideal; the lone, stoic figure standing for justice, a platform that placed him between passive black victims and the lower-class whites who persecuted them. Gregory Peck's moving portrayal of the hero stirred young moviegoers to Faulkner's view that better-blooded whites should assume the obligation of setting blacks free. It was a comforting thought, for who knew blacks better than their former masters; certainly not outsiders, northerners who knew nothing of the region's complex and darkly textured past. Not a few college men (it was still a young man's world to make), sitting in darkened picture shows, hand-in-hand with their girlfriends, thrilled to Atticus Finch's slow passage out of the courtroom as grateful Negroes, who had packed the gallery in hopes of seeing justice done, stood in silent tribute.

But these were images for the whites-only student body. One member of the class of '65 would be black. She saw the world differently. She would wish her classmates no less than they wished for themselves; only that whatever they wished not be purchased at her expense. In the fall of 1960, Vivian Juanita Malone had applied for admission to the university's Mobile branch campus, only to be denied. In 1965 she would graduate from the University of Alabama.

The campus

One could feel things stirring in the winter and spring of 1961. The student newspaper noted Kennedy's impassioned call to service and said that the silent generation was clearing its throat. But a disturbing current flowed beneath the surface. On March 22, a student reporter interviewed Robert Shelton, National Grand Wizard of the Knights of the Ku Klux Klan. Shelton bragged about the strength of the Klan on campus. He said that student Klansmen had infiltrated all campus organizations and were prepared to finger "liberals." Pressed to name "liberals," he demurred, but added that the activities of Iredell Jenkins, chair of the philosophy department, and Earl Whatley, editor of the *Mahout,* warranted attention. Shelton foresaw a major attempt to integrate the universty in the fall and promised that "student

Klansmen will play a part." Asked if that meant violence, Shelton said blacks had provoked violence already by breaking laws, thereby justifying the Klan in "bending the law."

Students were not the only ones concerned about Klansmen in their midst. The university administration received alarming reports from a "mole" planted in the Klan about plans to disrupt the campus. At a klavern gathering in January, Shelton told followers from Birmingham, Bessemer, Tuscaloosa, Montgomery, Centerville, and Pickens County that in the event of racial disturbances, they were "to create such chaos and disorder and trouble that the Governor of Alabama will close the university under his police powers." Shelton called it "their job to present the Governor with this opportunity." The Grand Wizard planned carefully, placing himself in command of operations. Four automobiles, equipped with two-way radios, would provide communication. He assigned task groups: one to disable fire trucks, another to pass through crowds increasing their "tempo" and preparing them "for action . . . as opposed to mere noise," still another to pour oil of mustard on the seats and steering wheels of police cars. Shelton warned Klansmen not to bring firearms or large knives on campus so as "to remove the possibility of arrest for carrying concealed weapons." Small knives, however, would be needed to cut fire hoses. Finally, "under the impression that the police will be wearing blue battle helmets," Shelton ordered Kluxers to buy military helmet liners from local Army-Navy stores and paint them blue to confuse the police. There was an allegation, which Jeff Bennett discounted, that the Klan had a plant in the Admissions Office. These chillingly specific plans brought an immediate response from Bennett. Within a week he worked out countermeasures with university, city, and state officials.[1] Secrecy presented Bennett his only difficulty, for tipping the Klan to what he knew would endanger the university's spy.

Bennett suggested to Rose that the trustees be informed. Following the board's February 15 meeting, Rose wrote Patterson one of his typically ingratiating letters. "It is my interpretation," he confided, "that the Board of Trustees gave both of us the authority to proceed with the handling of the problems of the University in the manner in which we think would serve the best interests of the institution." Rose expressed satisfaction "that we had a chance to talk it out as I have found the members of the Board of Trustees, previous to our meeting, extremely cautious and concerned about some of my activities relating to my handling the problem in cooperation with you."[2] Rose never articulated what concern the board had about his working with Patterson. University people always considered the Phi Beta Kappa Patterson one of the law school's brightest products. Only his connection with the Kennedys (he had supported the Massachusetts senator for vice president in 1956 and had come out early for him in 1960) would have alarmed some board members; but even there, Patterson's demagoguery on race made him safe on segregation.

There was always the reluctance to yield board control to any governor, but Rose wisely reached out to Patterson. The threat of violence made planning necessary. Riots that accompanied the admission of two blacks at the University of Georgia, sit-ins, and soon Freedom Riders threatened to push events beyond safe bounds. People worried. Benjamin Muse, who had just traveled the state for the Southern Regional Council, reported, "The danger of violence remains. Some mentioned Tuscaloosa as the next likely scene of disorder—in view of the possibility of application by a Negro . . . and the unrestrained activity of the Ku Klux Klan among the student body and the populace of the city." Buford Boone of the *Tuscaloosa News* "agreed with others that Tuscaloosa was simmering for another explosion when the next Negro should try to enter the University. . . ."[3]

But the violence that attended the Freedom Riders' foray into Alabama also made it less likely that any black students could summon sufficient support to go through with plans for desegregation in the fall. The only ripple on an otherwise calm surface in the summer of '61 came when the Auburn student newspaper declared that the Freedom Riders had "law and morality on their side" and predicted that integration would soon come to Auburn. The editor hoped that when the day came Auburn would not degrade its name "as the name of Alabama has already been degraded"—an obvious reference to the Lucy episode. The Auburn board of trustees swiftly censured and censored the paper. Though the *Crimson-White* regretted this assault on freedom of the press, the editors nonetheless understood the provocative nature of student commentary on so sensitive an issue.[4]

Progress and poverty

If the Rose administration took comfort in student passivity, other occurrences spelled trouble ahead. The trouble was a byproduct of good fortune. In reaction to *Sputnik,* the nation poured vast sums of money into higher education with emphasis on science and technology. Because the Tuscaloosa campus had never been a significant research center, the branch campuses threatened to usurp its research prerogatives, especially in Huntsville and Birmingham. Huntsville had proved its value to the nation within ninety days of *Sputnik* when Wernher von Braun converted one of General John B. Medaris's Redstone rockets into *Juno I,* the launcher for *Explorer I,* the nation's first satellite. Huntsville was winning the war over location of the nation's missile program. As von Braun's Germans settled in to conquer gravity, the little town grew exponentially, concentrating scientists and engineers in a technological village.[5]

To keep abreast of progress the technocrats required continuing education, especially at the graduate level, and naturally turned to the University Center. They were willing to pay. In addition to tuition for students, they provided a ready pool of highly skilled part-time faculty. Recognizing its good fortune, Huntsville in 1958 dedicated 83 acres of prime land for a campus, and in 1961 the state appropriated money for an Alabama Research Institute. A true "university-industry-government complex," as NASA administrator James Webb called it, had been established. Von Braun's dream of a "technological community" unfolded. The town that scarcely had 16,000 souls, many of them in agri-business when the Germans arrived in 1950, numbered over 60,000 in 1960. The campus blossomed with federal contracts and grants.

The growth of the Huntsville Center had two effects on the University of Alabama, and both caused Frank Rose much concern in the spring of 1961. First, the Tuscaloosa faculty worried publicly about the establishment of the Alabama Research Institute in Huntsville. Rose had raised money, built buildings, and improved faculty salaries, all of which he had promised and for which the faculty was grateful. However, he had no vision for making Tuscaloosa the true center of a statewide university system. He willingly gave up branches in Mobile, Dothan, Selma, and Montgomery, and by the end of his administration in 1969 had established autonomous administrative units in Birmingham and Huntsville, universities that would compete with Tuscaloosa for resources.

The apparent diminution of Tuscaloosa's role, especially in research, provoked the chair of the physics department, Charles Mandeville, to lash out at Rose for crippling the research mission of the main campus while allowing a Research Institute in far-off Huntsville. Mandeville warned that new buildings and faculty salaries alone did not make for a first-rate research institution. There would have to be state-of-the-art equipment, graduate fellowships, and reduced teaching loads. Mandeville, along with all other department heads in the sciences, complained about a new central computer Rose had acquired and took exception to Rose's boast that the new machine would be a major asset for scientific research. The scientists called it a "compiler" and complained that it had been bought without consultation.[6]

The Mandeville criticism came to a head in July, when at the invitation of Tuscaloosa's state representative, who was Mandeville's neighbor, the physicist appeared before the house Ways and Means Committee and challenged Rose on budgetary line items for scientific research. Mandeville long had the reputation of an eccentric, some said crackpot, but one who happened to be right about scientific research at Tuscaloosa. With Rose and his aides present, Mandeville shoved a stack of reports at the legislators and concluded his presentation by singing "Oh Susannah." As the physicist departed, Rose broke the tension for startled legislators saying, "Now you see what I have to put up with." He told the press he had never been so humiliated. On his return to Tuscaloosa, Rose obtained Mandeville's dismissal. Bennett had warned the physicist it would happen. In August a faculty committee sustained the president's action, and Mandeville went packing with a year's pay. Twelve years later Mandeville, his outrage unappeased, wrote a novel titled *The Chronicle of an Unscrupulous Administrator Who Brought Sin, Sex and Scandal to the Hallowed Halls of the . . . UNIVERSITY.*[7] As messenger, Mandeville killed the message about scientific research, but Rose's popularity was undiminished. Most faculty did not have strong research programs and had seen real improvement in their salaries. Still, those who wanted the institution to move into the first rank of southern universities were disheartened. One young chemist, a recent post-doctoral fellow at Yale, found laboratory facilities deplorable and wrote in support of Mandeville's complaint, though not his behavior.[8]

Huntsville's growth, fueled by federal dollars, occasioned a second worry for Rose—federal intrusion into the university's racial policy. The university stood to receive six million dollars in the 1960–61 school year, three million in research grants plus three more to construct dormitories. However, on March 6, Kennedy signed an executive order forcing equal employment opportunity on federal contracts. Alex Pow, Vice President for Contracts and Grants, urged Rose to bring the matter before the trustees. Pow argued that the university could refuse to sign the compliance agreement and lose the revenue, or sign with no intent to comply. The latter, Pow believed, risked popular outrage if word got out that the document had been signed. The board saw no course but for Rose to sign. Governor Patterson's reaction was the same, "expressing his belief that while President Kennedy would like to see this done he does not intend to push it." Patterson promised to bring the matter up at a luncheon with Kennedy on May 8.[9]

As Patterson and Kennedy lunched, Freedom Riders rolled through the Carolinas heading for their Mother's Day rendezvous in Alabama. The events that occurred over the next two weeks permanently ruptured the relationship between Patterson and the President. More seriously, they caused the Kennedy administration to draw back from the civil rights movement. The waves of Freedom Riders that followed spilled into Mississippi's Parchman Penitentiary until they numbered 200, abandoned by a federal administration that viewed them as extremists who embarrassed President Kennedy's efforts to negotiate peace with the Soviet Union and extend freedom to Latin America. Robert Kennedy's ire seemed especially directed at a group of riders with Ivy League credentials. They had ridden into Montgomery with a handful of SNCC organizers just as General Henry Graham of the Alabama National Guard, after tedious negotiations, escorted the first group of Riders into Mississippi. The President's brother could understand the misguided actions of Quakers and students from the South's poor, all-black colleges, but not the behavior of these people. In this second group were William Sloan Coffin, Jr., Yale's chaplain, and John D. Maguire, a young faculty member at Wesleyan University in Connecticut and a recent graduate of Yale. Coffin, Maguire, and company avoided the terrifying experience of Mississippi's Parchman Prison only by being arrested for violating Montgomery's segregation laws before they could board the bus for Jackson.[10]

Meanwhile, in Tuscaloosa, a young faculty member followed happenings in Montgomery with more than casual interest. He picked up a pen and wrote his friend, Maguire. "Would it be ludicrous of me," he mused, "to say 'welcome back to Alabama'? " Robert Garner, a native of Mobile, and Maguire, a native of Montgomery and later president of the Claremont Graduate School in California, met as young southerners through the Baptist church in New Haven. There they took comfort in the familiar sounds and cadences of their regional dialect and delight in the Sunday night suppers they shared with fellow exiles. It was not uncommon, of course, for young southerners to go North toward home. Garner wished Maguire and his companions "physical safety and the success (however remote) of doing something to do away with senseless laws and customs that thwart all our hopes of understanding. Only time," he said, "will be able to judge the value of your dramatic efforts, but it will surely praise your courage and dedication."[11] Garner was among those "liberal" professors the Klan watched. He was also the young chemist who wrote to endorse Mandeville's general complaint. His handwritten letter to Maguire was opened, photostatically copied, and wound up in the segregation files of Jeff Bennett. Garner never knew that. By whatever means Bennett got the copy, his intentions would not have been to admonish the young professor. In fact he never spoke with Garner about it. It is probable that a busybody on the staff with segregationist sympathies turned the copy over to Bennett, or even that the university's own Klan informant intercepted the mail.

Intentions aside, the effects of a more general surveillance of thought, even the felt need to protect faculty from the consequences of their thought, were palpable. Faculty of adventurous intellect continued to leave and promising talent refused to come because of the turmoil. A professor of engineering at the University of Texas declined a job offer and based his decision "rather strongly on what I believe to be a shaky future for the public school systems in the South." In passing this refusal along to Rose, Dean Cudworth's covering note disclosed, "This is the third letter we have had that has contained such statements." In February, the usual month for the most intensive faculty recruitment effort, Paul Siegel, head of the psychology department, wrote a similar letter.[12] By June, Buford Boone, resident friend of the university, said privately that the campus was having "some trouble getting and holding faculty members . . .—not only because of school and community uncertainty, but because of the stifling intellectual atmosphere."[13]

Rose was good for the university in many ways. His eventual length of tenure alone provided continuity. (In the twenty years since George Denny, there had been four presidents and three interim appointments. No one had held the office for more than five years.) Rose's ability to raise money and to ride a cresting tide of post-*Sputnik* dollars improved facilities and salaries. He was popular and a much sought speaker. But in the end, he could not or did not prevent the politics of race from robbing the Capstone of any university's most important currency, a commitment to values. Unable to defend the values of an old, discarded South, the university felt it could not embrace the values of a new racial order. The University of Alabama found itself at the table of a high stakes, value-driven poker game, but too spiritually impoverished to ante up.

A line in the dust

It was a tribute to Kennedy style that events in the spring and summer of 1961 did not plunge the nation into a morass of introspection, though perhaps they should have. Yuri Gagarin's space triumph on April 12, followed within the week by the Bay of Pigs debacle, the quixotic Freedom Rides in May, the disastrous Kennedy-Khrushchev summit in early June, and the Berlin Wall in August—each crisis provoked another rhetorical gesture. If cosmonauts trundled the upper atmosphere, America would go to the moon. Zapata Bay was Eisenhower's legacy, while Freedom Riders were commended to a silent martyrdom in Parchman. Though difficult to put a better face on failed summitry, it was not difficult to diabolize the Soviet Union for throwing up a wall of shame. So the nation kept its wits, if only by mirrors.

In Alabama it was time to prepare for another political season. Early speculation centered on Jim Folsom. A straw vote by the segregationist *South* magazine in January 1962, showed Folsom way out in front of a pack of eight gubernatorial candidates vying for the Democratic nomination. "Kissin' Jim," it was thought, still had his pull, and knowledgeable pundits said that unless the race issue heated up, the aging populist might recapture his old job. There were plenty enough candidates to fan the flames of segregation. MacDonald Gallion, the state's attorney general, was thought to be reckless, but it would have been hard to rank him ahead of Birmingham's Bull Connor. The smart and affable Sam Engelhardt, instrumental in the organization of Alabama's White Citizens Councils, was a non-candidate but was rumored to have the support of Governor Patterson. Lieutenant Governor Albert Boutwell, a strong but soft-spoken segregationist, stayed in the race until late, then withdrew. Of all the extremists, George Wallace commanded the most support. He placed second to Folsom in *South*'s January poll. Moderates held out hope for George Hawkins from Gadsden or Jimmy Faulkner in Baldwin County before settling on Tuscaloosa's Ryan de Graffenreid. However, the pundits who thought Folsom had a chance miscalculated on two points: one, race was already the central issue, and two, nobody, including Folsom, could out-campaign Wallace.

Standing in the cavernous lobby of Birmingham's Tutwiler Hotel, Wallace fielded questions from reporters sprinkled among a larger cast of admirers and well-wishers. Benjamin Muse, on an investigative tour for the Southern Regional Council, took notes. It was January, and the City Commission had already voted to close the parks rather than yield to Judge Grooms's order that they be desegregated. Many Birminghamians, including the city's business community and newspapers, had protested the park closings, but Wallace warned, "We have to make sacrifices and do without things if we want to stand up for what we believe in." He then expanded on the mistake of "appeasing niggers" both at home and abroad. When asked for his opinion of "Bobby" Kennedy, "his reply," Muse noted, "was too revoltingly obscene to be repeated."[14] Within days of Wallace's performance, Judge Grooms sentenced six men for the savage attack on the Freedom Riders bus in Anniston. They did not spend a single day in jail. Meanwhile, Fred Shuttlesworth was returning to Birmingham to serve ninety days for his 1958 attempt to sit in a city bus.

Wallace rode such a wave of resentment into office that twenty-five years later, exhaused and spent, his followers still would not let go of him even when he confessed the error of his earlier views. Racism was the muscle and sinew of white anger but it was so interlarded with traditional suspicions of power and privilege that race itself could not be separated from other themes of the powerless: localism versus federalism, the working many versus the monied few, southerners versus Yankees, farmers versus bankers, average citizen versus pointy-headed intellectual, individual ballot versus bloc voter, little/big, light/ dark, white/black, and so on until the majority race in Alabama was transformed into a conscious minority, victims of an overwhelming international colored majority. Having drummed on these tensions for four years, Wallace came to Montgomery on March 10, 1962, to kick off his primary campaign. His speech that day coalesced all these themes into one all-encompassing symbol, a symbol so powerful that it resonated with the fears and hopes of a nation. Not lunch counters for Wallace, buses, or even ballot boxes, but a schoolhouse door, the gateway through which all children must pass.

In a voice sounding like a cornet played out of tune, Wallace delighted his followers, promising to secure from the legislature personal authority to assign pupils. Rolling his shoulders like a prizefighter, he snapped, "And when the court order comes, I am going to place myself, your governor, in the position so that the federal court order must be directed against your governor"—his body for theirs, consubstantial with the faithful—"I shall refuse," he pledged, "to abide by any such illegal federal court order even to the point of standing in the schoolhouse door, if necessary." And there it was done, the course set straight into next year when the promissory note would have to be redeemed.

For now, however, it meant only that the racial theme in the campaign was heating up. Gone was the hypothetical talk of January about whether the campaign would turn to race, and if so what its consequences would be for "Kissin' Jim." In early April, Folsom rolled into Talladega still strumming his branchhead, populist themes. His singing wife, Jamelle, warmed up the crowd with strains of "Peace in the Valley," but there was no peace. Young blacks from the local college took up positions across from the courthouse square to protest segregation.[15] Had Wallace been there he would have known to taunt them, to turn their newfound initiative to his advantage, but Folsom ignored the students only to confirm white suspicions that he was soft on segregation. This oversized, goodhearted man would stumble on like a punchdrunk fighter, parrying blows from Wallace, while trying to land a few awkward punches for segregation himself. Even before Folsom's infamous election-eve appearance on Montgomery television, when he drunkenly pawed his wife and could not remember the names of his children, he was politically dead; the victim of booze, incompetence, and race, but even at that, his was a dearly departed corpse, a bumptious reminder of the state's better angels.

The summer of '62

In 1962 little happened for the cause of human rights, and if it did, it did not happen right. Martin Luther King's movement languished in Albany, Georgia, 200 miles from anything resembling a major media market, striking blows for freedom against what turned out to be segregations's soft pillow, Sheriff Laurie Pritchett. The NAACP happily scored King for his embarrassing performance, while the students of SNCC expressed open contempt for his bourgeois ways and for being more interested in the financial coffers of SCLC than the grassroots movement in southwest Georgia. The *New York Times* joined the denunciation. Segregationists no doubt smiled.

Actually things were happening, but against segregation's soft underbelly. Throughout the urban South, business communities, faced with boycotts and sit-ins, made arrangements with local blacks. In Birmingham, students at Miles College began a selective buying campaign against downtown businesses that Norman Jimerson, the new director of the Alabama Council on Human Relations, declared 95 percent effective. Though Jimerson exaggerated, the city's Chamber of Commerce president, Sydney Smyer, began negotiating with black leaders and eventually included Fred Shuttlesworth in his talks. Smyer's experience typified shifting attitudes. He had been a determined states' righter and segregationist in the fifties, but a business trip to Japan in May 1961 let him see his native city as others saw it. Even Japanese newspapers could not ignore the news value of photographs showing Freedom Riders being beaten unmercifully in the Birmingham bus station. Smyer knew then that Bull Connor had to go.[16]

A year later, Smyer and others devised a plan to get rid of Connor by launching a campaign to change the city's form of government from commission to mayor and council. It was an end-around move. The image-conscious city might continue to elect Connor as one of three commissioners, but never as mayor. This campaign and others such as the long battle to end the County Unit system in Georgia were never advertised as means to improved race relations, only to good government, but those in charge knew that the racial order would change in the wake. August primary elections in Georgia, for example, saw the triumph of Carl Sanders, a handsome New South progressive, over former governor Marvin Griffin whose jowls shook with "no-never" declarations. Concessions from the business community and the elections of progressive governors around the South helped to confirm the politics of gradualism. Moderates at last had proof that the South's response to civil rights might be something less than absolute intransigence. Time, it now seemed, was on the side of good government and gradual change.

Even in Birmingham, the SCLC was able to host its national meeting in September without disturbing the peace or setting back the movement to oust Bull Connor. Sobered by their experience in Albany and entreated by Birmingham's black community to be cautious, SCLC ministers stuck to sermonizing. Besides, Smyer and Birmingham's leading businessmen had met with Fred Shuttlesworth and pledged to remove the "Whites Only" signs from local department stores. For good measure, and much to their own surprise, the businessmen also agreed to desegregate restrooms. There was an additional reason for the unobtrusive arrival and departure of Martin Luther King and the SCLC: the nation's attention had switched to Oxford, Mississippi.

But for Mississippi

Carl Sandburg, teller of tales, once described a farmer working in a field. From the south the old man hears the rumble of the Southern Flyer barreling northward. He looks up to see in the distance the Northern Comet hurtling southward. He thinks, "What a helluva way to run a railroad." The collision between federal and state authority at Ole Miss in the fall of 1962 reverberated the farmer's declaration. What happened at Baxter Hall on the night of September 30 shocked the nation and raised questions about the Kennedys' highly touted flair for crisis management and about Governor Ross Barnett's sanity. It also jarred Alabama officialdom to its senses.

If there were a racial bog in America it was Mississippi. Throughout the rest of the South there existed counties and geographic belts to match the Magnolia State's passions, but there was no state where racial attitudes were so uniform. The Citizens' Council movement started first in Mississippi, and Governor Barnett was its creature. The state had no urban center that was more than an overgrown town, no outside penetration of its business community, and no aggressively organized black population. Overwhelmingly rural, it was structured around smalltown courthouse squares and beauty parlors. Mississippians maintained segregation through a fiercely rigid code of manners that issued from an intimate day-to-day, side-by-side contact between its near 50-50 white/black populations. To offend this code invited retribution of the most tribal sort, including mayhem, mutilation, and/or death. The brutal slaying of Emmett Till in 1955 was but one instance in a centuries-old tradition of maintaining racial order by castrating, literally and figuratively, the black male population. In this sense, Mississippi was Dixie's true heart, and nowhere did that heart beat more truly than at Ole Miss, otherwise referred to in some long forgotten state charter as the University of Mississippi.

To break the color line at Ole Miss required a true maverick. In 1958 Clennon King, scion of the distinguished C.W. King family in Albany, Georgia, was crazy enough to apply, whereupon the state committed him to

Whitfield, the colored asylum for the insane. A year later Clyde Kennard applied at Mississippi Southern College. Not long after, Hattiesburg authorities arrested Kennard and put him on a chain gang for seven years.[17] King and Kennard rolled like loose cannons on a pitching deck in part because they were backed by no organized support group. Unlike Alabama, where blacks had a history of organizing, Mississippi blacks operated in isolation. Medgar Evers might complain to national NAACP authorities about the treatment of King and Kennard, but he could not, even in Jackson, organize committees to assist them.

James Meredith, inspired by John Kennedy's rhetoric, applied to enter Ole Miss in the fall of 1961. Eventually, with support from the Legal Defense Fund and Constance Motley, he cleared the legal hurdles and was ordered admitted for the fall term of 1962. On September 25, the day Martin Luther King entered Birmingham for the SCLC convention, Meredith, accompanied by Chief U.S. Marshal James McShane and John Doar from the Justice Department, held a gothic exchange with Governor Barnett in the corridors of the Woolfolk State Office Building in Jackson. Its upshot was to force Meredith's registration to the Ole Miss campus rather than at the comparatively safer Federal Building in Jackson. For the next five days Robert Kennedy and Barnett played cat and mouse on the telephone. Barnett tried to orchestrate a scene in which he, standing at the head of a thin gray line of state troopers, would call a retreat in the face of overwhelming federal force with drawn guns. The conversations, however, were without common ground, except that both Barnett and Kennedy wanted to save political face. Some comments were off the wall. At the conclusion of one exchange with the President himself, Barnett thanked Kennedy for his "interest in our poultry program and all those things."

If the door between the White House and Governor Barnett often swung comically, the consequences at Ole Miss were disastrous. On the night of September 30, a contingent of 500 U.S. marshals—a motley assortment of prison guards, border patrolmen, FBI agents, and regular marshals—formed a cordon around Baxter Hall. Meredith had retreated there following his registration. Oxford was jammed with Klansmen, Citizens' Councilors, and students trying to register for the fall term while going through fraternity and sorority rush. That night President Kennedy went on the air announcing that the books on the Meredith case could be closed and offering a conciliatory message for the white citizens of Mississippi, compelled to abandon their folkways. As he spoke, a mob surged on Baxter Hall. One fraternity house emptied to join the melee when a rumor spread that a sorority girl had had her left breast shot away by gunfire.[18] State police drifted from their posts. Frightened marshals fired teargas. Exploding cannisters were followed by ballistic fire from shotguns and hunting rifles. By morning 160 marshals had been wounded or injured—28 by gunfire—and two men lay dead, a jukebox repairman hit in the head by a stray bullet and a correspondent for the *London Daily Sketch* shot in the back.

Assistant Attorney General Nicholas Katzenbach watched and prayed for the arrival of reinforcements. He was on the phone with President Kennedy when whoops and shouts announced the arrival of the army. Eventually 23,000 troops crowded into tiny Oxford to protect one black student.

Repercussions in Tuscaloosa

Ole Miss forced the University of Alabama to get serious. Despite Rose's bravura in February 1961 about taking charge and working things out with Governor Patterson, little had been accomplished. A Southern Regional Council official reported that while campus authorities said they were "prepared to accept qualified applications," he saw "little indication that any considerable efforts are being made to secure applicants".[19] In fact, the university discouraged applications. Twice in 1962 Rose ordered early termination for receipt of applications in order to avoid consideration of black applicants and abandoned the university's longstanding policy of automatically considering applications received after the deadline for the next term. The reason given was pressure of enrollment. Among the applications put on hold were those of Vivian Malone and James A. Hood.

Still, Ole Miss woke the university to reality. Talk of calming down George Wallace's rhetoric became more insistent, even as Wallace, reacting to Ole Miss, stepped up his attacks on federal intrusion. Recently appointed trustee Winton Blount suggested the establishment of a trustees' planning committee much like one Rose had set up in the university. Rose suggested that Blount meet with academic and business leaders at the November 6 homecoming meeting of the board.[20] There, seeds were sown for what became the strategy for dealing with Wallace; the use of prominent businessmen to urge him to back off. While Rose and Blount mulled this idea, George LeMaistre, a lawyer and president of Tuscaloosa's City National Bank, began preparing a speech that called for a businessman's peace.

He took the message first to local civic clubs and eventually around the state. Claude Sitton of the *New York Times* heard it and heaped praise on LeMaistre for the courage of his convictions. LeMaistre deserved it. He and his wife quietly contacted the Southern Regional Council for advice and help.[21] A quiet man by nature, educated in law and for a time a member of the law school's faculty, he crafted a message that would enable Tuscaloosa's business community to rally to the cause of law and order. He could expect howls of denunciation from the Klan and the West Alabama Citizens Council along with threats to boycott his bank. Now LeMaistre stood before a Tuscaloosa civic club, tall and strikingly handsome, to measure "The Price of Defiance." When he was done, a relieved audience stood in sustained applause. His words broke an impasse and provided each man with common economic grounds for yielding to the inevitable. A local photographer wrote to express his gratitude. "The grocer, the plumber, the banker, the average business man feels as you do. . . . But they dare not say it because the rubber workers might stop trade, the plumbers might not work, and the radicals might boy-cott."[22]

LeMaistre's calculus was strictly economic. He talked about image and the cost of defiance. He totaled the price paid by Little Rock for Orval Faubus's antics. The message struck a responsive chord. In September, the *Birmingham News* had argued that Carl Sander's victory in Georgia, as well as recent elections in South Carolina, Tennessee, and Arkansas, proved that racism was losing its appeal. Lieutenant Governor James B. Allen condemned the violence at Ole Miss and said that it should not be repeated in Alabama. Richmond Flowers, Attorney General-elect, joined the chorus. Ministerial associations also chimed in: Methodists, Baptists, the Montgomery Pastors' Conference all wrote letters to Governor Patterson urging an end to extremist rhetoric. Even the much maligned Rubber Workers Union in Tuscaloosa called for law and order, though it is doubtful that their statement was ever submitted to rank and file vote. For his part, Governor-elect Wallace said that these pleas for law and order were being sent to the wrong address; instead they should go to "Castro, Martin Luther King, and that crowd in Washington . . . because if there is any breakdown in law and order it will not be caused by Alabamians but from those outside our state."[23]

Rose decided to step forward with his own message. He got encouragement from Bennett, who interviewed senior staffers and advised a plan for "leadership maintenance and coordination on campus." Bennett became particularly insistent after Martin Luther King expressed personal interest in the Tuscaloosa situation. "The publicity surrounding King's announced intentions on Fri., Oct. 19 [to press for desegregation of the university], followed by actual receipt of application on Wednesday, Oct. 23, have presented the campus with an entirely different matter from the relatively abstract debate which ran through the campus on the heels of the U. of Miss. debacle. The question of 'timing,' " he now concluded, "is fast demanding a new answer. We [the senior staff] believe it is impossible to continue total restraint of faculty and students for this reason alone."[24]

Rose needed no further inducement. A draft for a state of the university address began, "From every corner of the state, from every walk of life, and by Governor Patterson and Governor Nominate Wallace, I have been assured that the state will use every force at its command to see that we not have mob violence and bloodshed." As finally published, references to Patterson and Wallace were dropped. Claims by the *Birmingham News* about the demise of racist politics notwithstanding, neither man felt it politic to be identified with talk of compliance. "I have been assured by our authorities," Rose now said, "that the state will use every force at its command to see that we do not have mob violence. I am most encouraged by the response from the citizens of our state. This is a great University, with more than 100 million dollars of facilities; and with a distinguished faculty it carries on some of the most critical work for the space program and our national defense. This University must never become the scene of mob violence, battle troops, or bloodshed."[25]

Mel Meyer

The hundred-million-dollar university had a problem—Mel Meyer. Meyer was a young sports editor who wrote happily in the spring of 1961 about the balance between athletics and academics at 'Bama. Now in the fall of 1962 he was editor of the *Crimson-White* and, by university estimates, a radical. Unfortunately his calls for racial justice came at a time when the university felt it could least afford disturbance. With winds of change making desgregation not only inevitable but imminent, university officials wished to batten the hatches. Throughout 1962 they tried to avoid anything that might draw unwanted attention. One professor was warned about his visible association with the Alabama Council on Human Relations. Students were forbidden to continue joint meetings with fellow students from Stillman College.[26] The curtain was drawn on black entertainers as well. Rose and Bennett convinced the president of the Cotillion Club to cancel an agreement with Ray Charles for a performance that fall, despite a poll in the spring that made Charles the students' choice.[27] Only Mel Meyer refused to be reasonable.

Meyer was the son of a newspaper editor from Starkville, Mississippi. He had followed his sister to Alabama. She debated; he took up journalism. He joined a Jewish fraternity and made friends readily. Nothing in his background (and yet probably everything, being a Jew from Mississippi) could have predicted what followed. It began innocently enough. An editorial in the first fall edition "thanked God for Mississippi" and lampooned Barnett. The column described a meeting of the West Alabama Citizens Council, attended almost entirely by "elderly men and women with a sprinkling of grandchildren" and presided over "by a 300 pound behemoth." The "behemoth" introduced the movie *Birth of a Nation*. Declaring it a true depiction of Reconstruction and of things to come, the fat man, by that time "sweating profusely," said the film was based on a novel by D.H. Lawrence, which led Meyer to suspect that he had just "gotten out of a meeting of the Anti-Sex league."[28]

Raillery turned serious the following week. Two days after Barnett stood Meredith down in the state office building in Jackson, one day after Lieutenant Governor Paul Johnson repeated the scene at Oxford, and three days before disaster struck the Ole Miss campus, the *Crimson-White* heard John Donne's bell. Under the headline "A Bell Rang . . ." the editorial said morality and law were on the side of Meredith. If he could be excluded, who next? Catholics? Jews? For support they turned to Petal, Mississippi's maverick editor and author of the *Magnolia Jungle*, P.D. East: "If I were a Catholic in Mississippi, I'd be worried. If I were Jew, I'd be scared stiff. If I were a Negro,

I would already be gone." Meyer and his friends strained toward eloquence. Touched by events, they gave expression to an awakening conscience, aching to soar. The allusions were close at hand; from Peter, Paul, and Mary's "If I had a hammer . . ." to Donne's oft quoted "No man is an island." "There was no need," the editors concluded, "to send to Oxford this week to see what bell rang. It wasn't the bell of justice and freedom: it rang for you and me."[29]

The closest the editorial came to saying anything about Alabama was to say that Barnett "literally 'stood in the schoolhouse door.'" The reference to Wallace was unmistakable, however, and the response immediate. Rose wrote Judge Lawson in Montgomery to say that he had been on the line with Patterson and with Wallace, "and both felt that we were handling the Crimson and White [*sic*] editorial in the right way." Rose noted that "the editor has received several threats from the Klan and I believe it has impressed upon him the seriousness of the situation." Rose added that the KKK's Bobby Shelton had been "on the campus every night trying to solicit students," but upon a warning from him had stayed away the last two evenings.[30]

Shelton and the Kluxers were but the beginning of Meyer's problems. Judge Lawson's "first reaction to the editorial was simply to say close the newspaper down." But he and other members of the board were sobered by reactions from the Klan and the Citizens' Councils. The Klan threatened to "bust up" a pep rally. James B. Laseter, whose West Alabama Council had been the recent butt of *Crimson-White* ridicule, warned Rose, "You remember we objected when the Crimson Tide went to Philadelphia and played an integrated team. At your request we did not publish our objection in the papers, we did not want to hurt the University. We do not want to hurt it now, but we cannot sit idly by and let Meyer remain as Editor." Soon even the most conservative board members were trying to keep heat off the university. Dixon wrote Walter Givhan, the state's chief Citizens' Councilor, to say, "Summary dismissal of this editor and perhaps one or two others connected with the Crimson and White [*sic*] would probably have resulted in making them martyrs." Rose tried to defuse the situation by spreading word that Meyer had not written the editorial. He told the newest trustee, Ehney Camp, that it had been written "by a young man from Tuscaloosa whose father is in public life and the students are shielding this young man for fear that his father might lose his job." Rose said that the Student Government Association "will allow us to check the editorials on the race issue. We hope this is a guarantee, but as you know the paper is printed in Birmingham and editorials can be changed rather easily."[31]

The paper's staff backed away. Though alert to the new morality, they were not firebrands. For the most part they were a fun-loving, convivial bunch much like their editor. For a month the *Crimson-White* fell silent on race, not even mentioning the riot at Ole Miss except as a reminder when the Klan threatened the October 4 pep rally. By October 25, however, it did report that three black students had applied to the university and noted that Martin Luther King endorsed their effort. In the same issue, emboldened by calls from newspapers around the state urging Wallace not to create trouble, the *C-W* expressed its own view. Observing that the Cuban missile crisis was bringing the nation to the brink of nuclear war, the students implored Wallace, "Let's not have it here in Tuscaloosa. If we've got to fight, let's not fight those damn Yankees again. . . . Let's fight somebody our own size like the courts."

Student acquiescence did not make the problem go away. Besides, what most irritated the Klan and their fellow-travelers—ridicule and raillery—did not stop altogether. The students poked fun at Birmingham's mayor, Arthur Hanes, calling him "Little Art" for his inability to stand up to Bull Connor. The October 25 issue ran an ad announcing "The Formation of the UA Chou Tong Society: Named for the Chinese Communist Who Killed Captain John Birch." Similar ads ran in student newspapers around the country, but the irreverence enraged local Birchers. The threats against Meyer from the Klan became so serious that the university hired a detective for his protection. The detective followed Meyer and his friends for the remainder of the semester, occasionally showing more interest in the hemline of Meyer's girlfriend than where carloads of Klansmen prowled. But in all it was a friendly vigil, with Meyer and his buddies often giving their hired shadow the slip.[32]

The effort to protect Meyer was not appreciated in all parts. In November, Bennett received a call that took him by surprise. The voice at the other end belonged to Seymore Trammell, soon to be Wallace's finance director. Trammell was calling at the governor-elect's request. Bennett grabbed a pencil and began scribbling notes as fast as he could. Trammell wanted to know who was paying for Meyer's guards. Bennett told him the university. Trammell said "that would go 'pretty bad with the Legislature—as bad as everybody needed money.' " Trammell did not stop. "George," he said, "wants Meyer expelled." In disbelief Bennett asked him to repeat it. He did. "Well I'll be goddamned," Bennett breathed audibly.[33]

Wallace, Patterson, and the university

The university never knew whether to be afraid of Wallace or embarrassed by him. Perhaps it was the way he curled his upper lip. Despite a solid, middle-class upbringing and an enormously successful political career, he always seemed to be sucking toothpicks. His style was too jaunty, too feisty, too twangy. He had too much energy and too much ambition. Worse still, Fort Payne mechanics, Northport rubber workers, and wiregrass dirt farmers loved him. A natural tenor, his voice flattened into that familiar nasal whine, especially when he spoke about the "average" citizen of Alabama. Average was his métier. Average man, average woman, average citizen, average voter—they became a mystical force in his speeches. Unfortunately for the university, it took pride in standing against the average.

If an individual symbolized the gulf between the university and Wallace, it was Bennett. Those close to the university and the governor believed that

Bennett's antipathy for Wallace poisoned the relationship between Rose and the governor. Bennett thought Wallace was a demagogue who would stop at nothing to improve his own fortunes. Wallace knew what Bennett thought. More generally he knew how his thinking represented substantial opinion within the university. It did not help, therefore, when on Wallace's election, Bennett wrote to him: "I am sitting here remembering our arrival on this campus as incoming freshmen on the same train and the evenings we spent in the old rooming house on Tenth Street, debating where our paths would lead in the future. You predicted yours right on the button and I am confident that the University of Alabama family is delighted to know that another alumnus will soon undertake the heavy responsibilities of the Governorship of the State."[34]

Wallace took it all in stride, even the "kissing up." He showed up for the 1962 homecoming pep rally and dutifully praised the football team and Bear Bryant. Ironically Governor Patterson carried the Wallace torch that day. The trustees had assembled for their annual homecoming meeting. The principal order of business was to show support for Rose in the desegegation crisis looming ahead. Everyone marched to the same drummer. Gessner McCorvey, the old Dixiecrat warhorse, introduced a resolution supporting Rose's leadership and his determination to maintain law and order in the event of court-ordered desegregation. McCorvey always had voted to do the lawful thing, even back in the dark days of 1956, and now he resigned himself to making the university's passage as easy as possible. His resolution was seconded by Dixon, who also was slipping quietly from the scene. Rose could not have had truer university men nor, for that matter, truer segregationists pointing the way, but John Patterson would buy none of it. He took the floor and began preaching. He recounted the struggle to maintain segregation and declared that the battle was not over—there were "legal tools yet unused." He begged them not to say anything. "No one questions that this Board is for law and order." The public, however, would interpret such a resolution as "backing off." "Why make an announcement that will only give aid and comfort to those who would integrate the university? What if there were a negro applicant out there who was still uncertain about coming here, this might give him comfort."[35]

Though surprising, Patterson's outpouring was consistent with his racial thinking. He had pressured Rose to hire a psychologist who was helping the state prepare its defense in desegregation cases. Rose refused. In defense of segregation, Patterson desperately sought out discarded scholars and discredited scientific views. He paid $3,000 to W.C. George, controversial professor emeritus of histology and embryology at the University of North Carolina, to do a study on Negro students in Alabama. Already George had concluded that because of brain size Negroes were inferior to whites and said that "integration is not Christian." When asked why it was not Christian, George said, "Integration is evil. Doing evil is not Christian."[36]

So Patterson's speech was based as much on conviction as politics. Moved by the governor's forthright talk, Dixon wavered. He asked Rose to consider a more propitious time for releasing the board's statement. Patterson said that no time was right. He told his astonished listeners that bad things occur when the state makes an announcement in advance. He said that Bobby Kennedy had asked him to make an advanced guarantee of law and order in the case of the Freedom Riders. Because he finally relented, he was accused of having welshed on the deal when violence broke out at the Montgomery bus station. Had he said no to the request, they simply would have sent in federal marshals in even greater numbers. It was best, he thought, simply to say nothing.

Caddell finally got the discussion on a more productive path. He thanked the governor and President Rose for having done so much to maintain segregation, but he insisted that the resolution was necessary to strengthen the university's hand in federal court. Rose thanked Caddell for his generous remarks about his personal efforts to maintain segregation. He said that his refusal of some 213 applicants over the past few years had not been meant to usurp the board's prerogative to regulate admissions, but was necessary to maintain segregation.[37] Having proved themselves stout, the discussion centered on points in the resolution, until Dixon, tiring of the process, said it was good as it stood and should be passed, which it was. Patterson abstained. Rose emerged the victor. He had one more weapon in his arsenal to achieve desegregation peaceably; one more vote for him and against Wallace's planned demonstration at the schoolhouse door.

Wallace left for Mississippi to speak before a White Citizens' Council rally, complete with FBI informants, and to the state legislature. Nine in ten students, reported the *Crimson-White,* rejected Wallace's assertion in Mississippi that "removing politics from the operation of schools is removing the people." On November 30 Rose ordered receipt of applications for the spring term closed, again freezing all applications from black students. The *Crimson-White* noted the official reason as "the pressure of enrollment." "Few faculty members and students," concluded the reporter, "were taking this as the real reason." (In fact, only 78 applications remained to be processed.) Nor did the national press greet the action as other than a quiet surrender to Wallace, recent resolutions for law and order notwithstanding.[38]

The Kennedy administration, however, understood Rose's pragmatism. Since Ole Miss, the Kennedy people had been keenly interested in what might develop at Tuscaloosa. When Rose told them of his plans to stop applications, thereby throwing any chance of desegregation into the summer or fall of 1963, they no doubt breathed a sigh of relief. The smoke of Ole Miss had scarcely dissipated and the Cuban missile crisis was only a month old. Rose talked to Burke Marshall, head of the Justice Department's Civil Rights Division, and assured him of the university's good intentions and the wisdom of delay. Marshall recorded the names of the four most promising black applicants and listed the telephone numbers for both Rose and Bennett.[39]

On the same day, Senator John Sparkman relayed a message to the President through presidential adviser and former Arkansas congressman Brooks

Hays. Whether Sparkman acted on instructions from Rose is not clear, but Sparkman purported to speak for both the university and Wallace. "This may sound strange," the senator confided to Hays, "but it's true that the situation in Alabama is going to be much better than in Mississippi." Sparkman said that the university would not give "Wallace the power that was handed to Ross Barnett." Moreover, Sparkman reported that Wallace had asked him to plead with the President for "breathing time," and that while Wallace had vowed to resist "bodily," "he now adds . . . that he is under no misapprehension as to his ability to accomplish anything."[40]

The telegrams and correspondence between the university and the White House illustrated two things: Rose and the Kennedys were ready to work together and both wanted more time. Sparkman's message illustrated one other thing: no one knew what Wallace thought.

A Spartan refrain

Having bought time, the university could enjoy its Orange Bowl invitation. President Kennedy would be there, and Governor-elect Wallace had been invited—but he declined, knowing better than to share a political stage on less than equal terms. The Bear had had another great season. Under the field generalship of sophomore Joe Namath, the Tide opened the season smashing Georgia 35-0. Only a one-point loss to archrival Georgia Tech marred the campaign. There was bad blood between the schools. The Tech game saw Bryant appear on the sidelines wearing a business suit and a football helmet for fear of bottles thrown by Tech students. Like Wallace, Bryant knew how to be the center of attention. And as with Wallace, school officials did not always know what to make of their coach's popular style. Bennett confessed that the university had been embarrassed by "the potato chip and Coca-Cola sponsorship of Bear Bryant's Sunday afternoon show."[41] Needless to say, Bryant made tearing open potato-chip bags and swigging Cokes a trademark of the most watched Sunday afternoon show in Alabama.

Football in the South was so enwrapped with state and regional pride that one could barely separate Saturday afternoon heroics from the politics of race. It was our little skinny boys against theirs, and if theirs happened to be from outside the South, then it was their big, well-fed boys. It was as if the strong young men dressed in Alabama crimson and white transformed magically into butternut-clad Confederates, gaunt, ghastly, readied for battle. The mystique of southern football was such that JFK tried to get Johnny Vaught, the Ole Miss coach, to help quell the riots on that campus. Now Kennedy planned to be at the Orange Bowl where the Crimson Tide would square off against Oklahoma's Sooners.

The bloody-but-unbowed dramas on football Saturdays amounted to guerrilla theater, each being a re-enactment of Thermopylae or, more recently, any Civil War engagement. No one followed these plays more enthusiastically than Gessner McCorvey, but for a month now he had been a troubled man. The resolution passed at the board's November 10 homecoming meeting ran counter to Spartan expectations. As if to make up for his apostasy, McCorvey wrote to Rose in December asking that Wallace be invited to the Orange Bowl. The trustees, he thought, may have "misjudged George in thinking he was criticizing our Board . . . in these 'broadsides' he has been giving to 'resolution writers' "—a reference to Wallace's suggestion that the board's resolution be sent to Castro or Martin Luther King. "We certainly did not intend," McCorvey assured himself, "to aim anything at [Wallace] in our resolution but, of course, there are a lot of people who insisted that our resolution was telling our colored friends that we are rolling out the red carpet and would welcome them to Alabama 'with open arms.' "[42]

For McCorvey the unkindest cut came when his old friend and longtime ally, J. Miller Bonner of Wilcox County, published an open letter in the *Alabama Journal*. Bonner, one of the Black Belt's most powerful planters, decried the action of the board of trustees in resigning themselves to integration. He compared their work unfavorably with Wallace's recently announced State Sovereignty Commission. "It is better to die a noble death fighting," he scolded the board, "even if we know we cannot win, rather than to surrender to an ignoble death." "If two must be killed and scores injured, as at Oxford, or if ten thousand times two must be killed, I say, deliberately and solemnly, we will pay that price to prevent integration, intermarriage, and mongrelization in Alabama." Rather than resolutions, Bonner concluded, Alabama needed a Leonides with 600 Spartans. McCorvey virtually wailed in reply. "I just did not think my old friend, Miller Bonner, could write such a letter about me." Could Bonner not see that the University of Alabama was the last holdout for segregation? "Don't get the idea," he chided, "that our task has been an easy one and that the reason we have been successful in our efforts to maintain the kind of university you would like for your Alma Mater to be is . . . simply because we were 'lucky.' "[43]

With President Kennedy watching, the Crimson Tide ushered in 1963 by shutting out Oklahoma 17-0.

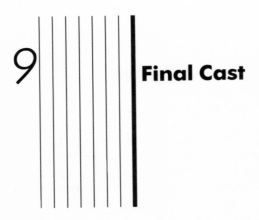

9 | Final Cast

In December 1962, the children at Verner Elementary School on the south side of the campus got the thrill of their lives. It was the last day of classes before Christmas holidays. A training jet from Air National Guard headquarters in Birmingham appeared over the campus, switched on its camera and began taking reconnaissance photographs. The pilot pulled up, surprised when an RF-101 Voodoo streaked in from the east also making photos. To the delight of the children, the two airmen decided to show their stuff and began a series of low level runs that included rollovers and thunderous sound effects. Another Voodoo and trainer arrived a bit later. The show lasted about two hours.

Despite one parent's concern that a flameout could have ruined "Christmas in Tuscaloosa," the display amounted to little more than a moment of slack-jawed amazement for the kids, and as it turned out, comic relief for the impending confrontation between federal and state authority in Tuscaloosa. The reconnaisance had been ordered by the Justice Department and channeled through the Vice Chief of Staff of the Air Force. The request went to the Air National Guard (ANG), and as a backup, to Shaw Air Force Base in Sumter, South Carolina, which accounted for the Voodoos and the confusion. ANG Colonel Bill Teas was told that "the request came from topside and the pictures must arrive at Langley Air Force Base by 1300 EST 20 December to be dispatched by a staff car to make the deadline of 1700 EST arrival in Washington." As comedy would have it, mechanical malfunction and clouds aborted the mapping, and Washington had to be "satisfied by copying the University of Alabama part of a mosaic previously taken of Tuscaloosa."[1] The episode gave the Wallace administration and the local press ammunitition for ridicule—more often than not related to recent overflights and misadventures in Cuba.

More seriously, the mapping represented the Kennedy administration's obsession with avoiding another Ole Miss. They did not know how much they could count on local officials, and Wallace remained the complete enigma. The early Wallace, reasoned some, the one who supported Kissin' Jim and the national Democratic party, was the real Wallace. Refusing to be "out-niggered" was strategic. An Anniston correspondent typified the intelligence Washington received on the governor: " . . .he does not believe in segregation as a Southern way of life. He was quite liberal when he was Folsom's main speech writer [an exaggeration]. . . . The last time he ran for Governor he looked for the Negro vote and was defeated. He made it this time with the old cry and it is a lesson he will never forget. He is convince[d] that the South will hold the balance of power in the race in 1964 and that he will hold a big place in it. He told me this himself."[2] The correspondent had the facts right. He erred in the thought that Wallace's words were more apparent than real. He ignored, as did many others, the self-hypnotic powers of Wallace's rhetoric.

For Wallace, his own 1962 campaign gave racism the imprimatur of democracy. How else could he have surrounded himself with John Kohn, a Montgomery lawyer whose elitist racism disdained all lower orders, or Ace Carter, who learned the power of language from Gerald L.K. Smith and who hated Catholics and Jews only slightly less than "niggers"? Indeed, it was Carter who penned the inaugural lines "segregation now, segregation tomorrow, segregation forever."[3] On that cold day in January 1963, Wallace believed every word of his speech. He was as thoroughgoing a racist as the civil rights movement could have invented. The only people more deceived than those who held out hope that Wallace did not mean what he said were those who believed his extremism would undo him. Charles Morgan, a young, liberal Birmingham lawyer, detested Wallace and all he stood for. After visiting Morgan, the Southern Regional Council's Paul Anthony said, "The Governor's inaugural address impressed everyone. Some surprised and discouraged. Chuck Morgan is delighted: 'we couldn't convert him, his defiance helps our chances of putting him in jail.' "[4] Morgan's was a fool's paradise. He underestimated the capacity of the privileged race, class, or gender to rationalize its advantages as "natural." Unlike Morgan, the Kennedys understood Wallace's power in the same way they understood Boston's ward heelers, often unsavory characters but essential to getting business done. They spent considerable time figuring some political calculus that would make Wallace less dangerous. In the end, it was the Kennedys, not Wallace, who earned the opprobrium of history for betraying their rhetoric.

In mid-February a triumphant Wallace took his first official out-of-state visit to Washington, home of the enemy. On arrival, he chortled, "I took some aerial views of Washington as I flew in. I can give assurance there are no nuclear launching pads on the University of Alabama campus." He told Drew Pearson "they will have to arrest me before they integrate the University of Alabama. We will not close the University, but we will not be pushed around by a court ruling that is not the law of the land. . . ."[5] Over the next months the governor proved the most consistent actor in the schoolhouse drama. Only those who wished to make more or less of his words were surprised.

Holding the university to account

Wallace had a way of making Frank Rose squirm. Returning from Washington, the governor called a meeting of the trustees in Montgomery. He wanted to know what the university was doing about mounting pressure to desegregate. Seymore Trammell, who prompted the governor's monologues with little asides, sat in, as did Cecil Jackson, 1957 Student Government Association president at Alabama and now Wallace's legal adviser. Anticipating a confrontation, Rose met with the executive committee to prepare for this first direct exchange with Governor Wallace. As the board filed into Wallace's office, they hoped for the best. Patterson's warning that the board's resolution in November would tie Wallace's hands was a worry the board need not have feared. Wallace knew what they had done and why, just as he knew what he was going to do even if nobody else could be sure. Wallace asked Andrew Thomas if the Autherine Lucy injunction still applied now that Hubert Mate had replaced Adams as dean of admissions. Thomas equivocated, and Wallace insisted that the university pursue this profitless legal avenue. Wallace wanted to know the status of Negro applications. University officials breathed easier. They boasted that over the past year they had received 236 letters of inquiry and applications but none had cleared the elaborate hurdles erected since 1957. Black applicants had to take additional tests and provide more detailed information about their transcripts on pretext that the university had no prior experience with the schools which they previously attended. University officials also discussed specific applications and what they were doing to discourage them.[6] Despite the tension, all went well. The governor did not rant and rave, and the university proved itself a worthy guardian of the customs and laws of Alabama.

University officials still felt uneasy. Since late 1962, they had prepared for desegregation but now had to buck a governor determined to resist. No one doubted who was the more powerful in any confrontation between Montgomery and Tuscaloosa. Rose's talent for managing what people thought of him worked, at least in the early going, with Wallace. The trustees helped. They knew they had a prize in Rose and worked hard to make him happy. In early February, Ernest Williams asked Wallace's good friend, McCorvey, to urge the governor to voice his support for the university's president. Williams feared that Rose might leave Alabama for the University of Kentucky, whose president had recently resigned. "I do not feel that he has any desire to leave the Capstone," Williams wrote, "and would do so only if he felt the current political climate would make the program of the administration intolerable." Williams had little to fear. There is no evidence that Rose made Kentucky's short list. Moreover, Wallace's performance at the February 23 meeting relieved anxieties. Wallace was not about to change the "political climate" for Rose's convenience, but he did not treat Rose and the university as the enemy.[7]

By nature most university presidents are political creatures. Every day is scheduled from morning until evening: letters by the gross to dictate, programs and policies to be talked up, interviews and conferences requiring the utmost diplomacy. Words in such profusion become a first line of defense, a way of extricating oneself from daily crises of no lasting importance, as when W. R. Jones, chief of the Investigative and Identification Division of the Department of Public Safety, turned up the screws about university efforts to maintain segregation. Rose replied testily, if evasively, "The closing of enrollment, as you know, has always been our practice and is one of our legal steps. You know us well enough to know we don't need any encouragement or advice on closing our registration."[8]

Rose's words placed himself at the center of drama, but also removed him from accountability. This aspect of his character came to be reflected in the way the two protagonists in the desegregation crisis, Wallace and Robert Kennedy, viewed him. When dealing with Wallace directly, Rose positively fawned. "It is impossible for me to convey to you fully," Rose wrote, "our personal gratitude for your participation on the occasion of our annual Governor's Day. . . . Your address [a speech the SGA president felt compelled to rebut] was the finest I have ever heard on this campus and was most widely accepted with favor by our faculty and students." Of course common courtesy dictated a response, but this effusion was beyond bounds. It typified Rose's correspondence with Wallace. Though Rose remembered himself as confrontational with the governor, the record does not support it. By the end of the desegregation crisis, Wallace regarded Rose as helpful yet he openly distrusted Jeff Bennett, even though Rose and Bennett dealt from the same hand. At the same time, Robert Kennedy admired Bennett while puzzling over Rose, their longstanding acquaintance notwithstanding. "[Dr. Rose] was on our side," Kennedy observed. "At least as he talked to us, he was on our side. . . ."[9] Enigmatic behavior, some thought it duplicitous, was a part of Rose's personality. It produced his success, even as it detracted from his leadership. He seemed to operate from no fixed compass.

In university councils Rose walked tall. However he exaggerated accomplishments or fudged the truth, he gave clear signals to trustees, faculty, staff, students, key alumni, and supporters that desegregation was coming and that the university must prepare for it. The same man who could boast to Wallace about steps taken to delay desegregation enlisted the university community in plans to prepare. Everyone worked on the planning, none harder than Bennett. By early March he was in steady contact with Burke Marshall to advise the Justice Department about the status of black applicants and university planning. Marshall came south to establish personal contact with the applicants. On March 6 he was at the Space Flight Center in Huntsville where he met Marvin Phillips Carroll and Dave Mack McGlathery separately in the office of Wernher von Braun.[10] It appeared that the Huntsville Center might be the best location for desegregating the university.

Dave Mack McGlathery

Dave McGlathery never wanted to leave Huntsville, but before 1962 the Redstone Arsenal did not hire graduates of Alabama A&M, unless they were willing to work in menial jobs.[11] McGlathery, the son of a sharecropper, was born in April 1936 at a place called Pond Beat, the present-day location of the Marshall Space Flight Center. When he was three or four, the family moved to Huntsville on Church Street near the depot, and Dave's father went to work for the Arsenal as a laborer. They stayed until 1944, when to avoid the draft, his father moved to Chapman Mountain to take another sharecropping arrangement. The children were happy on the farm and did not mind the three-mile walk to Winston Street School; the school year was short because of farm work. As Dave grew older, he had even less time for school. He would pass first semester exams by cramming just enough to get by. During the second, he would go full time and get his grades up. He never missed a year and graduated on time in 1954.

Life on the farm was happy. His mother, Anna Langford, was attentive to the children, and grandfather William lived with them. Dave always wanted to pick more cotton than anybody in the field. A few times he picked over 300 pounds. His reward was a pat on the back from his father, accompanied by the boast, "My son can pick more cotton than any of you guys." This family of six never made much money but never went hungry either. In 1948 they left the Chapman area and moved to another farm. Things were never so good again. One day Dave looked up to see the owner, Mr. Charlie, strike his father, then prevent him from retaliating by holding a gun on him. As a result, they moved to the Blue Springs Nursery where Dave's father again signed on as a tenant. In 1952 his parents divorced, and the children moved with their mother into a dilapidated house on the Bradford place. Soon Mr. Bradford built them a cement block house which they liked, and Dave put his mind back on school. After junior varsity, he gave up basketball because he was too small, 5'6" and 118 pounds at graduation, and took up snare drumming in the marching band. He made friends for the first time with children from the teaching middle class and knew that he would have to go to college if he were to join their ranks. He signed on with the Navy in 1954 to pay for his education.

Boot camp at Great Lakes gave him a black company commander. He was impressed, but on duty he saw all-black companies working as steward's mates. He decided never to accept service in one of those units. He spent most of his time in Southern California and in Hawaii where he learned hydraulic maintenance for F9F Grumman Fighters and took charge of a crew. Mustered out as a second-class petty officer in 1958, he had never made an error on his watch, and, equally important for a small lad, he attained a height of 5'10" and weighed a solid 175 pounds. GI bill in hand, he returned to the comfortable surroundings of Alabama A&M. The college on the hill offered a parochial education. He majored in math, his weakest subject, but the instructors never went into much depth. In his last two years, he made straight A's, was named "Mr. Brains" in the college yearbook, and graduated magna cum laude. It was a proud day for the McGlatherys and a beautiful day in Huntsville.

Wanting to stay in this happy valley, McGlathery applied for teaching positions. He filled out forms at NASA, but had little hope of success despite a 91 on the qualifying exam which, with a military service bonus, entitled him to a GS-7 rating. He also applied for graduate study at Vanderbilt and was told that the university did not accept blacks. Despairing of success at home, he went to the Naval Weapons Laboratory at Dahlgren, Virginia, as a GS-7 mathemetician. Within a year, he married, took his wife to Virginia, and returned to Huntsville when a position opened at NASA in October 1962. The excitement of McGlathery's new job did not blind him to his superficial background in math. So on December 8, 1962, he applied to take undergraduate courses at the university's Huntsville Center. He told no one.

Five days later Dr. Russell D. Shelton, Chief of the Ion Physics Branch of the Research Projects Division, called McGlathery into his office and told him that his application had "hit the main campus." Shelton, who held progressive views on racial matters, also was the liaison officer between NASA and the university. He told McGlathery that the university wanted him to withdraw his application for the winter quarter. McGlathery agreed. He did not want trouble, and he knew that even Shelton's cautious support was not matched throughout NASA. His immediate superior, Dr. Ernst Stuhlinger, strongly opposed his applying. He seemed excessively sensitive to "the environment that the German team was operating under"; knowing "they were in a region of the United State that was . . . racially volatile." Nevertheless, on February 9, he applied for admission to the graduate school for the spring term. Shelton advised the university that he could not get him to change his mind. McGlathery wished to avoid publicity. Other than his superiors, no one outside the university knew of his application. In fact another employee at NASA, Marvin Phillips Carroll, was the first to draw attention.

The Huntsville support group

Carroll was the same age as McGlathery.[12] Except for a rounder face and larger eyes, he could have passed for his brother, but neither knew the other. McGlathery's family ties to Huntsville drew him into church and social connections that lay outside the city's middle class. Carroll, a newcomer, readily made friends among the black elite because of his employment at the Arsenal. One of his earliest acquaintances was John Cashin, a dentist who with his wife Joan led efforts to end segregation in the Rocket City. Carroll and Cashin belonged to the same fraternity, Omega Psi Phi, and frequently found themselves at the same card table.

Carroll grew up in Atlanta known as Marvin Philips, later adopting his mother's last name.[13] He attended Morehouse for a year before transferring to Howard University to study engineering. Despite generally average grades, he made honor societies in engineering, mathematics, and ROTC. He was elected to the student council and named to *Who's Who in American Colleges and Universities*. He graduated in 1958 with a degree in electrical engineering and became the first black to participate in Allis-Chalmer's graduate training program. After a year in Milwaukee, he took positions at the Emerson Research Lab in Silver Spring, Maryland, and later at the National Bureau of Standards in Washington. Coming to Huntsville in November 1962, the Correlation Branch of the Electromagnetics Laboratory of the United States Army Missile Command employed him and gave him duties relating to the analysis and synthesis of passive missile homing devices.

Carroll and McGlathery shared a common motive for pursuing study: getting ahead. The Center offered some 260 undergraduate courses and 60 graduate classes per year. The graduate offerings concentrated on physics, math, and engineering. Moreover, the Center agreed with the Army to offer special courses for personnel at the Arsenal. At the very moment Dr. Shelton sought to dissuade McGlathery, the university got a $600,000 NASA grant for the Research Institute. Another $1.8 million was being considered. These facts were not lost on Carroll or McGlathery or on those who supported them.

Their support group came to include the Cashins and others in the Community Service Committee, the Alabama Council on Human Relations, and the Anti-Defamation League (ADL). Dr. Raymond Ettinger of the ADL corresponded with the Southern Regional Council to keep the Atlanta office current on developments in Huntsville.[14] Carroll was the first to be noticed by these activists because he frequented the Cashin home and let his intentions be known. Dr. Cashin saw value in desegregating the University Center before local groups attempted to integrate the public schools that fall. In late January he encouraged Carroll to apply for a graduate course. Carroll got permission from his supervisor and a tuition grant from the Army. The Cashin group sought to enlist the NAACP Legal Defense Fund, which promised to inform the Justice Department. However, the Fund's reaction seemed unenthusiastic, and on February 5, Dr. Ettinger encouraged the Southern Regional Council to "move rapidly [to contact the Justice Department] as Spring Quarter is fast approaching." Ettinger said "that a second Negro from Redstone Arsenal may have applied to the Huntsville extension center for the Spring Quarter—even before Marvin Carroll. (When it rains it pours.)"[15]

Meeting at the Cashins', the Huntsville group talked again of a desire for "a political lever to avoid a crisis over the public schools."[16] Each member took a specific task: one to talk with the chief of police, others to talk with political and financial leaders, another to witness the application process at the Center. They wanted to avoid publicity that would make the applications appear to be initiated by committee. Meanwhile the university gave signs of trying to sandbag the applications.[17] Getting the right transcripts proved tortuous. Two of McGlathery's reference letters failed to appear despite declarations from all writers that they had been mailed. McGlathery had to take an additional test, but could not find out what kind of test it was. He investigated irregular post-graduate status so that he might take courses without an entrance examination. In the end a lawyer at the Southern Regional Council advised him to take the test, and he did.

Through it all, officials were unfailingly courteous, if slow. University counsel cleared correspondence between officials in Huntsville and the applicants. At the same time the university began working with the Justice Department to effect the eventual change. Burke Marshall's visit with McGlathery and Carroll on March 6 was held with university cooperation. On information from Bennett, Marshall advised the two men that they would not be admitted without another court order. They therefore could no longer entertain the idea of enrolling in the spring. He gave them his personal number in Washington in case trouble developed. The delays, however, had other significance. All parties, except those in Huntsville, began to see Tuscaloosa as the more significant venue for desegregation. The Legal Defense Fund decided to press for victory there.

Thus it came to pass that the most logical site for quiet desegregation, Huntsville, was set aside in the interests of the movement and, as it turned out, in the interests of George Wallace. The University of Alabama had become a last-ditch symbol and would have to be defended and taken by show of force.

Three for Tuscaloosa

The shift to Tuscaloosa represented the broader interests of the movement but also reflected an amalgam of concerns within the state civil rights community. Since the Patterson-engineered demise of the NAACP in 1956, factionalism among black leaders in Alabama increased. The state convention of the NAACP had difficulty coalescing the interests of rival groups, and after 1956 the absence of a common tent made concerted action difficult. In Birmingham, Fred Shuttlesworth's Alabama Christian Movement for Human Rights affiliated with the SCLC. This was a constant source of irritation to the NAACP old guard which tried to rally under W.C. Patton's Alabama State Coordinating Association for Registration and Voting. In the capital city, the Montgomery Improvement Association, after giving birth to the SCLC, struggled to keep local issues alive in the face of harassing law suits brought by city officials. Huntsville's Community Service Committee chafed at having local concerns submerged; besides, its 14 percent black population believed its problems were different. In Mobile, John LeFlore, who long headed the state's second most active NAACP branch and who often resented Birmingham's favored status with the national office, formed the Non-partisan Voters League and Citizens Committee. With his friend the Reverend Joseph Echols

Lowery of Warren Street Methodist Church, LeFlore moved after 1956 to identify Mobile's organized black community with King and the SCLC.

As executive secretary of Mobile's NAACP branch for over thirty-five years, LeFlore fought job discrimination in government offices, segregated public transportation, police brutality, and disfranchisement. He raised money, corralled new members, and attended state and national meetings. More than anything else he wanted his branch to win the Thalheimer award, the NAACP's highest recognition for an individual chapter. Though occasionally churlish about not winning top honors, LeFlore nonetheless pressed human rights with unrelenting courage, even risking his own job at the post office. He encouraged young people to get involved. In 1955 he started a youth council with Cecilia A. Mitchell as its president. Mitchell had just become the first black to attend Mobile's all-white Spring Hill College, a Catholic school.[18] One student working with LeFlore was Vivian Malone, a sixteen-year-old attending Central High School. Malone would go to Room 5 of the Franklin Building on Davis Avenue to volunteer for secretarial tasks. The work not only imbued her with the spirit of the movement but brought her to the attention of community leaders. She needed little introduction, for one of her uncles, Joseph Malone, already had an active role in ward politics, especially in registering voters, and was a close associate of LeFlore.

Vivian Juanita Malone was born the fourth of eight children on July 15, 1942.[19] Her father worked in maintenance at Brookley Air Field, where her mother, a former domestic servant, also worked. The family insisted on education and her older brothers had already gone to Tuskegee. Malone's parents lent their support to all efforts to end segregation. Though not members of the Reverend Lowery's Warren Street Church, the Malones lived within walking distance and often attended meetings there. They always celebrated Emancipation Day as a family. While others played on the holiday, the Malones scrubbed, dressed, and went to church. The most memorable of these celebrations brought Martin Luther King to Warren Street where Vivian's youth choir sang. Such occasions made her feel that "she had to be involved some way." These activities and associations did not destine Vivian, as they had Pollie Anne Myers, to move to the forefront. She was by nature a more reserved person. They did make her susceptible to a call which came one evening as the 1960 fall term approached.

A neighbor, Lewis Koen, returned from a Non-partisan Voters League meeting that evening where discussion had to do with selecting four or five students to desegregate the university's Mobile Center. Koen knew about Vivian's exemplary record at Central High, and she was ripe for the suggestion. She had applied to Alabama A&M, but a mix-up in sending her transcripts delayed enrollment until February 1961. Moreover, she wanted to study accounting, an ambition duly noted in her high-school yearbook, and neither A&M nor Alabama State had accounting programs. She readily agreed to Koen's proposal and went with her compatriots to the Mobile Center where officials let them fill out forms but denied them admission.

Applications to the university and its campuses came from all parts of the state, some individually motivated, others as part of movement activities in places such as Gadsden, Huntsville, Mobile, Birmingham, Tuscaloosa, and Montgomery. From Alabama A&M alone, 86 students applied to the school of nursing in Birmingham on January 15, 1962. From wherever they applied most were told that quotas were filled or enrollments closed.[20] Despite president Rose's report to Governor Wallace that the university had turned aside more than 230 applicants, in later years he said the university actually sought qualified applicants, even asking for Martin Luther King's help.[21] No evidence corroborates that remembrance. Though the university could have become involved at any time, it played no part in selecting the candidates.

From all the private acts of heroism that produced the groundswell of applications, only a few would be there for the closing act, and not because the others were without qualifications. Many dropped out because of intimidation. All were being investigated by the Department of Public Safety. Shortly after going to the Mobile Center, Malone and her family had a visit from two white men purporting to represent the state. They told the Malones that their daughter's actions had disturbed violent elements in the community and that they could not protect the family from retaliation.[22] Some applicants dropped out because they could not afford the delay and the hassle. Some were not qualified. From the shortened list, movement leaders themselves picked the final group.

When Martin Luther King came to Birmingham in late September 1962 for the SCLC convention, he talked with local leaders about the University of Alabama case. Ole Miss was going forward, albeit with violence, and Alabama seemed ripe for the plucking in emancipation's centennial year. Besides, King had met James Alexander Hood. Hood was an attractive young man. When Florida A&M's Bob Hayes became the world's fastest human at 9.3 seconds in the hundred-yard dash, Hood was little more than a step behind, clocked at 9.6 while still a senior a Gadsden's Carver High. Hood also played halfback on the football team and presided over the student body. His popularity and good grades earned him a trip to Montgomery where he met then Judge George Wallace. "We had a coke in his office," Hood recalled. "He didn't use the word 'Negro' once." Hood preached his first sermon at age twelve and harbored some thoughts of going into the ministry. He was rather high strung and suffered bouts of anxiety, or what his mother called "nervous spells." He compensated by covering awkward moments with humor. When a reporter asked whether his family had pets, he first said no, then added: "We have a police dog imported from Birmingham."[23]

The oldest of six children, Jimmy was born in Gadsden on November 10, 1942. His father, Octavie Hood, operated a tractor at the city's Goodyear Tire Plant. His mother, the former Margaret Hughes, worked at home. The entire family was proud of Jimmy, and equally proud to be part of Gadsden's civil rights movement. When he applied to the university, Hood was in his second year at Atlanta's Clark College where he maintained an A-minus average. He regularly attended SCLC, SNCC, and CORE meetings, and during his first summer home helped organize Gadsden's Citizens Committee, an affiliate of the SCLC. In the spring of 1962, the *Atlanta Constitution* reported that Governor Patterson had paid a professor emeritus from the University of North Carolina $3,000 to study the intellectual ability of Negro students in Alabama. The professor, W.C. George, concluded that because of a thinner cortex "Negroes are about 200,000 years behind whites in developing brain structures associated with higher mentality." A light went on. Hood had taken an intelligence test in high school, which he now concluded must have been part of Professor George's study. Incensed, he went to his philosophy professor, who suggested that Hood research the subject and write a letter to Ralph McGill, editor of the *Constitution*. With the help of Julian Bond, a rising star in SNCC, among others, he did his research and wrote the letter. McGill did not publish it but suggested that Hood send it directly to Professor George, who responded that Hood had not "attained sufficient intellectual maturity to challenge his findings."[24]

Seeing that Hood's outrage was not appeased, Andrew Young, whose acquaintance Hood had made at SCLC headquarters, suggested that the young Alabamian take more direct action by applying for admission to his state's university. Hood's ministerial background inclined him toward a career in clinical psychology, and Clark College could not satisfy that interest. So Young, according to Hood's account, put him in touch with Legal Defense Fund lawyers, and Hood made application. He failed to designate race on his first application, and the form came back with a red checkmark by the blank space. He returned the form declaring himself to be "Negro, American Negro."

It is possible that Hood's application explains in part Martin Luther King's passing interest in the Alabama case and his declaration of support in October. Meanwhile the state's Legal Defense Fund's lawyers, Arthur Shores and Fred Gray, checked to see which other applicants were still available and most likely to succeed in a final push. When Malone received her call from Shores, she did not have to be persuaded. She longed for an opportunity to get an accounting degree from an accredited institution and to strike a blow against segregation. Her parents pledged their support. A call also went out to Sandy English of Birmingham, who attended Stillman. Others were called but these three seemed most promising.

Sandy English would not make the final cut, but in the fall of 1962 he too showed promise. Part of his problem was Stillman College. Located in the shadows of the university, Stillman did not want to risk the displeasure of its big brother. In English's case, the college chaplain called to say that his credentials were questionable and that English along with two other applicants "had been advised by the Negro leadership not to make application to the University of Alabama for February, 1963."[25] Undeterred, he remained an active candidate until late in the process. The university received English's application on November 20 and those of Hood and Malone six days later. This was still a Legal Defense Fund operation, but King, who even then had designs on Birmingham, found value in identifying with the struggle to topple segregation at the University of Alabama.

Setting the stage

By mid-March no one could be sure that the Tuscaloosa applicants would go first, but it looked more and more as if they would. John Cashin was dissatisfied with the treatment his group received from the Legal Defense Fund and had a positive dislike for Jack Greenberg, who recently had replaced Thurgood Marshall at the Fund. The New York staff itself was not very happy with Greenberg's succession, vexed that Constance Motley had been passed over for a white man.[26] By the end of the month, Cashin called Charles Morgan in Birmingham to get what he considered more effective counsel for Carroll and McGlathery. In the meantime Governor Wallace called a meeting of the trustees for Monday, March 18. Trustees and university officials met at the Jeff Davis Hotel in Montgomery, few realizing that Wallace was about to give stage directions for the final act.[27]

Wallace declared the meeting in executive session. The trustees had more than desegregation on their minds. On that same day, the March 23 edition of the *Saturday Evening Post* hit the streets with the story of a college football "fix" between Bear Bryant and Georgia coach Wally Butts. Though later proved untrue in court, the story was sensational and, according to the case's legal historian, was more than smoke.[28] Rose told the trustees that Bryant had "admitted to two telephone conversations & many others but did not get game plans from Butts." Bryant also denied a rumor that he had placed a "$10,000 bet" on the Georgia Tech game and called a risky "pass play" that resulted in a loss. Rose reported the results of a lie detector test that Bryant had taken the day before in Birmingham. Rose's presentation satisfied trustees that Bryant had done no wrong. Breathing easier, the board moved to the issue that most concerned Wallace.

The discussion began with talk about the status of the Huntsville applications. Trustees wondered whether the university could require McGlathery to take the School and College Ability Test if he applied for graduate study (the test the Huntsville Center told McGlathery the previous Friday that he must take), and they discussed the absence of Carroll's Morehouse transcript. McGlathery's statement that he would not litigate bothered the trustees, for it removed an additional reason to delay admission pending a court order. The trustees wondered how far to press the scope of the Autherine Lucy injunction. Andrew Thomas had told the governor that the court likely would construe the injunction as still in effect against Dean Mate, but there was additional worry. Thomas said "that any construction of the scope of the injunction would probably place in issue the reason for the resignation of Dean Adams, in order to test the bona fide action of the Univ[ersity]."

Normally the question would have caused no concern, but Memphis police arrested Adams in October of 1960 and charged him with making "obscene suggestions to a vice squad lieutenant in Adams's room at a downtown hotel." Though the charge of "soliciting males" was dropped when the officer failed to appear, Adams pled guilty through his attorney to a lesser charge of disorderly conduct. The board met in November and received his resignation.[29] The trustees feared damaging publicity for they were in the uncomfortable position of having denied admission to Pollie Anne Myers on moral grounds. It was Dean Adams who had given the university's version of her motives to the *New York Times*. Moreover, they were even then digging into the backgrounds of the Tuscaloosa and Huntsville applicants for damaging information. They would succeed in the instance of Marvin Carroll. After discussion, Wallace and the board felt that the risk of embarrassment was worth taking if any chance remained of forcing the Legal Defense Fund into the time-consuming task of reopening the whole constitutional question, as opposed to the simpler expedient of coming in under the Lucy injunction.

Wallace then took the floor to make clear his position about the schoolhouse door. He said that it was time to set the record straight. He knew that board members were nervous about his pronouncements but strategically refused to lump them with those criticizing his threat to stop integration by physically barring the entrance of black students. Wallace declared himself on four points. First, he was against integration of any school. Second, he would take the action promised in his campaign. Third, he would take all steps necessary to prevent mob violence. Finally, if desegregation were attempted in Huntsville first, he would mount the same show of resistance there that he had pledged for Tuscaloosa. The governor spoke with such conviction that Judge Lawson agreed "in principle" and pledged to vote against admission of the students. Wallace thanked Lawson but replied, "If the court orders admission, the board must comply." With that, the trustees breathed a collective sigh of relief. There was more than generosity in Wallace's method. He had studied the Ole Miss debacle. He knew that one of the problems there had been a university administration and trustees that did not prepare adequately for the possibility of violence; indeed they had talked as defiantly as Governor Barnett. If the "stand" were to work, Wallace knew he must do it alone; a fact that coincided with his natural instincts.

At that point Wallace launched into a speech. Notes taken as he spoke bear repeating as recorded:

> I will help wake country up—I will be there in person—people will follow me—If I leave [the situation] alone, Univ. will be wrecked—[Negro students] can't be admitted without great group of people—Supreme Court case forbids dispersal before violence—I will ask people to stay away from campus—I will stand at U of A & make them present claim—Move all those bricks chains posts bottles, etc. Not to bring in

Gd [National Guard] unless nec. Will be alerted. People may peaceably assemble but on campus those without business will be banned.—Gd for 1 purpose only—to keep peace—Brd of Tr. under injunction rightfully used law & order resolution [reference to November 10, 1962, resolution] but others were wrong.—Do not want marshals—4 or 500 people needed—can't use patrolmen or Gd with state funds—Let them [federal government] come in with force and protect us.—I will let them [Negro students] stay as long as you [federal government] stay. Then [as soon as federal forces leave] I will take them [the students] to another school.—As Gov. & chief law enf. off. responsible for peace of state under police power will refuse admission.—Now [after refusing admission] will not let civil authority breach line—[Will force the federal government to] rely on troops.

Wallace could not have been clearer. His stump speeches had fanned racial antagonism, but he believed nothing could be gained from violence at the university and that he had a plan to avoid it. He laid out every particular, almost every detail of what he was going to do. All the guessing about Wallace remained a function of those who doubted his word. It was precisely the plan Ross Barnett had hoped to use in Mississippi, only Barnett had failed to cultivate either his followers or the university and lost control. Wallace prepared both, and consulted Citizens' Council chiefs in Jackson to avoid the mistakes Barnett had made.[30]

The day after the board meeting, Bennett called Burke Marshall to report the results. He said the trustees had concluded "that the University should desegregate by having these two students [McGlathery and Carroll] admitted to graduate school in Huntsville in June." It was a late hour to reach such a sensible conclusion. Events, of course, were rushing past the board to place the final site in Tuscaloosa. Bennett also told Marshall that Wallace intended to "take over" once the university voted to comply. "He will go to Huntsville personally," Bennett elaborated, "with the State Highway Patrol, a group of sheriffs, and citizens serving as sheriffs' posses. He will preserve law and order but will turn away the students. He will force the federal government to come back with troops in order that he can announce that he has yielded only to force. He will then also say that it is up to the federal government to protect the students." Marshall copied these notes to the Attorney General.[31] Thus, it came to pass on March 19 that the Justice Department learned exactly what Wallace planned to do; only they could never be sure they had just learned it.

RFK and Wallace

Robert Kennedy was thirty-seven years old and still had the body of a twenty-year-old. Wallace was forty-four, and though a bit rounder, had not developed the after-forty paunch. They were young men, successful beyond their years—Bobby in the wake of his brother's own youthful triumph and George by dint of extraordinary political energy. Both played presidential politics, RFK from the certainty of the eastern establishment, Wallace from the equal certainty that no southerner could be President. They were on a collision course that Kennedy did not want and that Wallace sought. They met on April 25 to talk things over.

Kennedy wanted a settled understanding of how to proceed. Poor communication with Ross Barnett, not all of it the governor's fault, was in part to blame for the Ole Miss explosion. Kennedy did not want a repeat. The Attorney General had thought about coming to Alabama in early January, but his stock was so low among Alabamians that friends persuaded him to cancel his trip.[32] It was a far cry from the warm reception he received on two previous visits to Alabama: one when Frank Rose invited him to speak at the university's Law Day in 1958 and the other at the invitation of Red Blount to address the state Chamber of Commerce. Ed Reid, longtime director of the Alabama League of Municipalities, set up the meeting between Kennedy and Wallace. When Reid gave the sign, Kennedy telegraphed Wallace to say, "I will be in Montgomery, April 24–25, and would be pleased to pay you a visit if it would be convenient for you." Wallace replied, "I will be in the governor's office Thursday morning, April 25, and will see you at 9:00 a.m."[33] The governor turned this exchange over to his segregationist friends as protection against any suspicion of collusion.

Wallace did not want the meeting. He did not want to appear to be cutting a deal. He believed the Kennedys had leaked information on conversations with Barnett that made the Mississippi governor look bad. Moreover, he feared for Bobby's safety and ringed the capital with state troopers.[34] The Kennedy name set off powerful emotions. On his arrival in Montgomery, placards greeted Kennedy: "Koon Kissing Kennedy Go Home" and "Alabama Will Resist with Vigah." Two defiant women placed a floral wreath on the spot where Jefferson Davis swore his oath of office and explained primly that their purpose was "to keep the enemy off sacred ground."[35]

As Kennedy approached the governor's office, he drew a different conclusion about the reason for state troopers. One shoved a stick in his stomach, "the point [being] to show me that my life was in danger in coming to Alabama because people hated me so much."[36] Al Lingo's troopers could be as mean as their commander. Having walked the gauntlet, Kennedy entered the governor's suite of offices. Bill Jones, Wallace's press secretary, greeted him and took him to Earl Morgan's office. Morgan was executive secretary. Kennedy and Reid next went into Wallace's office where the governor welcomed the Attorney General to Alabama and said that their conference would be taped to avoid misunderstandings. The possibility of any real bargaining now went aglimmering. Having dealt with Barnett, Kennedy had been prepared for even the most bizarre of conversations. He now braced for the possibility of no conversation.

Asked if he objected to the taping, Kennedy replied, "Whatever you like, Governor."[37] It did not get better quickly. Kennedy had called his trip a "courtesy visit"; Wallace expostulated on Alabama as the "courtesy capital of the Nation." The conversation drifted from a discourse on the flags that stood in one corner to how nice the governor's office was, which allowed Wallace to drone on about mansions in Georgia and Mississippi and how much had been spent on them. Finally Kennedy interrupted, "I didn't know whether we might discuss the problem we are perhaps facing here in the State." Wallace answered, "I don't hear good." "I said," Kennedy spoke up, "I don't know whether you care to discuss the problem that we might face here in the State in connection with integration of the university." With this opening, there followed a pointless discussion of whether the guest or the host determined the agenda. Kennedy finally said he hoped "the problem" of desegregating the university could be handled locally and that his responsibility was simply "to enforce the law of the land." Wallace replied that he interpreted the law differently. As a governor he had the right to raise constitutional questions under the Eleventh Amendment, the reserved powers clause of the Constitution. He also reiterated his pledge to the people of Alabama.

The exchange began to heat up. When Kennedy fell back on his theme of following court orders, Wallace shot back, "Well, Martin Luther King said you have a right to disobey unjust laws, and he also says that this injunction [issued by a state judge] is lawless in Birmingham. Now, do you believe with Mr. King?" (In fact, the decision to violate a court order in Birmingham troubled King, but he excused it as disobedience to an unjust state order.) The Attorney General stiffened: "I believe that he [King] should follow, like everybody else, Governor, should follow the orders of the court, and I believe it for federal courts and I believe it for state courts. I think that applies to an ordinary citizen. I think it applies to a minister. I think it applies to a Governor of a state." Kennedy jabbed with another debating point, but it brushed harmlessly against Wallace's ever-moving talk. Wallace slid away to his duty to provide for the "peace and protect the safety and health and morals of the people of the state, . . . and you just can't have any peace in Alabama with an integrated school system."

Soon they were lecturing each other on abstract principals, such as whether the Supreme Court, which had made segregation lawful in 1896, had the power to "all of a sudden, for political reasons, . . . pull the rug from under us." Wallace became so heated that he declared, "I will never myself submit voluntarily to any integration in a school system in Alabama, [and] . . . in fact, there is no time in my judgment when we would be ready for it in my lifetime." Kennedy reminded him that he was also a citizen of the United States and that he had taken "an oath under the Constitution" to uphold the law. "I think it transcends," Kennedy pressed, "the question of segregation or integration or anything like that. I think it just goes to the integrity of our whole system. If the orders of the court can be disobeyed by you, Governor, with all respect to you and your position, then they can be disobeyed by anybody throughout the United States who does not happen to think that the particular laws of the federal court applies to them or feels that it is not the kind of law that could be good for them." Wallace expatiated at length on the peace and tranquility of Alabama as compared with Washington, D.C., when fortunately coffee arrived.

And so it went, a gulf separating them as wide as the channel that separated Duke William from King Harold in 1066. Kennedy tried once again to press the issue of local responsibility, suggesting South Carolina as a model; Wallace replied *never* in Alabama. Wallace complained that Mississippi had not been given a chance to exhaust its remedies under the Eleventh Amendment before being forced to accept Meredith; Kennedy questioned what the governor would do if the Eleventh Amendment actually were "litigated: Governor, would you [then] follow the orders of the court?" Wallace evaded. Seeing the impasse, Reid interrupted, "Governor, one thing that Mr. Kennedy was talking about, he doesn't know whether he wants to say anything to the press. He thinks it ought to be left up to you to say what is said to the press." Even this reminder that both were in a political goldfish bowl did not stop the digressions. The Attorney General said his brother would be coming into Alabama soon, which led Wallace to reminisce about his support for JFK in 1956 and for the party in 1960. Kennedy grinned, "I heard you were very good. I have had indications that might not be true in 1964." The move to presidential politics launched Wallace into the shabby treatment the South received from both parties.

Kennedy swung to voter registration and pressed the attack. On this issue the governor could not use worn maxims. So Kennedy demanded to know whether Wallace, if he were the U.S. Attorney General, would enforce the constitutional right to vote. Wallace said he would protect that right consistent with state laws and local prerogatives, giving Burke Marshall an opening. Normally reticent, Marshall interjected, "Governor, I don't know whether you know this or not, but it is a fact that there is not a case that has been brought in Alabama or another state in the voting field that was not first put to the local registration board." When Kennedy reinforced his civil rights chief's point, Wallace pounced, "Has the Justice Department ever taken any action in the cities of Dearborn, and Owatta and Wyandotte, Michigan, where they don't let colored people live? You know that exists, don't you?" There ensued a lengthy discussion of housing and President Kennedy's assertion that he could end discrimination in housing with "the stroke of a pen." Why not in Michigan? Wallace asked.

Finally Wallace said, "Of course, you mentioned a while ago about the integration matter. Of course I understand your position, and I am sure you understand mine, and it looks like it may wind up in court." Kennedy relented, saying, "as long as we wind up in court, I will be happy, Governor. . . . I just don't want it to get in the streets. I don't want to have another Oxford, Mississippi. That is all I ask." The conversation seemed to have ended in exhaustion, but Wallace was not done. Governor Barnett's experience made one item imperative, and Wallace saw an opening. "I do not want another Mississippi myself," he agreed, "but you folks are the ones that will control that matter, because you have control of the troops." Kennedy did not see instantly where this was going and simply agreed that he had a responsibility to enforce the orders of the court and would use "all of the force behind the Federal Government to that end." Wallace got more specific. "I know you are going to use all the force of the Federal Government," he repeated. "In fact, that is what you are telling me today, if it is necessary you are going to bring troops into Alabama."

KENNEDY	"No, I didn't say that, Governor."
WALLACE	"Well, you said all the force of the Federal Government."
KENNEDY	"To make sure that the orders of the court are obeyed."
WALLACE	"I know, but all the force includes troops doesn't it?"
KENNEDY	"Well, I would hope that that would stay in the courts. . . ."
WALLACE	"But it does involve troops if the law is disobeyed."
KENNEDY	"Well, I am planning and hoping that the law will be obeyed."
WALLACE	"But suppose it is not in your interpretation of obedience, you will use troops?"
KENNEDY	"I am planning and hoping that the orders of the court will be followed and that everybody will live up to their oath of office."
WALLACE	"But you are going to use all of the power of the Federal Government and that involves troops."
KENNEDY	"I would hope that that was not necessary." And then it dawned on the Attorney General: "Maybe somebody wants us to use troops, but we are not anxious to."

It was just one more sortie in a war of words that had no end. As they argued over what to say to the press, Kennedy tried to make it clear that troops were not his idea. "I have no plans to use troops," he reasserted. "You seem to want me to say I am going to use troops." Wallace could not let that go. "No," he declared, "we don't want troops," but he did want to know what all the "U-2" (Voodoo) flights over the campus were about if not to plan for the "ingress and egress of troops." Finally, after twenty minutes of trying to decide what to tell newsmen, they agreed to say that their positions remained unchanged.

Underlying themes

The Wallace-Kennedy confrontation added nothing to unfolding events, but as an exchange of words it said much. For Wallace there was a moral issue, an issue of public order without which there could be no law. The enforced subordination of Negroes through segregation made Montgomery's streets safe to walk. A free, and in many ways an equally impoverished black underclass in Washington, D.C., made that city unsafe. By all the content of his culture and by observation, Wallace believed blacks to be inferior. As James

Hood pointed out, he did not use the term Negro, the then respectful way to refer to blacks. In polite company he called them colored people. Among cronies or when he wanted to twit university officials he called them "niggers." He knew Alabama was the last state holding out against integration, but he believed that even where desegregation had been coerced or voluntarily accomplished there would be an eventual revulsion of feeling that would return the South to segregation.

Wallace knew people. He knew more white Alabamians by name than any living person, he knew them in sickness and in health, and sent them reminders that he cared. He knew racism and knew that practically speaking northern transgressions exonerated segregation even as they embodied the greater sin of egalitarian hypocrisy. He hated cruelty but knew it to be in human nature. As he listened to the soft, downwardly inflected Boston staccato that accented Kennedy's words, he sensed a cool detachment that failed to comprehend the visceral hold of racism. Only the day before, a quixotic white postal worker who had set out on a Freedom Walk from Chattanooga to Jackson was shot dead just outside Gadsden. Wallace knew the kind of folk who murdered the poor man, and the sympathy the killers drew from their communities. So when he pleaded with Kennedy to go slow, or better still to use his considerable persuasive powers against integration, he did not so much believe that he could change the Attorney General's mind as he despaired of the consequences that would attend his personally giving up and giving in. "[T]hey [the same people who killed on roadsides] will wreck the university," he had told the trustees.

An edge to Wallace's talk distanced him from racists like John Kohn and Ace Carter. Wallace cared whether blacks suffered. He was uncomfortable when Kennedy chided him about voting rights and refused to say, as would a Carter or a Kohn, that blacks should be denied the ballot. He talked about making life better for blacks. He pleaded with Kennedy for help in bringing industry to Alabama so that the vocational programs he sponsored for colored people could succeed. Wallace clung to the mythic morality of separate but equal (the hypocrisy in a commitment to that principle no greater than lip service to equality). In the end, Wallace's conviction that people "ought" to be treated equally made it possible for him to turn his back on segregation. But as he talked to Kennedy, he was a man pledged and covenanted with the white people of Alabama.

Had Kennedy not encountered Ross Barnett, Wallace's performance might have rendered him incredulous. Compared with Barnett, however, Wallace was a model of coherence; only the strange dance over the use of troops reminded him of the Mississippi governor. Kennedy was practical. He did not try, as Wallace did, to get a conversion on the social issues, though he sparred effectively on those points. He wanted two things: to establish common ground on law and order and to establish communication with the governor. Kennedy believed that communication was of paramount importance because of the garbled exchanges between the Justice Department and state officials in

Mississippi. But even here Wallace was not forthcoming, believing that he had already said all that needed saying. Wallace treated Kennedy as he did university officials; they were all superfluous. They need only play their parts, and he (Wallace) would do the rest. Other than a close-up view of the governor, Kennedy got nothing.

Kennedy's performance, however, was credible, even passionate at times. Though he put the Justice Department behind voting rights as a way of ducking more volatile civil rights issues, he spoke about these rights with a conviction that put Wallace on the defensive. When Wallace talked about the incapacity of blacks to govern the District of Columbia, Kennedy applauded their accomplishments in the cities of Africa. With the recorder on, he feared what Wallace might later do with the tapes. "At least for me," he recalled, "I couldn't let anything he'd say go by as if it had been unanswered."[38]

If Kennedy felt a common bond with Wallace, it was politics. He understood Wallace's sense of covenant, but insisted that, as governor, Wallace also had a responsibility to uphold the Constitution. Kennedy worked hard on the Alabama case and understood the trap that ensnared his brother between the morality of civil rights and the politics of holding a Democratic South. Already the Kennedys were in the process of abandoning the South in favor of an electoral strategy that could use civil rights to win re-election. The difference between Kennedy and Wallace that day was that Wallace had moved farther toward wedding politics with segregation's peculiar morality than Kennedy had in joining presidential ambition with civil rights' higher calling.

The meeting accomplished little, but Judge Frank M. Johnson, Wallace's old nemesis, believed the Kennedy visit "served a beneficial purpose insofar as the general attitude of the open-minded citizens in the area are concerned." Kennedy flew off that afternoon for a speech before the University of South Carolina Law School where he declared that the integration of Clemson University represented "states' rights at its best." Wallace went on statewide television that night and gave a masterful speech on roads, industry, and education. Because he knew the audience wanted to hear about the meeting with "Bobby," he opened with a brief mention of the conference and closed saying that Kennedy had not changed his mind or he Kennedy's.[39] He played no tricks with Kennedy's visit, perhaps out of respect for the Attorney General's sincerity and certainly in anticipation of the President's impending visit to Muscle Shoals. No "Bobby" baiting this time. The schoolhouse-door drama had moved beyond rhetorical flourishes to center stage.

Other April happenings

Spring had come to Tuscaloosa and the azaleas and dogwoods were doing their riotous best. It was the tenth spring since Myers and Lucy had made application. This time, however, the university moved full speed ahead to do what Wallace wanted to stop. During the first week of April, a group of Tuscaloosa's leading businessmen and civic leaders, George LeMaistre, Harry Pritchett, and Jack Warner, went to South Carolina to learn how the business community there had handled so successfully the admission of Harvey Gant to Clemson. The South Carolinians urged them to come out early and strong "*in advance* of a showdown." They advised nine steps for law and order, beginning with an opening statement from the state Chamber of Commerce and continuing with a series of pronouncements designed to box Wallace in. Anticipating a sharp reaction from Wallace, the Carolinians advised, ". . . then your Lieutenant Governor [Jim Allen] and your Attorney General [Richmond Flowers] (separately or together) should back up your joint statement."[40] This became the master plan in business's effort to set backfires against the firestorm they feared from Wallace.

On April 8, Wallace met again with the board in Birmingham. Wallace said that he intended to call Senators Hill and Sparkman to request that McGlathery and Carroll "be transferred away from Huntsville." He promised a thorough investigation of the two Huntsville applicants and a reminder to Wernher von Braun "that an incident in Huntsville would be bad for the space program." Bennett duly reported all this to Burke Marshall the next day. Marshall learned that Wallace, ever true to his word, had already made the calls.[41]

The university mounted roadblocks to normal consideration of the five applicants and awaited the litigation that would force final consideration. McGlathery and Carroll were put off at least until summer. On April 19, the Legal Defense Fund filed suit in behalf of English and Malone for the fall term and for Hood's request to start in summer. That same day, McGlathery, with great apprehension, walked into the Huntsville Center and took the School and College Ability Test, which he passed easily. The test also included an essay on an "interesting book" that he had "read during the past year." He wrote on *The Universe and Dr. Einstein,* a popular book in the fifties that appealed to technocrats and laity alike as an introduction to the great physicist's thought. No one marked the occasion, but it represented the first time a black applicant wrote anything to which anyone associated with the university responded with a positive "Yes."[42]

On the last day of April, Wallace met the trustees in Tuscaloosa for the annual Governor's Day, a time for marching, celebrating, and honoring. He told the students that the South's problems could be traced to Reconstruction. Following the festivities, he asked Rose to report on legal steps being taken to block the applications and again insisted that the scope of the Lucy injunction be pursued. He reminded the board that it must not move without a court order. Wallace knew the university was maneuvering behind his back, but he held all the trumps. He put it simply, "If we go in ahead [of a court order], you are asking for it."[43] Wallace got one other thing, an agreement that the lawyers petition for a delay beyond the summer term because of racial tension in Alabama. Martin Luther King was in Birmingham, and the city simmered. Three days later, Bull Connor savaged the demonstrators with police dogs and fire hoses.

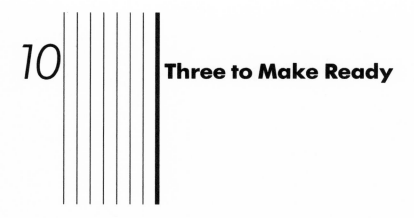

10 Three to Make Ready

Like spiny-finned pilot fish swimming with sharks, certain people can always be found around trouble. On March 8, 1963, former Major General Edwin A. Walker and his spiritual accomplice, the Reverend Billy James Hargis, came to Birmingham's municipal auditorium to rally white citizens behind the mayoral candidacy of Bull Connor. General Walker's mind worked tortuously with all the demons of America's radical right. Under duress, he had commanded the paratroopers who executed Eisenhower's orders at Little Rock. Repentant, he came five years later to Oxford, Mississippi, to rally the resistance forces behind Ross Barnett. Calling on Mississippians to "bring your flags, your tents and your skillets," he sounded the alarm. "It is time! Now or Never!"[1] Following the Ole Miss riot, the government charged him with insurrection and seditious conspiracy and shipped him off to Springfield, Missouri, for psychiatric observation. His brush with federal authority only enhanced the general's standing in Dallas, his adopted hometown.

Walker and Hargis came to Birmingham as drummers for "Operation Midnight Ride," a coast-to-coast speaking tour for God and country. Among those filling the auditorium were Klansmen from Birmingham's violent Eastview Klavern 13; Dr. Edward Fields, the chiropractor who headed the 200-member Alabama branch of J.B. Stoner's National States' Rights party; Bobby Shelton and Ace Carter. A local radio station, WIXI, with which Carter affiliated after being fired by WILD when he linked the National Conference of Christians and Jews with the international communist conspiracy, sponsored the event. WIXI also carried Hargis's daily program. Enough anger filled the auditorium that evening to fuel a Midianite war of extermination. The immediate purpose, however, was to boost the fortunes of the Wallace-endorsed Bull Connor over the moderate segregationist Albert Boutwell. Hargis fired them to their purpose, and the very sight of Walker, "the general they couldn't muzzle," inspired them.

Bull Connor was in trouble. Moderate forces in Birmingham grabbed the upper hand when in November they won the right to reorganize city government into a mayor and council system. On March 5, Boutwell polled 30,000 more votes than Connor despite the presence of another attractive moderate in the field. Now all that remained was to defeat Connor in the mayoralty run-off set for April 2. To be classified a moderate in Birmingham meant pledging to reopen the closed city parks. Moderation also meant the possibility of negotiating with the city's black leadership, but no one, least of all Boutwell, admitted as much. It remained one of those things that might happen in the future.

On April 3, the *Birmingham News*, festooned with an editorial cartoonist's rendering of a sunburst rising above the Magic City, carried Boutwell's jubilant message of "a new day for Birmingham." On that same day, just three blocks from city hall, the first band of Negro marchers sallied forth from the 16th Street Baptist Church. Whites were appalled. Norman Jimerson of the Alabama Council on Human Relations quoted "one white liberal leader" as saying that this was "the most cruel and vicious thing that has ever happened in Birmingham." Jimerson felt betrayed. From his vantage point, Martin Luther King and Fred Shuttlesworth, without consulting the real black leadership, had forced the movement on Birmingham whether it needed it or not.[2] He was wrong about consultation. King had spent the better part of a week in February cultivating all the preachers and leaders who opposed him.[3] Though the more conservative black leaders continued to withhold endorsements, King had attained broad-based support when the first disappointingly small band of marchers set out on April 3.

Jimerson's misgivings amounted to one more instance of whites, no matter their connections to the black community, not knowing what Negroes thought. They could not fathom how deeply Birmingham's blacks resented the cancellation of the limited agreements reached back in September, a cancellation to which blacks acquiesced in the interest of moderate electoral politics. On Good Friday, April 12, sixty-three "Negro Leaders of Metropolitan Birmingham" explained why they could wait no longer: "As a result of conferences with leaders in the community, we waited past the gubernatorial elections last year. We waited in September. We waited in March through the first election. We waited in April through the run-off. We feel compelled at this time to lay our case before the general public." That same day Bull Connor's men grabbed Martin Luther King by the back of his overalls, forcing an undignified tiptoe into an awaiting paddy wagon. Four days later King completed the famous "Letter from the Birmingham City Jail," though it remained unpublished for more than a month.

Expressions of surprise and dismay deepened to questions about King's sanity when on May 2 children began to empty from 16th Street Baptist Church. Cries of "King is our leader!" which issued from the mass meetings now gave white liberals serious pause: Was he still the Black Moses? Had he become a latter-day Peter the Hermit? Who was this man who suffered little children to do the work of adults? Then on May 3, everything changed. Bull Connor opened the fire hoses and unleashed the dogs. Shutters clicked and reporters rushed from other stories to catch the action. With anxious parents wringing their hands in prayer, still more children poured into Connor's jail until their numbers reached over 2,000. The pressure became so great that within the week Birmingham's business community and black leadership, Burke Marshall serving as mediator, reached a desegregation agreement—an agreement that brought out Birmingham's violent men.

On Saturday night, May 11, the home of King's brother exploded at 10:45 p.m. The Reverend A.D. King scrambled to get his dazed wife and their five children out the back door, just as another, larger blast tore a gaping hole in the front and sent debris flying like shrapnel. Angry blacks, many in their bedclothes, filled the streets and threatened revenge. King's brother had just begun to calm the crowd when another explosion rent the night air from the direction of downtown. This time it was the A.G. Gaston Motel, where dynamiters set off a blast in the room beneath Martin Luther King's headquarters—he was in Atlanta. Rock-throwing blacks could not be contained. Police Chief Jamie Moore, a graduate of the FBI National Academy and a far less extreme segregationist than Connor, rushed to the scene with his K-9 corps and other units. Near 2 a.m. the riot began to subside, but it was not over.

Thirty minutes later, 250 of Al Lingo's troopers returned to the scene they mysteriously left just hours before the bombing. Before him, Lingo saw the orgiastic fulfillment of his darkest fantasies, buildings burning and Negroes milling. "Oh, was he [paranoid]," recalled Department of Public Safety Investigator Ben Allen. "Paranoid, period. Paranoid, period. Takin' pills by the handful. Just a whole handful of tranquilizers."[4] Despite protests from Chief Moore that the situation was under control, Lingo sent his men charging into the streets, clubs swinging. Blacks were beaten in a wanton display of force. As the smoke cleared on Sunday, Martin Luther King returned to work the bars and billiard halls to restore calm. Robert Kennedy, alerted by John Doar at 2:30 Sunday morning, met with his brother throughout the day in an atmosphere of crisis. Everyone figured that Wallace had gotten what he wanted, a chance to intervene in behalf of Bull Connor and possibly to declare martial law. Relieved that King did not ask for federal troops, President Kennedy nonetheless authorized a standby call-up and ordered General Earle Wheeler to move a contingent to Fort McClellan in nearby Anniston. By Monday, however, calm returned and King announced that the deal with Birmingham's business community was still on.

Almost certainly Wallace was not as darkly motivated as moderates and liberals feared. Al Lingo was a force to himself, and Wallace was loath to control him. When Jefferson County (Birmingham) Sheriff Mel Bailey beseeched Wallace to get Lingo to pull his troopers out, Wallace shrugged and left Bailey to deal with Lingo himself. Wallace occasionally winced at Lingo's excesses, but as in most of his dealings with loyal subordinates, he seldom had the stomach for controlling him directly. Albert Jennings Lingo of Eufaula (Barbour County), near Wallace's hometown of Clio, was an original 1935 member of the Alabama Highway Patrol, now designated "Troopers" by the "Colonel." A tough, pot-bellied, bug-eyed archetype of southern law enforcement, Lingo moved comfortably among the likes of Bull Connor, Dallas County Sheriff Jim Clark, and their allies in the klaverns and White Citizens' Councils.

Wallace's other motive in turning Lingo loose was political. Birmingham's Big Mules never warmed to Wallace, at least not then, nor he to them. He knew that cautious whites who wanted things calm in their own city in expiation might vote for a strong show of resistance at the state level. Even if they did not, Wallace's staunchest supporters would let out a good yahoo on learning that state troopers had truncheoned Negro rioters. Law and order meant keeping blacks in their place, nothing more. Blacks provoked violence; whites reacted. It was the logic of General Walker, Billy Hargis, Bobby Shelton, Ed Fields, and Ace Carter; it was also the logic of the Wallace electorate—the same tautologous connection that leads abusive parents to conclude that they have unruly children. Subtly this logic of law and order worked to transform Wallace's solemn vows about preserving law and order at the University of Alabama into a pledge to keep blacks in their place, not an acquiescence to federal authority and desegregation.

Lingo's headbusting charge into Birmingham's near west side had one other effect: it made the Kennedy administration and university officials fear that Wallace's troopers might be a destabilizing force in Tuscaloosa. Five days after General Walker's March 8 appearance for Bull Connor, Buford Boone, the Pulitzer Prize-winning editorialist from the Lucy crisis, spoke the worry of all moderates. In a private letter to Wallace, he wrote, "I have feared and I still fear, that inciting statements will encourage people to violence because they think it is expected of them." Speaking like the prophets of old, Boone warned, "You are going to lose, as John Brown did, as Lee did and as Barnett did. But are you going to lose with dignity, with intelligent courage, and with proper regard for the long-range welfare of our people in Alabama? Or, are you going to take the low road and work the situation for all it is worth in current popular support and with too little thought of the tomorrows that always come?"[5]

Wallace no doubt respected the Tuscaloosa editor, but words like "dignity," "intelligent courage," and "proper regard," in the governor's parlance, had little connection with the reality of politics. Edwin Walker, not Boone, spoke the language that reverberated in the union halls, an apocalyptic language laden with images of righteous death and destruction. Two days after General Walker warned Bull Connor's supporters about the communist conspiracy, Lee Harvey Oswald photographed and sketched the general's home on Turtle Creek Boulevard in Dallas. Two days later he clipped a coupon and sent $21.45 to a Chicago mail-order house for a Mannlicher-Carcano rifle. One month later, under cover of dark, Oswald tried to kill General Walker from a distance of 120 feet and missed.[6]

The month of May

As bombs exploded in Birmingham, only one calendar month remained before the showdown at the schoolhouse door. On Thursday, May 16, with "Gordo" Cooper soaring high above on his historic twenty-two-orbit mission, Judge Grooms made short shrift of the effort to narrow the scope of the Lucy injunction, binding Dean Hubert Mate to his 1955 ruling. That afternoon, Mate, a Latin Americanist who had been with the university since 1937, sat uneasily while Constance Motley took his deposition. Motley tried repeatedly to get Mate to say whether the plaintiffs, Hood, English, or Malone, would be admitted if their credentials were in order and met standards. Andrew Thomas refused to let him answer. Motley took the refusal to Grooms, who sustained Thomas. The maneuvering continued as Thomas played for time to allow the board to make this final, momentous decision. Motley also assailed the proposition that students applying from institutions with which the university had no previous experience, that is, black, should meet additional qualifications. She asked if that had ever applied to a white student who transferred from, say, Harvard.[7]

The following day, the ninth anniversary of *Brown*, Charles Morgan entered a "show cause" petition to force the university's hand. All the cases had been consolidated in the May 16 ruling, and the court would respond to Morgan's brief. On that same day Air Force One lifted off for Nashville, Tennessee. Kennedy had been looking for an occasion to show the flag in hostile territory. His fortunes had slipped in national polls, and though still high at a 64 approval rating, he was down from a still higher 76 in January. Vanderbilt University's 90th Anniversary Celebration and ceremonies for the 30th anniversary of the Tennessee Valley Authority, set for Muscle Shoals the following day, proved the right combination. Nashville had made considerable progress since the 1960 sit-in demonstrations, and the Shoals, with its dependence on TVA, represented the only congenial area for the national party in Alabama. At Vanderbilt the President did not use the word "Negro" once in his speech but did chide rebellious southern leaders for failing to recognize the inevitable expansion of human rights. He also "blasted lawbreakers" generally, which for most of his southern listeners meant riotous Negroes in Birmingham.[8]

The ticklish situation about visiting Alabama was how to get the President in and out without being served a subpoena. When Kennedy ordered 3,000 troops into the state on May 12, Wallace retaliated with a state's sovereignty petition to enjoin the President. Lee White, special counsel on civil rights, had briefed the President on ways to avoid the subpoena, recommending that the Secret Service block any who attempted service. If that failed, White suggested that the President "dismiss it with humor."[9] In fact, things went smoothly. The President extolled the value of TVA and proclaimed the federal government to be no adversary of the South, whereupon Wallace, along with members of

Alabama's congressional delegation, boarded the President's helicopter for the short trip back to Huntsville. Kennedy asked Wallace where he wanted to sit, and Wallace chuckled, "I'll sit on the right, Mr. President, and you may sit on the left." Wallace pointed out the sights along the beautiful Tennessee River Valley and Kennedy said a few words about having Birmingham's business establishments hire more blacks in responsible positions, but to Wallace's surprise, the President never mentioned the University of Alabama.[10]

Wallace returned to Montgomery where on Sunday, the 19th, he met with the trustees. They decided what they were going to do. Malone had sent a telegram on Friday requesting summer admission, and her application along with McGlathery's was complete. Wallace asked once more about the thoroughness of the character investigations, and Rose assured the governor that the students were qualified in all respects. Wallace said it was his constitutional duty to vote against the applicants but reiterated the trustees's obligation to vote for admission. Wallace asked only that their statement be in terms "the average man can read." Excepting Wallace, the vote was unanimous to admit Malone and McGlathery. McCorvey, absent due to illness, pledged his support from Mobile. They instructed Andrew Thomas to tell Judge Grooms of their decision but to insist on a delay until racial tensions eased. Any announcement would await Tuesday's hearing.[11]

On Tuesday, Grooms took judicial notice of recent racial unrest but dismissed the university's petition, observing with some irony that Wallace himself had pledged to maintain law and order. He also refused to stay the injunction pending appeal, noting that when he stayed the Lucy injunction in 1955, though sustained by Judge Rives, "the Supreme Court promptly slapped both of us down." That afternoon the board announced its decision to admit Malone and McGlathery, and Wallace called a press conference to say that he was interposing himself between an intrusive federal government and the people of Alabama. To let the trustees off the hook, he said that a federal court "would not hesitate to jail, imprison, and inflict severe punishment against any lesser official than the governor of this state, and this, of course, includes trustees and other officials of the University of Alabama."[12]

That same afternoon Grooms ventured the opinion that attitudes on race were "softening." The 6′3″, sixty-two-year-old judge was in a position to know. He attended church on Easter Sunday when blacks fanned out over Birmingham to "kneel-in" at white churches. Some worshiped in his congregation, even as Martin Luther King sat on jailhouse bedsprings, hunched over a newspaper, scrawling his famous letter in the margins. Despite recent turmoil, Grooms noted that his reinstatement of the Lucy injunction had drawn "considerably less protest mail." He attributed this "to greater moderation and a feeling of inevitability."[13] Martin Luther King also concluded that things had changed. In summing up the Birmingham campaign, he detected a paralysis in the white community. "Strangely enough," he observed, "the masses of white citizens in Birmingham were not fighting us." A year before and "Bull Connor would have had his job done for him by murderously angry white citizens." In 1963 they stayed away. They were not "in sympathy with our cause." "I simply suggest," he concluded, "that it is powerfully symbolic of shifting attitudes in the South that the majority of white citizens in Birmingham remained neutral through our campaign."[14]

If Grooms and King were right, the larger lesson for Wallace was that he was right too. Birmingham's violent explosion in May did not terminate the agreement between its white business community and black citizens, nor would Wallace's Tuscaloosa show stop school desegregation. For the moment, things did not look good for the forces of segregation. With expectations of immediate success lowering, angry and perplexed whites looked for a sign, if only symbolic, that the fight had just begun, a struggle that eventually would end in success—victory made inevitable by the self-fulfilling logic of white supremacy.

The business of government

The pace now quickened at the Justice Department. A memo circulated that spelled out the essentials of Wallace's plans "as told to friends": he would "surround [the] campus with troopers to keep mob out"; he would "meet federal marshals & tell them students can't enter"; "if [federal] troops sent, [he] will say in effect 'I can't buck the troops'"; and finally, he "will tell federal authorities 'the negro student can remain as long as you are here. As soon as you leave, the negro student will come out of the school.' "[15] With small details added at the schoolhouse door itself, it was the plan outlined in March and now reiterated. So, why all the fuss? The reason was that Wallace would not communicate and those who purported to speak for or about him spoke with different voices. In the end, information overload exaggerated the mystery.

Some of the newer trustees pitched in to help the Justice Department. At Robert Kennedy's request, John Caddell flew to Washington for a conference with the Attorney General. The flight, as it turned out, was at his own expense, but he bore the cost cheerfully because it helped ease tension.[16] Red Blount played a key role in devising what the Justice Department believed to be its most useful tool in defanging the governor: a campaign to get Alabama's leading businessmen to pressure Wallace for a peaceful resolution. With Blount's help the Justice Department constructed a list of 128 business leaders of whom 108 were contacted, and 80 agreed to call Wallace. Cabinet officers and other high appointed officials received the original list of 128 and marked names of people they knew personally or professionally. The remainder was assigned randomly. On making contact, the Cabinet officer not only asked that Wallace be called but also inquired about the governor's intentions and state of mind.[17]

Most businessmen supported the initiative. Most, though not all, thought Wallace was determined to have his "stand" and placed little confidence in their ability to change his mind. Lewis Jeffers of Hayes International Corporation, a firm that did contract work for the government, told Secretary of the Treasury Douglas Dillon that Wallace "was aware that the business men of the state are against him and that the University of Alabama is against him, [and that he personally] was doubtful that Wallace could be controlled." Alto Lee, a Dothan lawyer and past president of the National Alumni Association, told Dillon that the pledge to stand in the door had "become an obsession with the Governor and that 'he is hellbent on us putting him in jail.' "[18] Still, Justice remained confident that the cumulative effect of these calls would be beneficial.

Alto Lee's letter, however, indicated the down side of the strategy: a tendency for those who knew Wallace to exaggerate what he might do, thereby heightening the crisis atmosphere. Trustees contributed to the image of Wallace as dangerously volatile. On reflection, Robert Kennedy came to understand the negative effect of that kind of information. "The trustees [not all of course] were opposed to the Governor," he remembered. "So they would relate what the Governor said at the trustees' meeting and what his position was going to be. But this gave us great concern, because they always reported that he was crazy, that he was scared. Inevitably they'd say he was scared; inevitably they'd report that he was out of his mind; inevitably they would report that he was acting like a raving maniac." Though exaggerating himself, Kennedy realized that it frustrated his ability to "make plans on the basis of what a reasonable man would do if he were the Governor of the state of Alabama."[19]

Kennedy wanted to deal on a reasonable basis. He wanted to orchestrate a resolution, only he was playing with half the orchestra and his half was all strings. Wallace controlled the brass but refused to rehearse. The reports about his erratic behavior, though enlarged, were not without foundation. Physically Wallace moved with an enormous amount of energy. He worked crowds aggressively. He walked quickly. He ate out of necessity. He shredded paper into little piles during conversations. He lived politics with an intensity that made the Kennedys appear diffident. He thrived on adulation to the point of addiction. His enemies feared him. The "over-the-mountain" crowd in Birmingham, the monied interests, still show scars to any who will look, blaming the tardy completion of their interstate system on Wallace-styled vendetta.

Many who wrote about Wallace traced his emotional state to the disability that sent him home after eight or nine missions over Japan in a B-29. Oregon's Republican-turned-Democrat Wayne Morse trotted out the record: he was hospitalized in 1943 for acute cerebral spinal meningitis, and in September 1945 "for severe anxiety state, chronic, manifested by tension states, anxiety attacks, anorexia, and loss of weight." In 1946 he filed for and received a 10 percent, service-connected disability for psychoneurosis. In 1956, he stood examination again. "He was tense," wrote the physician, "restless, and ill at ease, frequently drumming the desk with his fingers, changed position frequently, sighed occasionally, and showed a tendency to stammer, resulting in the diagnosis of anxiety reaction." The psychiatrist recommended a continuation of the disability. With Senate heads wagging, Morse concluded that Wallace needed another examination.[20]

Emerging from the Freudian fifties, these conclusions about Wallace weighed heavily with readers of the *New York Times*. Among the mothers and grandmothers who flocked to Wallace's rallies, however, Morse's attack was akin to one schoolboy making fun of another for being a nervous child. Modern-day readers of the *Times* would know to put greater emphasis on the neuro-physiological damage done by cerebral meningitis. Whatever, Wallace was not crazy, but for those who feared him, the governor's excessive nervous energy confirmed the diagnosis. Elements of this same fear accounted for a tendency by some trustees and university officials to enlarge reports about Wallace to the Kennedy administration. They moved around Wallace with the trepidation of an unfaithful wife, fearful that perfidy would show through their wan smiles. Never knowing when he might lash out, they watched him closely, impressing on each gesture their wilder imaginations. Wallace knew their infidelity and used their anxiety to exact private concessions for his plans and outward pledges of support.

In response to what it heard, Justice redoubled efforts. On Wednesday, May 23, the day the new mayor and council took office in Birmingham, President Kennedy told a news conference that federal troops would be used if necessary. The *Birmingham News* felt all along that the May 12 troop call-up was intended more for Tuscaloosa than Birmingham. On cue, Attorney General Richmond Flowers, on whose support the business strategy counted, declared, "As dark as these days are, defiance that would provoke violence would only make the days darker, and upon those who resort to these measures must lie the blame if federal troops are used in the state of Alabama." The *Birmingham News,* with inside tracks to the Justice Department, continued to say that troops would be sent only as a counter-measure to Wallace's vow to use state troopers.[21] Then on Thursday, the 24th, Justice lawyers sought an injunction to stop Wallace from standing in the schoolhouse door. Judge Seybourne Lynne ordered the governor to appear in Birmingham for a hearing on June 3. On the 25th Rose and Bennett flew to Washington for their last visit before the confrontation. That night as Rose spoke to the D.C. chapter of the National Alumni Association pledging that the university would "meet the crisis with dignity," Wallace spoke at a high school in Northport, just across the river from the university, renewing his vow to stand up for Alabama.

Over the weekend Rose and Bennett visited the Attorney General at his home in McLean where at breakfast Bennett urged Kennedy not to use federal troops. If troops became necessary, they should be local boys in the national guard. They took the issue to the President himself, who endorsed Bennett's suggestion. Standing in the Oval Office, feeling the thick carpet underneath and seeing all the trappings of power, Bennett was overwhelmed by the realization that "here was the leader of the Free World enlisting the whole weight and might of the United States government to support two black students." Returning to the Justice Department, Rose wisely exacted a promise from Robert Kennedy to send twenty-five FBI agents into Tuscaloosa to harass Klansmen by interviewing them and checking hotels and motels to see if Klan members had registered.[22]

In the camera's eye

Just when the media came to realize they were sitting on top of made-for-television drama is uncertain, but once the lights went on in corporate headquarters, producers scrambled like the scribes who rushed to catch up with Grant at Appomattox. In December at the annual Blue-Gray all-star football game in Montgomery, NBC switched off its field cameras rather than show Wallace at halftime. Now the network wanted him in New York for Lawrence Spivak's *Meet the Press,* an invitation that Oliver Carmichael had spurned in 1956, but one that Wallace was eager to accept.[23] Most of New York's journalistic establishment figured that Wallace's Gothic style of race-baiting, judge-goading politics could not survive national television exposure. This caricature of one-gallused ignorance who delighted in telling his thin-lipped audiences that he had "called Judge Johnson publicly in 200 speeches . . . a 'integrating, scallywagging, carpetbagging liar' " and who bullied crowds into jostling New York reporters would have his come-uppance.

The governor flew out of Birmingham under the alias Jack Bailey. He wheedled Grover Hall, Jr., editor of the *Montgomery Advertiser,* into going with him. Hall, a bachelor who dressed to the nines and invariably wore a rose in his lapel, served alternately as close confidant and journalistic nemesis of the man who had become king of Alabama and whose sights were on the White House. Hall's father had won a Pulitzer in the thirties for attacking the Klan and some believed that the son had served Wallace poorly in 1958 by doing the same. Hall's considerable intellect burned brightest when belaboring northern racial hypocrisy, and in Wallace he found the perfect champion. Too smart not to find the mote in Wallace's eye, Hall occasionally used a journalistic beam to blacken it. He chided Wallace for the political folly of endorsing Bull Connor. The *Advertiser* even editorialized against the pledge to stand in the schoolhouse door, but that owed more to the paper's owner. Hall admired Wallace and winked at the governor's eccentricities. Wallace in turn leaned heavily on Hall's loyalty as well as his talented pen.[24]

Wallace especially wanted Hall along on this first venture before a national panel of journalists. On the flight up Wallace told people where to sit and twisted nervously in his seat. The unflappable Hall sat across from the governor and all went smoothly until Wallace suddenly leaned toward Hall and said, "I've got to have a foreign policy." For one of the few times in his life, Hall was nonplussed and stared at Wallace in disbelief. "Those boys," Wallace insisted, "are going to question me about my beliefs tomorrow. They're going to want to know about what I think about things in this nation. And they're going to want to know about my foreign policy. If I'm going to run for the presidency next year I've got to have a foreign policy." Wallace's obsession about foreign policy continued through the evening and into the next morning. By that time Hall had ripped a column from the *Wall Street Journal* and suggested that Wallace read it, which the governor did, chomping incessantly on his cigar.[25]

Wallace and entourage landed in Newark, greeted by a sprinkling of reporters and a security force equal to that assembled for Premier Khrushchev on his 1958 visit to the United Nations. Security problems dictated that the program, normally produced in Washington, be moved to New York. Pickets promised to be out in force. The prospect of tumultuous demonstrations gave Wallace a chance to sound his familar themes about race relations in the North. He reminded University of Alabama alumnus Gay Talese that Bobby Kennedy could walk the streets of Montgomery in safety while he [Wallace] could not even appear in Washington. He declared cities like New York to be "citadels of hypocrisy." "I've been to these northern hotels," he continued, "and I hear clerks telling Negroes, 'Now, let me see. Oh, yes, we're full up' or 'Doesn't seem your reservation is here.' " Wallace charged that Alabama had 10,000 Negro teachers while the state of New York had only 4,000 to teach 207,000 Negro children. Michigan had 450. "We pulled ourselves up," he concluded, "and pulled the Negro up with us. He lived. He ate. Who else pulled him up? We did. Or else he'd have starved to death."[26] It was vintage Grover Hall served up Wallace-style.

At the studio Wallace worked the camera crew as if they were constituents. Waiting for the red light to blink on, he drummed his fingers idly and shifted to a more comfortable position. Across the way sat an imposing bank of reporters: Anthony Lewis for the *New York Times,* NBC's Frank McGee (formerly of Montgomery's WSAF), Vermont Royster from the *Wall Street Journal,* and, of course Spivak. The journalists knew a successful politician could not be dismantled by thirty minutes of questioning. Still, it was difficult not to hope that this time the gnarled and twisted logic of segregation could be exposed—George Wallace, at last, forced to stand to reason. When the red light went on, Wallace stopped fidgeting and his fingers relaxed. He was cool and resolute. Spivak probed his plans for the schoolhouse door. Wallace said there would be no violence, that he had demonstrated his opposition to violence by sending state troopers into Birmingham. "I am not hoping to have myself arrested," he explained. "This is a dramatic way to express to the American people the omnipotent march of centralized government to destroy the rights and freedom and liberty of the people of this country."[27]

Anthony Lewis, whose close relationship to Bobby Kennedy worried his bosses at the *Times,*[28] reminded Wallace that the authority of the Supreme Court had been tested "dozens of times since the country was founded and the supremacy of the Constitution has always prevailed, most recently in Little Rock, Arkansas, and Oxford, Mississippi." Wallace parried with his own version of constitutional history. "*Plessy vs. Ferguson* many years ago was decided in favor of the separate but equal facility doctrine but constant efforts by those who opposed this interpretation resulted in the *Brown* case, so I see no reason why we . . . should not continue to raise questions, and the court might decide to change its mind as it did in the *Brown* case." Shifting easily from history to sociology, Wallace contrasted the safety of any street in Alabama with Central Park. Newspapers back home crowed with satisfaction over Wallace's performance. Their champion had bearded the lion. Moderate papers felt reassured by his pledges of law and order. Segregationist weeklies delighted in his ability to "turn the table" on northern hypocrisy. Wallace relished his own performance. Striding into the lobby when it was done, he saw Hall and broke into a big grin. Crumpling the article from the Wall Street *Journal* and tossing it into a trash can, he laughed, "All they wanted to know about was niggers, and I'm the expert."[29]

The racial theme aside, Wallace's New York performance signaled a serious turn on the governor's part to ensure against violence. Mounting pressure from business, and university claims that its accreditation was threatened, had some effect. In the legislature a handful of brave senators diluted a resolution in support of Wallace's stand. One of those leading the rebellion was the son of Judge Horton of "Scottsboro boys" fame, and, like his father, he would lose office, only the son's failure came not so much for bucking the racial tide as for opposing Wallace.[30] (In 1963 no one yet knew how fatal it was to cross Wallace.) The pleas of businessmen, educators, and legislators, however, only paralleled Wallace's own determination to prevent violence. The success of his stand depended on maintaining law and order, which by his constituents' definition meant not letting "niggers" and outside agitators provoke violence. As Wallace put it to a Birmingham reporter, "The integrating crowd is just looking for trouble, looking for violence. Well, they're not going to get it. There's going to be peace, there's going to be tranquility. . . ."[31]

In the days ahead Wallace would go on statewide television twice and publish a strongly worded proclamation in every state newspaper. Reminding his followers of his solemn pledge to stand in for them, he said, "I need your prayers and moral support, but I do not need your presence on the campus of the University of Alabama at Tuscaloosa or at the University Extension Center in Huntsville." Vowing to have sufficient troops in place to maintain order, he repeated, "I, as governor of the State of Alabama, ask you to stay away from the campus at Tuscaloosa and at Huntsville." "If we have a breakdown of law and order," he concluded, "our cause is hopelessly lost."[32] The last assertion was the closest he ever came to suggesting that a breakdown in order might be traceable to his own constituency.

Cinema verité

The Kennedy reputation for media exposure went without question. What was not so well established was the art of made-for-television documentary film, but Robert Drew and his associates worked to redress the imbalance. Most documentaries had the feel of walking through an art gallery accompanied by a talkative docent. Drew wanted action. A film on NASA experiments with weightlessness made the *Ed Sullivan Show* and CBS news. A big breakthrough occurred when Drew and filmmaker Ricky Leacock talked Jack Kennedy into letting them film the Wisconsin primary, including the intimacy of watching primary returns with his family. The resulting documentary, *Primary,* put politics on the black-and-white tube.[33] Drew was still not satisfied. Technological breakthroughs made it possible to move in on subjects with minimal intrusion. Drew wanted to do presidential decision-making in action, preferably in a crisis where the outcome was uncertain. In May 1963, Greg Shuker, who was on a story-finding mission for Drew, stumbled across a column about the impending drama in Tuscaloosa. As news, it was still back-page stuff, but it leapt at Shuker. He got in touch with Don Wilson at the United States Information Agency. "I can set it up for you to meet Bob Kennedy," Wilson said, "because he's going to handle the integration thing with his deputy Katzenbach. If you can sell it to him, you can sell it to the President because he already knows you guys."[34]

Initially reluctant, Robert Kennedy took it to the President, who said, " 'It's up to you Bob, it's your show.' "[35] With enthusiasm building, Drew Associates reached an agreement. The focus would be on the Justice Department, though if decision-making spilled over to the White House they could set up there. One problem remained: how to get Governor Wallace involved. To show only the Kennedy side would tag the project as promotional at best. Wallace talked to the filmmaker and agreed to participate. He did not ask for the right of preview, nor was it offered, even though the Kennedys had been extended that courtesy. Wallace had been burned often by media that talked about such things as the way he "curled his lip" or the "sucking noises" he made while eating, but he knew the value of publicity. With film he had some measure of control over what he said and how he looked saying it. The filmmakers also got agreements from Vivian Malone and, after June 3, Jimmy Hood. With the cast set, camera crews spread out to their assignments in teams of two. Drew remained in New York to edit. For the first and last time, the nation was about to get an inside look at presidential crisis management as it happened.

Back to court

Judge Lynne's order that Wallace show cause why he should not be enjoined from blocking the admission of Malone and McGlathery set off another comic round of cat and mouse. Wallace's advisers were as determined that he not be subpoenaed as Lee White had been for President Kennedy. On the first attempt a marshal mistook Earl Morgan, the executive secretary, for Wallace. The second attempt was worse. Showing up at the governor's mansion, a befuddled marshal handed the summons to one of the maids, a trusty from the state prison. Wallace needled the marshals for serving papers on someone "civilly dead." Just as Kennedy was urged to surround himself with secret service agents, Wallace encircled himself with twelve state troopers. Finally, a marshal who had worked the case continuously for four days dropped the papers at Wallace's feet as he boarded a commercial flight in Birmingham for his *Meet the Press* appearance. Even then Wallace would not admit he had been served.[36]

Wallace would be represented by counsel. As part of his strategy and for good public consumption, Wallace had appointed a State Sovereignty Commission to advise him in resisting desegregation. It was another trick he learned from the Mississippi Citizens' Councils. Leading lawyers, most of a rigidly conservative bent, formed the commission: James Simpson, J. Kirkman Jackson, and Reid Barnes from Birmingham; John Kohn and T.B. Hill, Jr., an Eisenhower Democrat who had the added value of being both a cousin and a nemesis of Senator Hill, represented the Montgomery bar. The commission had a few more progressive members such as John Caddell, who accepted appointment primarily to protect the university's interests. Barnes and Jackson represented Wallace.

As the hearing opened, Burke Marshall, assisted by Harold Greene and Doar, opened by playing a tape of Wallace's May 21 speech in which he pledged defiance. (The FBI was good at tape recording and taking pictures.) Marshall sought the injunction to resolve in advance the question of sovereignty that Wallace said he wished to test: Did he as governor of a sovereign state have the constitutional authority to challenge an order of a United States court by blocking its implementation? Marshall observed a more practical reason for the injunction, noting the violence that had taken place in Mississippi when Barnett claimed a similar right. "We think the Governor has every right," Marshall urged in his high-pitched, Texas-flavored accent, "to raise legal questions, but he does not have the right to bring himself in conflict with officers enforcing the orders of this court."[37] The governor's lawyers stuck to a narrow construction of the court's injunctive powers. They argued that Wallace could not be enjoined until he had taken a specific action. Lynne could have ruled immediately but instead set Wednesday the 5th for his decision. Part of the reason may have been a call from Jeff Bennett recommending that Lynne order university officials to meet with the two students and their advisers in his chambers. Bennett remembered the disaster of zero communication during the Lucy try. Lynne directed that all parties meet following his Wednesday ruling.[38]

In the meantime, the trustees announced that Jimmy Hood would be admitted for the summer term. Hood heard the news over the radio while on his way to South Carolina for a summer job as a minister. He called his parents, who confirmed the report, whereupon he turned back and headed for Gadsden to await further instructions. That afternoon Wallace met in Montgomery for 90 minutes with Rose, Bennett, and Alex Pow, Director of Contracts and Grant Development. The discussion centered on security arrangements for the campus. Following that meeting, the governor huddled with Sovereignty Commission lawyers for a report on the day's proceedings in Lynne's court.[39]

Wednesday dawned hotter than usual for early June. Judge Lynne spoke for a patrician order that wanted to go gently into an uncertain future. Yet part of the patrician ideal required that a senior judge take some occasion to address the future. Judge Lynne now prepared to use his bench as a platform and for the first time employed the personal pronoun in a judicial opinion. "I love the people of Alabama," he started. "I know that many of both races are troubled and, like Jonah of old, are 'angry even unto death' as a result of distortions of affairs within the state, practiced in the name of sensationalism." Then pointing his remarks toward Montgomery and Governor Wallace, he declared, "Thoughtful people, if they can free themselves from tensions produced by established principles with which they violently disagree, must concede that the governor of a sovereign state has no authority to obstruct or prevent the execution of the lawful orders of a court of the United States." Lynne barred Wallace from obstructing the entrance of the two students but did not prevent him from being in Tuscaloosa.[40]

Bill Jones released a statement from Montgomery that heightened the suspense: "The govenor says his stand is exactly the same as it has always been." What did Wallace mean? Did Wallace mean what he said? What had Wallace said? Judge Lynne's hoary wisdom contrasted sharply with Wallace's stump rhetoric, and the state's dailies used the contrast to suggest that moderation was more southern than rebellion—a thesis difficult to sell. "There is drama before us," wrote the *Birmingham News*. "It involves courage, professional honor and integrity, love for and devotion to Alabama." Wallace's defiance was no longer against far-off Washington. "For the opposition declared by Gov. Wallace is against the legal integrity and performance of Decatur-born, Auburn-graduated, University of Alabama law-trained Seybourne H. Lynne." The *Alabama Journal* intoned, "Judge Seybourne Lynne's ruling enjoining Gov. Wallace from the University door Monday embodied both his sworn duty as a federal judge and his natural anguish as an Alabamian."[41]

Wallace went on statewide television urging all Alabamians to stay away from the university. For what must have seemed to him the umpteenth time, he said, "I appreciate your prayers and I cherish your support. But I do not need your physical presence at the university. I implore you to stay away and let the governor of your state handle this all-important matter as your representative." Leaving the studio he met with a group of more than 100 "Women for Wallace" who pleaded to let them stand with him. He thanked them, but urged them to stay away.[42] Wallace never said he would resist to the bitter end. In fact he was resisting for principle, not jail. The Klan understood and said it would stay away. "The Klan is standing 100 per cent behind Gov. George Wallace," proclaimed Shelton, "and whatever legal means that he might deem necessary." (Shelton never could decide whether he wanted to be respectable.) The National States' Rights party accepted Wallace's leadership and pledged the same. One by one, radical groups fell in line. Even Bull

Connor came to the Holt Citizens' Council on the night of June 7 and urged acquiescence. Leonard Wilson was also there.[43]

Still, Wallace would not communicate with the Justice Department, and no one in Washington could say with confidence what he would do. Finally, on Saturday the 8th, Robert Kennedy put through a call to Wallace. Kennedy wanted one more chance at open communication. His hopes were quickly dashed. Cecil Jackson took the call at 4:40 in the afternoon, 5:40 Washington time. "Mr. Jackson," Kennedy opened, "I was trying to get hold of the Governor. Is that possible—to talk to him." Jackson answered, "Mr. Kennedy the Governor is not available and we have some question as to whether or not it would be proper for you to talk with him since you, as the Chief Law Enforcement Officer of the United States, have filed suit against him. I have here with me Mr. John Kohn who is one of the Governor's leading counsel and he will be happy to talk with you." Kohn, whose worldview was even more darkly etched than that of General Walker, was not about to say anything, and Kennedy dangled. Kohn wanted to establish that Kennedy and not Wallace initiated the call. Finally, in exasperation Kennedy asked for Jackson again and said, "I thought we were going to handle it as civilized people and gentlemen—but I am just—what I called for was to try to get some information—I am not trying to force anybody and not trying to stop anybody—but I can't get that obviously." Amidst protestations of mutual civility the conversation sputtered to a halt.[44]

Specters

The question lingered: Could Wallace bank the fires he stoked? Everyone knew that Shelton was notorious for talking peace in the valley while the Eastview Klavern 13 in Birmingham dealt mayhem. Even a call to Frank Rose from Al Sisk, Shelton's assistant, pledging that the Klan would stay off campus did not dispel anxiety.[45] On Friday night, June 7, some 750 state troopers poured onto the campus, took up residence in empty dormitories, and cordoned the area. Because of a Klan meeting set for 9 o'clock Saturday night, authorities imposed a curfew on the few students already on campus.

On Saturday, at the very moment Kennedy tried to open communication with Wallace, FBI informant and Klansman Gary Thomas Rowe loaded his car with five carbines, four twelve-gauge pump shotguns with the clip unattached so extra shells could be added, five boxes of .00 buckshot, four bayonets, a dozen or so fragmentation grenades, six tear-gas hand grenades, a .45 caliber machine gun, a dozen dynamite sticks with caps, and a bazooka with six rounds of ammunition. Rowe earned his reputation among Klansmen by bashing Freedom Riders in Birmingham. To his "handlers" at the Bureau, Rowe often seemed to practice more violence than necessary to protect his cover. Two years later, he rode in the car that shotgunned civil rights worker Viola Liuzzo to death near Selma.

His mission on Saturday, the 8th, was Tuscaloosa, where according to his Klan buddies their task was "to tear the school apart" if necessary, Shelton's assurances notwithstanding.[46] When Rowe called his FBI handler about the arsenal he had gathered, the handler balked at the bazooka but simply told Rowe to use his own car so the police would know who to arrest. However, at Butch Bell's barbeque and beer joint, they switched. A whip aerial adorned the car they took, part of the Klan's infatuation with police and military gadgetry. Bellies full of beer and sipping whisky to boot, they roared down Highway 11, headed for the Klan rally just south of town. Nearing Tuscaloosa they stopped at a roadblock manned by a dozen patrolmen. Normally fearing little from the law, they found things different this time. Shoving and cursing followed and the patrolmen confiscated their weapons, save for a switchblade in Rowe's hip pocket and .25 beretta concealed in another Klansman's trousers. Once the shouting and pushing stopped, Rowe's friends shared some whisky with the patrolmen as they pulled out for the county jail. "Jesus Christ," volunteered one trooper, "we sure hate to bust you when you came down here to help us keep the goddamn niggers away from the school!"

Rather than going to jail, the Klansmen were driven onto campus where Ben Allen, "one of the best investigators Alabama has ever had," interrogated them. Among Allen's more recent assignments had been a background check on Vivian Malone, about whom he concluded, "Vivian was good people."[47] Allen was tough and made the sweat pop out on the Klansmen. After determining their purpose, he sent them to the county jail. By 3 o'clock the next morning, Shelton had them out. The arrest and interrogation, however, convinced them that Shelton and Al Lingo meant business when they said no violence this time. By getting arrested they missed the throng of 3,000 that gathered to see a sixty- by forty-foot cross go up in flames and to hear Calvin Craig, head of the Atlanta Klan, offer thanks "for the greatest man in Alabama, Governor George C. Wallace."[48]

The cool reception in Tuscaloosa for Rowe and his friends was further evidence of Wallace's determination to keep the peace. Still, the fact that they descended eager to do the governor's work indicated how fine a line separated violence from the racism that permeated Wallace's speeches. The following morning Wallace wired President Kennedy to say that "out of an abundance of caution" he had called up 500 National Guardsmen and ordered them to stand by at the Fort Brandon Armory just off campus. "My sole purpose," he concluded, ". . . is to fulfill my pledge to preserve the peace. These guardsmen will be used for no other purpose."[49]

Local arrangements

For the university, this was Rose's finest hour. On the Tuesday following his return from Washington, Rose called Tuscaloosa's resident FBI agents to his office. He told them that he was "a personal friend of Attorney General

Kennedy," that he had requested twenty-five additional agents to harass the Klan, and that he expected "to be advised by the Bureau of developments in the situation." As the agents entered Rose's office, they saw Circuit Judge Wright and civic leaders LeMaistre and Pritchett leaving. Rose told the agents that the three had just interviewed Robert Shelton and told him "that if he created any difficulty in Tuscaloosa, they intended to run him and his Klan friends out of Tuscaloosa, seeing to it that they would not be able to get jobs." They warned him not to show up for a meeting of business and union leaders scheduled that night. (In fact, an unsubdued Shelton attended the meeting and got into a shouting match with union leader Aaron Waldrop. He called anyone signing a petition against Wallace's stand "an integrationist." Nonetheless, 212 people signed.) According to the FBI agents, Rose remained "considerably perturbed about the [Klan] situation."[50]

Rose, Bennett, and Pow worked hard. Their deans and staffs joined in the crusade to keep 'Bama peaceful. No stone was left unturned, literally, as squads of maintenance workers combed the campus for rocks or sticks large enough to be used as missiles. Paper cup dispensers replaced bottle vending machines. Meetings by the dozens with faculty, student leaders, and staff led to calls for dignity and decorum. Officials invoked a 10 p.m. to 6 a.m. curfew. The campus was to be cordoned off by Saturday, June 8, using 26 checkpoints. No one would move without identification. Students received cards at the old Northington campus, a collection of mostly abandoned Navy barracks about two miles east of the main campus. From Northington, students were shuttled by bus to Foster Auditorium for registration. Rather than have parents unload their children at the dorms, a traditional rite of passage, special trucks carried student luggage. Faculty drove to the edge of campus and left their cars before walking or taking shuttles.

The plan paralleled the one drawn up by Bennett when the university still expected the court to order Lucy's return in February 1956. Planners this time included city, county, university officials, and Al Lingo. Meeting at the Hotel Stafford in Tuscaloosa on May 24, they hammered out details. Lingo underestimated the number of troopers that would eventually be ordered to campus, indicating that 200 would be sufficient. He did ask for a command post on the third floor of the Administration Building with desks for ten lieutenants and four phone lines. Unlike Bennett's plan in 1956 which called for a National Guard officer to be in command, Lingo assumed that responsibility. In fact the planners did not envisage calling the Guard. Still, they laid out the basic details for isolating the campus by controlled access. They prepared a document that Wallace reviewed while Bennett and Rose were in Washington. Similar, if less elaborate, plans were worked out for Huntsville.[51]

The university also prepared for the national and international press corps. Unlike Carmichael, Rose tried to put his best foot forward for the media. The university leased space in an empty building owned by the *Tuscaloosa News*. Telephones, typewriters, teletype machines, outlets for sound-on-film cameras, and air-conditioned darkrooms filled the building. The university issued press passes from the news center and provided shuttle transportation to campus. All campus buildings were off-limits except the Old Union Building and Foster Auditorium. For the confrontation itself, eight media representatives were to be allowed on the floor of Foster Auditorium and 34 outside. No live broadcasts were permitted, though sound-on-film equipment was allowed at the site of the confrontation, and, in a nod to Drew Associates, transistorized sound-on-film could be used at locations other than at "the door." The rules forbade students to be stringers or photographers. Bill Dorr of *Publishers' Auxiliary* called it "the most thorough and elaborate press set-up the South has seen in its chain of racial incidents."[52]

Rose looked like the man in charge. *Life* took a picture of him sitting on a bench beneath a spreading oak; well creased, well cuffed, contemplative, even serene, as news of the impending confrontation swirled about. No one doubted Frank Rose's leadership at that moment. Reporters jockeyed for his words, and student leaders yearned some day to be like him. When he strode before the cameras at 2 o'clock on Sunday afternoon, June 9, reporters hung on each word of this calm, collected, easy giant. "His appearance was immaculate," observed one, "dark gray suit, matching tie, shiny black shoes, four pointed white handerchief." He exuded control, so much so that when he announced George Wallace's presence as "essential to insuring law and order," it all seemed natural.

Federal plans

Rose had no other choice. As hard as university officials worked, they exercised no meaningful control. By laboring behind scenes with the Kennedys and by holding the sheet music while business leaders orchestrated pressure on Wallace, university officials helped. In reality they were stage hands while Wallace produced and directed the play. President Kennedy did not accept the necessity of Wallace's presence. In response to the governor's telegram calling out the National Guard, Kennedy wired back, "I am gratified by the dedication to law and order expressed in your telegram. . . . The only announced threat to orderly compliance with the law, however, is your plan to bar physically the admission of Negro students . . . in violation of accepted standards of public conduct. State, city and University officials have reported that, if you were to stay away from the campus, thus fulfilling your legal duty, there is little danger of any disorder being incited which the local town and campus authorities could not adequately handle."[53]

The weeks leading up to the enactment at Foster Auditorium revealed that the Kennedy administration had Wallace trapped: the court order of May 16, followed by Judge Grooms's ruling of the 21st against further delay; the suit to enjoin Wallace filed on May 24, heard June 3, and followed by the actual injunction issued on June 5—all had the effect of encircling the governor.

Like Grant chasing Lee westward from Petersburg with the only question being how Lee would stand down, Robert Kennedy's call to Wallace on the 8th amounted to a last-hour attempt at getting Wallace to say how he would surrender. Wallace, however, insisted on an inscrutable silence, and the federals were forced to plan based on what others told them the governor would do. On the last day of May, they learned from Bennett that Wallace "intends to stand in the door and that members of Alabama Highway Patrol will assist governor in barring entry of Negroes even though Negro students will be assisted by marshals, and that it will therefore be necessary to call in federal troops." This report heightened fears of another Ole Miss. After June 3, however, the news got better. Ed Ball, director of news operations in Tuscaloosa, "said whole thing is greased. Wallace is going to make a gesture then step aside." Rose reported much the same on the 5th, but gave it an unusual twist, saying that "Wallace has attempted to forbid campus and Tuscaloosa police from protecting Negroes—[but that he] Rose refused ultimatum."[54]

The Justice Department had to plan for all contingencies. The use of federal troops had been a commitment from the beginning. In a memorandum to Robert Kennedy dated May 31, Katzenbach outlined the necessary steps for showing force. The assistant attorney general spent considerable time in Alabama during May and worked closely with General Creighton Abrams, who was on the scene. Katzenbach did not take sides on whether the troops should be National Guard or regular army, but he did say that even if Justice decided to deploy regular army troops "it would probably be necessary to federalize the Guard to take it away from Wallace's control." Katzenbach urged that the necessary proclamations and papers be drawn up in advance. He suggested that troops be "prepositioned" at Redstone Arsenal for use in Huntsville and at Columbus (Mississippi) Air Force Base for Tuscaloosa—his estimated deployment time being one hour in the case of Huntsville and a generous five hours to convoy troops sixty miles from Columbus.[55]

Justice had plenty of people on the scene but did not put together a final plan until the last moment. At various times Marshall, Joseph Dolan, Doar, Louis Oberdorfer, Katzenbach, and General Abrams entered the state to work with local officials, the Legal Defense Fund, and the U.S. Attorney's office for the Northern District of Alabama. Still, there was confusion. On Sunday before the Tuesday registration, U.S. Marshall Peyton Norville told reporters that only a few marshals would be used and that he personally had the responsibility of relaying the court order to the governor at the schoolhouse door. If the governor made more than token resistence, he was to withdraw the students and await instructions.[56] Norville's understanding was off the mark.

Fortunately, the uncertainty within Justice did not spill over into the relationships with the three students. Doar personally escorted Vivian Malone from Mobile to the June 5 meeting in Judge Lynne's chambers. Also present were Lynne, Hood, U.S. Attorney Macon Weaver, Shores, Katzenbach,

Doar, and Bennett. The purpose was to assure the students that the university and the federal government would be at their side at all times. Bennett remembered too well the failure to communicate in the Lucy case and talked at length about everything from dress, to registration procedures, to press relations.[57] It was a friendly meeting that relaxed the students. Ever mindful of Wallace, Bennett assured news reporters that no deals had been made to enroll the students secretly. The conference centered on plans for the blacks' arrival on campus but not for the governor's promise to block their entry.[58]

From Birmingham, Malone and Hood flew to New York where they got further reassurances about the NAACP's continued support through registration and beyond, including the payment of all fees. They took in all the famous attractions, including a visit to the NBC studios at Rockefeller Center where they met new singing sensation Leslie Uggams. For Malone it was her first trip on an airplane and her first memorable trip out of state—a family visit to Chicago when she was young was simply not the same.[59] They flew back into Birmingham on Sunday night to prepare for Tuesday.

The university influenced two important decisions. First, Hood and Malone were to enroll on Tuesday to avoid the crush of Monday registration when the majority of summer students would be in line. Second, McGlathery would enroll two days later. This decision, made for the convenience of Governor Wallace, distressed Robert Kennedy and made him wonder about Rose. Justice always favored the Huntsville site because of its proximity to the Arsenal, and if it could not get that, it at least wanted to make it impossible for Wallace to be in two places at once. "No matter what he did in one place," Kennedy reasoned, "he couldn't get to the second place at the same time. At least, there would be a student registered and going to the university. We thought that that would get us over the hump." Of course, Rose had no option other than to play along with the governor, and Kennedy understood. "In the end, we took different positions," Kennedy reflected. "By the last day, the relationship with Dr. Rose was not as close. The general feeling was that Rose . . . was under tremendous pressure. It was a difficult situation for him. . . ."[60] Wallace ran the 1963 show with even more authority than the board of trustees exercised in 1956. In neither instance was the university president more than a figurehead, though to Rose's credit and unlike Carmichael, he appeared to be a man in control.

The lowest common denominator

In fact no one could rise above the racism of Wallace and his henchmen at the Department of Public Safety without paying an enormous price. As a gutter instinct, racism drags all civil thought to its level. When state investigator Ben Allen gave Al Lingo the bad news that Vivian Malone's record was spotless, Lingo replied in terms not "fit to print." "Let's just say," Allen said, "he was very much upset, because I think his intentions were to keep them all out. He thought all of 'ems backgrounds would prevent any of 'em from attendin' college."[61] Wallace believed it too. Too many speeches before too many Klan and Citizens' Council crowds had convinced him that no black person's record was spotless, a conclusion reinforced by having white supremacists Carter and Kohn at his ear. Besides, segregation had its own cruel ways of reinforcing the assumption; one of which was to make the jailhouse or the penitentiary the schoolhouse for vast numbers of black males. So the investigators dug deeper, sure that "all of 'ems backgrounds would prevent any of 'em from attendin'." They found out, or thought they had, why Dave McGlathery's father could not attend his son's 1961 graduation from Alabama A&M; he was in the state penitentiary.[62] Fortunately someone, perhaps it was the university's lawyers, refused to allow the Department of Public Safety to visit the sins of the father upon the son. However, Marvin Phillips Carroll they could get.

In April 1953, at age seventeen, he was charged with bastardy. A jury found him not guilty. Three years later he was charged by another woman with the same offense, and after a brief, pro forma hearing in Fulton County, Georgia, Carroll got a twelve months' suspended sentence and was ordered to pay $6.50 a week in child support. On his application for admission to the Huntsville Center, dated April 12, 1963, Carroll checked *no* to the question: "Have you ever been officially charged with a criminal offense other than a minor traffic violation or been accused in a Court Martial proceeding?" The university's lawyers asked Dean Harrison of the Law School if bastardy amounted to a criminal offense. He replied that under Section 74-302 of the Georgia Criminal Code a "father can be brought before a Justice of the Peace to give bond. If he refuses to give bond he may be found guilty of a misdemeanor under Sec. 74-9901. This latter section is under Chapter on Crimes."[63]

How long the university held this information is not known. Sometime between June 1 and June 5, Charles Morgan was told and he in turn told his client. In fact, the university could have done no less. Had the university not sabotaged Carroll's application privately, Lingo, Wallace, and radical racists would have had a field day with him publicly—or so the reasoning likely went. Wallace's henchmen worked in the shadow of General Walker's bushy eyebrows, convinced that no "nigger" could pass muster unless practicing the subterfuge of a communist. So the university pulled the trigger on a young man who by the procedures of his day took responsibility for a child.

More puzzling was a conversation between Carroll and Dean Mate that took place in January 1965. For some time, Carroll had been trying to get Dean Mate to talk to him about taking a calculus review course at the Huntsville Center. Mate refused his calls until he could get to a phone with a tape-recorder attachment. When Carroll finally reached Dean Mate, his tone apologetic and his voice nervous, he asked about taking the course. Mate stood firm. He reminded Carroll that he had lied about prior criminal convictions and therefore could not be enrolled. In case Carroll had forgotten Mate added that his attorney, Morgan, had been notified of that fact in 1963. Mate suggested that Carroll try to find a correspondence course, but not at the Huntsville Center. Carroll thanked Mate for his explanation and hung up.[64]

Maybe Al Lingo would have found out about Carroll, even in 1965, but not likely. By then he was too busy with the impending crisis at Selma. The university had traveled the path of resistance for too long to be flexible so soon.

11 A Moral Issue

In late May, Brewer Dixon wrote Rufus Bealle to say that his Spanish-born wife Loba and her son, Francisco, wanted to enroll for English lessons during the summer term. "I realize," the veteran trustee observed, "this might be a very interesting time to be there, should the Negro girl be actually enrolled. I would rather like to be on hand at that time to see just [what] happens, although I would not care to have Loba and Francisco there at such time, and I probably would prefer to be with you in an armored car."[1] Fiery crosses and night callers made everyone nervous. Alex Pow, the university's man in charge of security, left his teenaged sons at home one evening armed with shotguns and explicit instructions. One stood at the bottom of the staircase, the other at the top. If anyone attempted entry, the son at the top was to shoot a window out. A nearby FBI agent would come running. If anyone forced an entry, the boys must shoot to kill. Pow was not alone in his forebodings. Earlier, Jeff Bennett took more direct action. When the Klan threatened his wife and others who met at Canterbury Chapel, Bennett called Bobby Shelton and made him personally responsible for his wife's safety.[2]

Personal fear did not make officials plan better, but it did underscore the necessity of planning. The plans leading to Tuesday's confrontation at Foster Auditorium could not have been better drawn. From midnight Friday, the campus was sealed. The shuttle system got people where they needed to go. On Monday, the main day for registration, the Northington operation went smoothly, save for an amusing glitch. A sophomore idly punched a hole in his identification card with a pencil. Escorted back to the identification center, the student apologized to Miss Jean Warren, who rolled another card into the typewriter. When she asked the young man's name, he replied sheepishly, "Joe Namath."[3] To have things go as well as they did reflected the work that John Blackburn, Sarah Healy, Alex Pow, and Jeff Bennett put into it. Besides, Lingo's men were on hand to enforce order. When Frank Rose's manservant, Otis, failed to explain quickly enough his business on campus, he took a crack on the head from one of Lingo's troopers,[4] demonstrating the persistence of old patterns in the midst of new plans.

Meanwhile Jimmy Hood and Vivian Malone flew back from New York, arriving in Birmingham Sunday evening. The Drew Associates camera team recorded their weekend conferences, most of which involved rehearsing possible scenarios for Tuesday, including responses for the press. Hood joked about being governor, saying that since whites believed blacks would take a mile if given an inch, it would suit him fine to be governor after attending the university. Malone concentrated on the lack of accreditation at A&M and the opportunity to study accounting at the university. The Legal Defense Fund could not have found two more attractive candidates—Hood, bespectacled, sporting a snap-brim hat complete with feather, articulate, and quick-witted; Malone, model-like in appearance, yet completely comfortable in the modest Villager styles of the early sixties. Though more reserved than Hood, Malone was equally well spoken, her soft inflections made all the more attractive for having retained a trace of the flat, country sounds of her native Monroeville. The cameras liked them both, but showed a slight preference for Malone's introspective beauty.

The cameras also preferred Robert Kennedy over Wallace. The thirty-hour documentary countdown to crisis began around a tumultuous breakfast table at Kennedy's Virginia estate. Three-year-old Kerry Kennedy is being coaxed into drinking her milk by taking a sip each for brothers and sisters and friends and grandparents. Robert Kennedy presides over the table in a no-nonsense style, but the presence of so many children and his attention to Kerry's milk warms the viewer. Amidst parting shouts from the children, the Attorney General of the United States is soon on his way to deal with the crisis. As Wallace no doubt expected, the cameras see him less favorably. The opening shot scans a cavernous mansion ballroom with sweeping staircase out of *Gone with the Wind*. In this empty museum Wallace's youngest daughter, Janie Lee (named for Robert E.), attended by her black nurse, bangs out discordant notes on a grand piano. The governor descends the stairs, coat buttoned, picks his daughter up and baby-talks her into a kiss, after which she strains to get down.

Combatively Wallace urges the cameras to take in the portraits of honor. Jeb Stuart stares solemnly down. Wallace points to William Lowndes Yancey, the fire-eating secessionist, and quotes him to the effect that "to live is not all of life and to die not all of death. I'd rather live a short life of principle than a long life of compromise." He turns to the film crew and says, "Of course that may not mean much to you folks." Walking out to a waiting car, he pitches a ball to his son but complains about being too old to be doing that sort of thing, and waves to prison trusties lined up to greet the governor before returning to their chores. A trooper opens the car door and welcomes the camera crew. Riding in the front seat Wallace looks stern, almost mean; contrasting sharply with a similar scene of Kennedy riding in the back of his car looking somber but thoughtful. Wallace acquits himself on the morality of segregation, saying that he did not think it immoral to separate people. "Separation," he observes, "has been good for the Nigra citizen and the white citizen." Back in New York, working his way through piles of film in his editing room, Drew sees the morality play exactly as the Kennedys defined it. To close the scene, he finds footage of the Alabama governor's unattractive habit of sucking air through his teeth, as if cleaning away food particles after a meal but looking more like a snarling dog.

The Montgomery camera crew, Jim Lipscomb and Mort Lund, also saw Seymore Trammell as Robert Kennedy had seen him—an "awful little man," an "impossible figure." The film shows Trammell, characteristically at Wallace's elbow, punctuating each of the governor's thoughts with his own running commentary. Wallace is shown with Trammell and others saying there will be peace in Tuscaloosa, that they will not allow "agitators and provocateurs to desecrate that beautiful campus." Correspondingly the Kennedys are shown preparing for all eventualities, especially how they will handle the governor—will he have to be removed physically and if so how; will they have to pick him up or shove past him? The film shows continuing debate over how to use the National Guard. Drew works these fragments into an account that is accurate but collapses the decision-making into a brief span of time, June 10 and 11, thereby creating a more dramatic effect of last-minute planning.

The university and the Kennedy administration did plan more earnestly and with greater urgency than did Wallace. They planned as if in a game where the opponent is perpetually on offense; a game in which they can claim a moral victory if the opposition fails to score, or possibly even win, but only if the offense makes a mistake. The Kennedys and the university never figured Wallace could win, but Wallace never acted as if he were going to lose—in part because he never made clear what winning meant—would he have to stop Malone and Hood to win or would a simple show of resistance suffice? In fact Wallace had done everything conceivable to make the confrontation symbolic, not physical. Unlike Barnett he did not usurp the university's authority to act as registrar. He told the board that it must comply with the court order. He told his constituency to stay away. He got pledges from all the hate groups to leave the confrontation to him. He sent in Lingo's troopers and called up the National Guard. The encounter between Rose's butler and Lingo's men showed their determination to keep outsiders off campus. Lingo even tried to keep the press off, save for the Alabama Capitol Press Corps. When Wallace's press aide, Bill Jones, arrived to find troopers cowing reporters, he stormed up to the colonel demanding an explanation. Jones reminded Lingo that the whole purpose of the confrontation was national publicity. Thereafter troopers cooperated in a friendly manner.[5]

Wallace's plan had two flaws. First, it contained no provisions for the day after. To Wallace, once the federals took over, law and order became their problem. The *Alabama Journal,* which kept a Cassandra-like watch on the governor, spoke the fears of most observers outside the South and many within. "Wallace cannot disavow all responsibility for what happens on D-plus 1 or 30. In defying federal courts in his way he is inviting those who cannot comprehend the niceties of the constitutional issue to defy the law of the state and the nation in their way." Declaring that "defiance begets defiance, even if delayed," the *Journal* characteristically laid some blame on agitators. "The outrageous spectacle King put on in Birmingham," opined the editors, "has galvanized the idiot brains of men who understand no law but the law of the jungle. . . . thugs are waiting in the wings." As if to underscore the *Journal's* warning, Ed Fields of the National States' Rights party said that his organization would honor the governor's request, but what happened after "was another matter." He also announced that J.B. Stoner, whose fulminations on race reeked of violence, would speak at their Sunday meeting before the big event.[6]

Also absent from Wallace's reckoning and not so apparent in the early days of June 1963 as it would become for the remainder of the decade, was assassination. To be sure, uppity blacks and kooks like the Baltimore postal worker had been executed, but that was the old business of lynch law. As yet no major civil rights leader or public official had been slain. Wallace no doubt knew the dark possibility, but a misfire against General Walker or a roadside execution did not make assassination as probable as events later in the decade made it seem. On May 28, the university and the FBI received a reliable report that a young man from Homewood, a comfortable middle-class neighborhood just south of Birmingham, took up "a collection for the purpose of buying a rifle with a telescopic lens" to be used in "taking care of Vivian Malone and Jimmy Hood."[7]

Trustees

University trustees began arriving on Sunday. They were as superfluous as gridiron alumni back to see new gladiators play. Having fought desegregation for ten years, they appeared almost relieved to have the ball in Wallace's hands, especially the older trustees. Wallace flew to Tuscaloosa on Monday afternoon, toured the campus, visited National Guardsmen, and checked into his headquarters at the Stafford Hotel. Wherever he went, well-wishers greeted him as a champion. At 7:45, accompanied by Cecil Jackson, troopers drove him to the Administration Building where he called the trustees to order at 8 p.m. Wallace chatted amiably before telling them how he planned to handle affairs the next day. Again, he pledged no violence and detailed the steps taken to insure law and order. Frank Rose explained cooperative arrangements between the university and state authorities. Wallace turned the meeting over to Gessner McCorvey and retired to the Stafford. Several trustees prepared a resolution of support for the governor. Though the board and Rose had made it clear that they wanted to avoid the confrontation Wallace sought, the board now unanimously said that the "presence of Governor

Wallace is necessary to preserve peace and order."[8] The tepid preamble indicated that some trustees had to struggle before resigning themselves to this paradox: the governor's presence made his presence necessary—the abyss calling forth the abyss.

While the board prepared the resolution, Winton Blount and Jeff Bennett drove to a downtown pay telephone and called Robert Kennedy. Even though the boardroom had pickup telephones to Washington and local command centers, they wanted to avoid the possibility of a tapped conversation or, as had often happened with Wallace, an operator eavesdropping and relaying information to the governor. Blount and Bennett described the governor as scared and nervous but said he had pledged law and order. Neither Blount nor Bennett could assure the Attorney General that Wallace would step aside without some use of physical force; however, they did expect him to yield in the presence of troops. Their description of the governor as agitated continued the suspense over Wallace's mental state. Others saw him differently. John Caddell remembered Wallace as relaxed, in control, and amiable.[9]

But Wallace had reason to be anxious. He faced the prospect of jail. If the Kennedys did not know how much the governor would resist, Wallace did not know how much Washington would tolerate. Judge Lynne warned Wallace's lawyers that a jail term would be more than token. Another factor surely played across the mind of this consummate politician—could he make the show work? Would all the elements of force-induced surrender be present? What if the federals with the cooperation of university people outflanked him? What if the drama unraveled and people laughed? It was enough to make any politician nervous, most of all George Wallace.

The deputy

Katzenbach also arrived in Tuscaloosa on Monday to take charge of federal operations. Plans were still in flux and improvisation the order of the day. The deputy attorney general had plenty of experience. He resourcefully commanded civilian forces during the Ole Miss crisis, at one point dropping coins in a campus pay telephone to communicate with the Commander in Chief when the army communications system went down. Ole Miss was a harrowing experience better forgotten. Yet Katzenbach would never forget a quick exchange with Robert Kennedy before boarding the plane for Oxford. Only the passage of time made it amusing.

"Hey, Nick," Kennedy called out.

"Yeah?" Katzenbach replied.

"Don't worry . . . if you get shot," Bobby laughed, " 'cause the President needs a moral issue."[10]

Katzenbach was vintage New Frontier. The son of a prominent New Jersey family (his mother presided over the state's board of education and his father was a former attorney general), he left Princeton and privilege in his junior year to become a combat veteran in World War II. Shot down over the Mediterranean, he spent two years in Italian and German prisoner-of-war camps, an experience he found "boring." He twice escaped only to be recaptured. The Harvard-slanted Kennedy administration did not place much store in his Yale law degree but it did like his Rhodes scholarship and his solid 6′2″, 200-pound frame on the touch football playing field. Nicholas deBienville Katzenbach had the look of a Roman or perhaps Gallic general: square jaw, long angular nose, dark penetrating eyes, high balding dome. Despite an imposing presence, made all the more so by a mellifluent, baritone voice, he looked rumpled. Rounded shoulders, a slight lean to his walk, and a casual, chain-smoking approach to the most important deliberations made him appear unprepossessing. In fact it made him the ideal commander, and if any doubted his authority, as some pushy reporters did on the day of the encounter in Tuscaloosa, his eyes cut like a fishknife on flesh.

From his command post in the Army Reserve Building about half a mile west of campus, he talked to the marshals about their duties. Protection of the students was the bottom line, even if it meant shooting any who threatened them. He cooperated fully with General Abrams, who was in overall military command. The 500 National Guardsmen Wallace called up were at the Fort W.W. Brandon Armory, half a mile east of Foster Auditorium where Malone and Hood were to register, and at nearby armories in Holt and Northport. The guardsmen spent the day lounging, jostling each other like young bucks, telling racial jokes, and practicing techniques for lifting the governor and physically removing him from the schoolhouse door. It was all great fun. Still, Katzenbach had no direct contact with the Guard or state authorities. Lingo's men patrolled the campus, and the deputy attorney general could not know what role they might play: would they assist the governor in barring the door? Would they form a line between the governor and federal troops? Would they drift away from their posts if violence erupted as the Mississippi patrolmen had done in Oxford?

Planning continued through the night. Doar, on leave from Mississippi, stayed in Birmingham to reassure the students and their lawyers that nothing was being overlooked. Still, details remained unsettled. Planners considered a scenario in which "the two students [would] be escorted to the point of confrontation by [U.S. Attorney] Macon Weaver and Marshal Norville." Katzenbach, Doar, and another marshal would follow. They wondered whether they should "shake hands with the Governor or not?" They figured that Wallace would speak first, but if not, Norville should be ready with a proclamation. If the governor refused to move, Weaver would remind Wallace of his constitutional duty. They planned for Weaver to "point out that he and Norville are graduates of this University and that the University is ready, able and willing to register these students in accordance with Court Order." He would "appeal to the Governor on behalf of the alumni . . . that this

University . . . not be sacrificed for politics." If that failed, Katzenbach would say that the governor left the federal government no choice.[11]

The plan reflected Monday morning's state of thought. Throughout the day Katzenbach simplified the plan, part of which involved surreptitiously getting keys to the students' dormitory rooms—not even university officials knew he had them. From General Abrams's headquarters at the Army Reserve Center, the deputy attorney general kept Robert Kennedy and Burke Marshall fully informed. Together they agreed that Katzenbach, accompanied by Marshal Norville and U.S. Attorney Weaver, should be the only ones to approach Wallace. In this way the students would be spared any indignity. Technically, it removed the Justice Department from the hook of having to arrest the governor for barring them. If the governor refused this overture, Katzenbach said that Malone and Hood would be taken to their dorms. Overhearing this, Abrams looked puzzled. Katzenbach, cupped the receiver and said, "Why not? I've got the keys to their rooms."[12]

The Attorney General and Katzenbach also discussed federalization of the guard. Convinced that Wallace would resist to the point of being forcibly removed, Katzenbach urged ignoring the governor altogether by simply having the students start attending classes. When Burke Marshall balked, Katzenbach replied testily that everyone on the scene in Tuscaloosa understood how it would work. Kennedy cut in saying that politically it would be impossible to deny the governor his platform without risking a deterioration of the situation. (Katzenbach's inclination not to retreat by taking the students to their dorms was brilliant. His desire to bypass the confrontation altogether risked disaster. Wallace would not tolerate losing face. His most probable option would have been stationing troopers at the entrances to Malone's and Hood's classroom buildings. To have substituted Lingo's men for the governor's singular symbolic act would have amounted to tossing a match in gasoline. As Katzenbach later reflected, "Bobby, I think, made the right decision. He said, 'Look, it's too dangerous a situation, and I don't know what that man'll do if he's crossed. He wants his show; you're gonna have to give him his show.' ")[13]

This last bit of improvisation was relieved when young Kerry Kennedy burst into her father's office. The Attorney General asked Katzenbach if he would like to talk with her. Startled, then beaming, Katzenbach chatted playfully. "Hi, Kerry, how are you, dear?" "Nick," she asked, "are you at our house?" He replied, "No, I'm way down in the Southland and it's awfully hot here." He asked Kerry to tell her father that "they were all going to get hardship pay," which she did her best to convey before having the phone wrested from her in the midst of "goodbyes." Sensing that Katzenbach wearied of Wallace's game, Kennedy suggested that his deputy use a dismissive tone when addressing the governor. " 'Well, he's really a second-rate figure to you. He's wasting your time. He's wasting the students' time. And, let's not make a big deal out of it. I don't want to put the students through that indignity. I don't want the man to stand there and say things to them.'. . . . 'Don't let him say anything to the students.' "[14]

The big picture

President Kennedy had to decide how events in far-off Tuscaloosa fit the larger picture. Movement leaders had come to view his administration as all talk and no show. Martin Luther King despaired. A presidential conversation with Coretta King on the occasion of her husband's incarceration in the Birmingham jail echoed only faintly the grand gesture of Kennedy's telephone call to Atlanta during the 1960 campaign. The Justice Department counseled patience throughout the Birmingham demonstrations and criticized King for violating the court-ordered ban on marches and later for the use of children. Now King wondered whether the march on Washington should be directed against the President as well as Congress.

On a more personal level, Robert Kennedy took fire directly. He accepted an invitation to visit with black cultural leaders in the New York apartment of Harry Belafonte. Gathering on the evening of May 24, Kennedy soon learned that the spirit of the meeting was not congratulatory or even friendly. James Baldwin decided the fire must come this time and refused any concessions to Kennedy gradualism. Baldwin's ignorance of public policy convinced Kennedy that Baldwin and his compatriots were uptown intellectuals blowing off steam because they had "complexes" about failing to put more than words and money on the line in Mississippi or anywhere else. The Attorney General felt uncomfortable, then angry. He left irritated at the lack of respect, if not gratitude. The fact that Baldwin and others rushed the story into print cinched Kennedy's belief that Belafonte's parlor had really been a grandstand.[15]

Robert Kennedy, however, was a man whose commitment to civil rights smoldered like a spark in old wadding, ready to explode when exposed to fresh air. Within a week of the unsettling New York encounter, he convened a summit of the administration's top legal and political strategists and proposed strong legislation to end discrimination in public accommodations. Politically shrewd advisers in the room said no, but the President and Lyndon Johnson stood solidly behind the Attorney General.[16] White House councils remained private, and Martin Luther King continued to beg for some sign, any sign, that the administration was prepared to move. He even grabbed at the idea of having JFK personally escort the Negro students to the schoolhouse door.[17] Still, the White House was silent, even refusing King's request to meet with the President on the same day, June 1, that Kennedy ordered drafts of the new public accommodations bill. The administration had moved up its timetable for civil rights legislation but wanted it to appear to be its own table. Time and circumstance now collided in Tuscaloosa to give the administration that precise opportunity: an event not created by King, an event that matched segregation's most defiant champion against the President—no intermediaries this time, no one else to claim credit, a media event to showcase Kennedy style.

The countdown to crisis had begun. King, still not privy to the President's plans, kept firing blindly, saying that Mr. Kennedy succeeded only in substituting "an inadequate approach" for Eisenhower's "miserable one." While King squeezed off round after round, the President stayed on the move, like a duck in a shooting gallery gliding unharmed toward oblivion. On Saturday the 8th, JFK was in San Diego for a speech. The next day he addressed the National Conference of Mayors in Honolulu, saying with characteristic moderation: "I do not say that all men are equal in their ability, character and motivation. I do say that every American should be given a fair chance to develop in full whatever talents he has and to share equally in the American dream." The vision was Lincolnesque. One did not have to believe in racial equality to argue for equal opportunity. Still, Kennedy pressed forward, trying to enlist more than the federal government in the struggle. "The time for token moves and talk is past," he warned the mayors.[18] Then on Monday morning, June 10, Kennedy gave what many considered the most remarkable foreign policy address of his presidency.

At commencement exercises for Washington's American University, he told the graduates, most of whom were born in the year of Pearl Harbor, that the assumptions of the Cold War must be reversed. Peace, they were told, was "the necessary rational end of rational men." He called for a re-examination of attitudes toward the Soviet Union. "No government or social system is so evil that its people must be considered as lacking in virtue." Abhorrence of war must be our common bond. "No nation in the history of battle ever suffered more than the Soviet Union suffered in the course of the Second World War." With so much to gain and so much to lose, Kennedy proposed negotiations leading to a nuclear test ban treaty and immediate unilateral cessation of our own atmospheric testing. "If we cannot end now all our differences, at least we can help make the world safe for diversity. For, in the final analysis, our most basic common link is that we all inhabit this small planet. We all breathe the same air. We all cherish our children's future. And we are all mortal."[19]

Spreading over the nation's radio waves and aired on television news, Kennedy eloquence had seldom been nobler. The applause was loud and sustained. The wordsmiths had worked overtime. Any attempt to follow immediately such a rhetorical triumph would be foolhardy. And yet at 5:30 that afternoon, at his brother's insistence, the President ordered draft ideas for a major speech on civil rights, a speech to follow the confrontation in Tuscaloosa. (In Drew's documentary, Bobby can be overheard mentioning that a draft was already in the works.) Larry O'Brien, special assistant to the President for congressional relations, objected. The president risked permanent rupture with the southern bloc and therefore his entire legislative program, not just civil rights. Another time and a less visible public demonstration would be better. Robert Kennedy felt the speech could operate as a double-edged sword. The President could get out in front of legislation he must eventually support anyway and could use the occasion to remind blacks of their responsibilities.[20] As the White House conference broke up around 6:30 Monday evening, the President thought his speech would be delivered in the midst of a continuing crisis. Thus the speech would be designed, at least in part, to explain the government's progress in securing the admission of the students. As events turned out, Tuesday gave the President a better platform than expected.

Tuesday morning

Jeff Bennett moved into the president's mansion Monday night and slept like a baby. Katzenbach got no sleep. Vivian Malone and Jimmy Hood, who had their final meeting for last-minute details at 11 p.m., spent the evening in the home of Arthur Shores's secretary, Agnes Studemeyer, and her husband, whose hobby of making and repairing guitars drew some of the evening's attention. Malone had already put herself into a calm, almost trance-like state as she concentrated on events beyond the next day's dramatics—what her classes would be like, life on the campus, even graduation. The air of calm resignation served her well through the days ahead. More immediately, she and Hood woke to the pungent smell of Agnes Studemeyer's famous hot rolls.[21]

Tuesday broke beautifully over the campus. The trees already wore their lush summer green, while the grass retained the tender growth of May's first warm evenings. The forecast called for unusually hot weather, but before 9 o'clock in the morning, one could still hope for a pleasant late spring day. Tommye Rose fixed breakfast for the mansion guests that included Bennett and a couple of the president's closest friends. One was Wyatt Cooper, husband of Gloria Vanderbilt and a budding writer, a friend and cousin from Rose's Mississippi youth. Several of the trustees also came for breakfast, including Caddell and McCorvey. All the trustees were to assemble at the mansion by 9 o'clock, then go to Little Hall, a short block away. Little Hall housed the physical education gymnasium and most of the athletic offices. The president and the trustees would assemble in Bear Bryant's office on the ground floor at the southwest corner of the building. The windows offered a good view of the north entrance to Foster Auditorium.

The schoolhouse door was neither little nor red. Built in part with Public Works Administration funds and completed in 1939, Foster served as a multi-purpose facility with a seating capacity of 5,400. Everything from graduation exercises to basketball games were held there. It was the site of President Carmichael's grave convocations during the Lucy crisis. University people remembered it more fondly for an event that took place a year later. With the University of North Carolina's national championship team visiting, an Alabama player took a rebound, spun, and hurled what remained until recent times the longest shot in college basketball. The building itself was stolid,

1930s' architecture, but above its north entrance rose six columns to give the otherwise drab structure a classical façade. Beneath the colonnade three doors admitted students into the gymnasium, doors that looked across the way to Farrah Hall, the university's law school. The afternoon before the confrontation, Bill Jones and Bennett directed workers to paint a semicircle in front of the center door. Behind that line, Wallace, facing the law school he had attended, would defy the federal court order.

At 8 o'clock Eastern time, 7 o'clock Alabama time, Robert Kennedy left McLean for the drive into the capital. Before leaving, he called the Pentagon and learned that the Guard could be federalized and deployed in less time than originally thought. (Since Guardsmen were already located just off campus, the only difficulty would come in getting General Henry V. Graham, who was on maneuvers at Fort McClellan, onto campus in time to command the units.) Kennedy arrived at the Justice Department thirty-five minutes later with three of his children in tow. Burke Marshall was waiting. In Birmingham, Katzenbach and General Abrams prepared to escort Hood and Malone to Tuscaloosa.

First curtain

At 9:15 Central Time, 10:15 Washington time, George Wallace emerged from the Stafford to a chorus of "Bless your heart; Bless your heart" and pats on the back from a cluster of women on hand to express affection and to wish him well. Before ducking into an awaiting car, Kohn approached Wallace and said, "You have Divine blessing, today, Governor. There is absolute peace here. It is a great tribute to this city and to this state—the people have shown great dignity. Good luck and may God bless you." Though overblown, Kohn's words genuinely moved Wallace toward what he considered his noble purpose. On arrival, Wallace saw 150 patrolmen in a cordon around the entrance to Foster. He waved as he approached the door where Bill Jones waited to give him last-minute instructions about the podium and the microphone. The governor wore a cool blue shirt for television. He jauntily strode the gymnasium floor joking and shaking hands with students. By 9:50 he settled himself in a comfortable air-conditioned office just inside the entrance. Ben Allen, the state investigator, was among those with him. "He was highly nervous," Allen remembered, "highly nervous, wonderin' if in fact they were gonna arrest him. Wonderin' if the federal authorities were gonna put him in jail. And I know it was awfully hot. We had men stationed on top of these buildings in this hundred-degree weather, stationed up there with rifles, and I don't see how they stood the heat." On several occasions Wallace turned to Allen and asked, "Ben, do you think they'll actually arrest me?"[22]

(Whether Wallace was nervous, scared, or acting irrationally is important. It goes to the question of how much threat Wallace posed to the American liberal-democratic tradition. If unstable, he could be dismissed as another in a long line of gargoylish demagogues. If not, he represented something more. His enemies, and even some allies, seized upon his nervous behavior as evidence of an irrational streak. However, Ben Allen, like all the others, probably misread the governor. Wallace often used question-begging to turn the table on his enemies. Later, when an opponent blamed Wallace with indirect responsibility for JFK's assassination, the governor went around for days asking people if they thought he had helped kill the President.[23] Put another way, the question Wallace asked Allen was: "You don't think they'll be fool enough to arrest me, do you?" Later Wallace dismissed the prospect of jail, saying "that if he were brought to trial on contempt charges, he was going to ask 250,000 Alabamians to come to his trial"—not an improbable figure.[24] Had he wanted them in Tuscaloosa, they would have been there by the busloads. Moreover, if Wallace was nervous while waiting in the anteroom, it never showed on camera.)

A phone conversation between Katzenbach and Robert Kennedy delayed the departure for the schoolhouse door by about fifteen minutes. Shortly after eight the caravan headed down U.S. 11 with Guardsmen, state troopers, and marshals making up the van. Katzenbach and Doar rode with Malone and Hood. Like everybody that day, they made light talk to relieve the tension. As they rolled past the small community of Vance, about twenty miles east of Tuscaloosa, a message reached the convoy that the Attorney General wanted to speak to Katzenbach. The radio-phone hookup, however, malfunctioned. Spotting a little church up ahead, the car carrying Malone and Hood along with another carload of marshals pulled into a cemetary on a hill behind the church. Another car wheeled in to take Katzenbach to a nearby grocery store to place his call.[25] Katzenbach was away for what seemed like an hour but actually was closer to thirty minutes. During this time the students and their escorts walked up under the shade of an old cedar tree. About them lay stone reminders of Tuscaloosa's early white settlers and their descendants. Under one weathered marble slab rested Zachariah Weaver, Pvt Co F 50 Ala Rgt CS Army. Off to another side, a crude limestone marker pushed up from the clay to announce in hand-scratched letters the birth and death of Billy Nabors in 1865, the year of black America's jubilee. There they were, Hood and Malone, the descendants of slaves, traveling toward a new plateau in the struggle for freedom, stopping to rest under a tree planted to shade their forebears' masters.

Of immediate concern for Katzenbach and Kennedy was difficulty in locating General Graham and what effect it might have on the day's planning. Kennedy also wanted to know once more what Katzenbach planned to say to the governor. Shortly after ten Katzenbach returned, conferred with Doar, and the caravan resumed its course, rolling into Tuscaloosa's east side around 10:20, heading toward the Army Reserve Center about a mile west of campus in the downtown area. At 10:29 Hood and Malone arrived at the Center with a four-car escort of marshals. They stayed in their car while a repairman worked on the malfunctioning radio transmitter. As they waited, Secretary of

State Dean Rusk, in Washington, signed the presidential "cease and desist" proclamation, a copy of which Katzenbach would present to the governor. At 10:41 the motorcade (now three cars) left for campus. At 10:45 they passed the first barricade at Tenth Avenue and Tenth Street where they encountered no delay. Small groups lined the route, trying to see the much-publicized students. At Denny Stadium they turned left to University Boulevard and continued east just past the president's mansion, where they turned right, circling behind Little Hall. At 10:48, in roiling heat (the thermometer had already climbed passed 95 degrees), the three cars pulled up in front of Foster. From the shadows of the door, Wallace watched Katzenbach, Weaver, and Norville approach. A year of pledges and promises, followed by months of planning, had come to this.

The first confrontation

Across the way, Rose and the trustees strained for a glimpse. The crush of reporters partially blocked their view, and Alex Pow climbed a radiator for a better vantage from which to call the action. At Foster, Jim Lipscomb, the cameraman for Drew Associates, desperately searched for his own perch. Finally, he took his belt and strapped one leg to the iron grating that covered a low window to Wallace's right. Leaning out he got an excellent camera angle, slightly above and to the side of the confrontation.[26] With a copy of the President's proclamation in his coat pocket, Katzenbach strode forward, flanked by Weaver and Norville. Standing behind a shellacked-wood podium, with a mike slung around his neck, Wallace raised his left hand like a traffic cop to stop them. He said nothing. The silence caught Katzenbach off guard. All the things he had planned to say rushed in at once, canceling each other out. He decided to push past the line Jones and Bennett had drawn for newsmen, a line he believed to have been placed there for the show. The theatrical trappings positively angered him. Katzenbach had one other reason for standing close to the governor. He wanted to get out of the sun. Finally, he identified himself and said, "I have here President Kennedy's proclamation. I have come to ask you for unequivocal assurance that you or anyone under your control will not bar these students." Wallace said, "No." Then warily Katzenbach pushed the presidential proclamation toward Wallace, who received it. Wallace said nothing. Trying to recover the initiative, Katzenbach began to speak. A lack of sleep, coupled with Wallace's unnerving abruptness, caused the deputy attorney general's voice to quaver ever so slightly as he searched for the right words and the right tone to convey the administration's message.[27]

Katzenbach folded his arms across his chest to avoid awkwardness of gesture or signs of anxiety. "I have come here," he stated more confidently, "to ask you now for unequivocal assurance that you will permit these students who, after all, merely want an education in the great University. . . ."

On hearing the editorial digression, Wallace stopped him. "Now you make your statement," he interrupted, "because we don't need your speech." Katzenbach said that he was in the process of making his statement, then repeated his demand for "an unequivocal assurance" that Wallace would do his constitutional duty and step aside. By now Katzenbach's pant legs showed sweat from the knees down. Wallace interrupted again, saying, "I have a statement to read." Wallace pulled out the statement he had prepared with the help of Kohn. He had rolled it nervously in his hands while waiting in the anteroom, so it was necessary to smooth it out on the podium before beginning. Then Wallace launched into a five-minute denunciation of the central government. After tracing his version of constitutional history and reviewing the tenets of sovereignty from a states' rights perspective, he said, "I stand before you today in place of thousands of other Alabamians whose presence would have confronted you had I been derelict and neglected to fulfill the responsibilities of my office." He declared the action of the central government (he seldom called it "federal" in official statements) to be an "unwelcomed, unwanted, unwarranted, and force-induced intrusion upon the campus of the University of Alabama." He closed with a proclamation of his own: "I . . . hereby denounce and forbid this illegal and unwarranted action by the central government."

Having finished, Wallace cleared his throat, took a funny little "skip step backward" and "hopped into the doorway" as two burly patrolmen closed in beside him. Katzenbach tried once more. "I take it from the statement that you are going to stand in the door and that you are not going to carry out the orders of the court, and that you are going to resist us from doing so. Is that so?" Wallace replied flatly, "I stand according to my statement." Katzenbach started again. "I'm not interested in this show," he declared. Then turning slightly toward the cameras, "I do not know what the purpose of the show is." With mounting exasperation Katzenbach grew more confident. "It is a simple problem, scarcely worth this kind of attention. . . . From the outset, Governor, all of us have known that the final chapter of this history will be the admission of these students. . . . I ask you once again to reconsider. . . ." Wallace stared straight ahead, chin thrust forward, refusing to say anything. Katzenbach tried a fourth time only to be greeted by silence. Thus denied, Katzenbach wheeled and walked toward the waiting cars. In all, Wallace's part of the show lasted about fifteen minutes, and it was a stand-off. The next move was Katzenbach's.

Katzenbach poked his head in the car and said, "We're going to the dorm." With that, Vivian Malone got out and proceeded with Katzenbach and Norville through the police cordon toward the west side of Foster Auditorium, in full view of the trustees in Little Hall. The car, carrying Hood, Doar, and the driver, pulled out for Palmer Hall, about a mile away. Katzenbach's end-around was in progress. When Pow announced from his perch atop the radiator the movement of Katzenbach and Malone toward the dorm, Rose was furious. Pow could not figure it: maybe Rose was angry because he did not want the board to know how closely they had worked with the Justice Department, or maybe he resented Pow's making the announcement directly to the board without going through him first.[28] As it turned out, Rose's anger owed more to Katzenbach's failure to let him in on the plan. Relations between the university and the Justice Department, especially those on the scene, had cooled. They feared that anything said to Rose and his staff would get to Wallace. When the university asked for general-area two-way radio communication with the Katzenbach team, they were refused. Instead, at the last minute they were given straight-line walkie-talkies, which meant that Katzenbach's men could have complete privacy and confidentiality by the simple expedient of ducking behind a tree.[29] Now Rose and his people, without advance warning, saw Katzenbach's surprise move toward Mary Burke Hall, just across the south parking lot from Foster Auditorium. As Malone and her party entered the brand-new dormitory, the house mother, having no reason to believe their appearance was outside the plan, walked up and said warmly, "You must be Vivian."

When Rose finally caught up with the deputy attorney general, he "just went absolutely crazy," Katzenbach recalled. "There's gonna be a riot if you do this." Katzenbach had told Vivian "to go up to her room and come down and go eat by herself in the dining room." He now turned to Rose and said, "Look, she's gonna be eating in that dormitory, she might as well start with lunch." In fact, things went better than expected. "She went [into the dining room] by herself, sat down at a table all by herself, and within thirty seconds, six or eight kids had joined her. And that was that problem. . . ." Rose later called Katzenbach to say, "You were right."[30]

While this was going on, Robert Kennedy tried to reach his deputy. Katzenbach could call from his car, but messages from Washington had to be relayed by radio signal. Katzenbach snapped at some reporters to move away from the car, slid into the front seat, and picked up the transmitter. Burke Marshall asked if they expected trouble. Katzenbach, perhaps from exhaustion, gave a garbled reply: "I don't expect any trouble, but they're thinking." While Marshall tried to decipher the response, Kennedy took the phone from Marshall and waited for a better connection. Marshall urged that they go ahead and federalize the Guard, but Kennedy insisted on hearing from his deputy. Getting a better signal, he asked Katzenbach if the governor had done anything that would make it unnecessary to federalize the Guard. Katzenbach said, "No." The Attorney General then called the White House to say it was time to authorize federalization.[31]

The President had been meeting with congressional leaders all morning to discuss his civil rights legislation. When the call came through shortly after noon Alabama time, he was in conference with the minority leadership and gave his assent over the phone; whereupon Robert Kennedy called Cyrus Vance, Secretary of the Army. The President did not sign the authorization until after his meeting with the Republicans, 1:35 p.m. Washington time, and by then the wheels were already in motion. The administration's procedure paralleled that used by Eisenhower at Little Rock—proclamation followed by executive order. At 12:05 p.m. Alabama time, the Chief of the National Guard Bureau in the Pentagon called Adjutant General Alfred C. Harrison of the Alabama Guard directing him to report immediately to headquarters in Montgomery. There he would be designated Commanding General, Alabama Area Command. Brigadier General Henry V. Graham of the Alabama contingent of the 31st Dixie Division had been directed "to proceed to Tuscaloosa to assume command of operations in that city." The Pentagon ordered all other units to remain at their home stations or field training sites, "prepared to move on four (4) hour notice" and ordered to initiate "training in civil disturbances and riot control with particular emphasis on the use of chemical munitions and gas masks."[32] Threats of Klan disturbances in other cities necessitated these statewide precautions.

General Graham's sad duty

General Graham was inspecting troops on summer training maneuvers at Fort McClellan. He knew nothing of the role assigned him. As he helicoptered back toward the air field, he got an urgent call to contact his chief of staff, which he did on landing. Graham's aide held a phone to connect him with Colonel W.G. Johnson, the senior regular army adviser to the Alabama Guard who had the Army Operations Plan for the entire state. Colonel Johnson told Graham to report to General Abrams in Tuscaloosa immediately. Not yet appreciating the urgency, Graham said that it would be evening before he could get there. Johnson said, "General, there's a helicopter right outside your headquarters building waiting for you to take you to Tuscaloosa—now! You are to report forthwith." Graham said, "Yes sir," but still took a quick shower to remove sweat and dust from the morning's inspection.[33] Within minutes he lifted off for Tuscaloosa, 120 miles by road but less than an hour by air. With him was army Colonel Gene Cook, the only officer at Fort McClellan who knew the plan.

General Graham was straight out of central casting. A Birmingham real-estate executive, he was tall, already silver-haired at forty-seven, with a commanding but not overbearing presence. He joined the National Guard in 1934 and entered regular army service in 1940. Early enlistment meant quick advancement. He rose to brigadier general during the African, Italian, and French campaigns of World War II. He regretted not having led a brigade in combat (his assignment had been "as safe as eating a barbecue sandwich in Ollie's"), and almost got his wish in Korea, but again wound up at headquarters, Tenth Corps.[34] Though third in line of Guard command, Graham's selection to confront George Wallace meant that he could he trusted. It also helped that he was the senior line officer. Walter J. "Crack" Hannah, the senior commanding officer in Alabama and one of Birmingham's Big Mules, was an ardent states' righter and supporter of Bull Connor. No one could predict what General Hannah might do under pressure of facing down his governor to effect an outcome he passionately opposed. Thus, Hannah, whose nickname came from crack shot, had been dispatched to Fort McPherson in Atlanta and made conveniently unavailable. General Harrison, number two in command, was ruled out for the practical reason that he served as the governor's adjutant. In fact, when Harrison received the call federalizing the Guard, he was in Tuscaloosa assisting the governor and dressed in civilian clothes. Happily, that left Graham a man of moderation and even disposition.

At 1:40 p.m., Graham's helicopter set down on an apron adjacent to Fort Brandon Armory. He proceeded immediately to the Army Reserve Center where General Abrams briefed him on the morning's developments and gave him instructions for confronting the governor that afternoon. Graham did not welcome the idea of forcefully removing the governor from the door. If it came to that, his business career almost certainly would be ruined. Nonetheless he understood his duty. Fortunately a call came through from General Taylor Hardin, a close friend of the governor's and later his finance director. Hardin wanted to come to the Reserve Center and speak with General Graham privately. The two men met in a small conference room. Hardin told Graham that Wallace would step aside peaceably if allowed to make a statement. While Hardin waited, Graham conveyed the message to General Abrams, who in turn consulted with Katzenbach. Though reluctant to give Wallace another platform, the deal was cut.[35] So at approximately 2:30 on the afternoon of the 11th, the Justice Department at last received what it had sought for two weeks: a direct assurance from Wallace that he would go without being forcibly removed or arrested. Earlier in the day Wallace's press secretary told reporters that the governor had "never said he will oppose the armed might of the federal government," and another aide, following the morning standoff, said, "The governor is waiting for the troops. There will be no shoving or pushing. There may be a minute's word battle and then he will fly back to Montgomery or to Huntsville if necessary."[36] These were comforting words, and in line with what the Justice Department had heard since Jeff Bennett's call to Burke Marshall on March 19, but indirect assurances were no substitute in Katzenbach's mind for the kind of confirmation provided by Hardin.

Shortly after noon, Wallace ordered troopers to bar all windows and entrances to Foster Auditorium, except the north entrance. Katzenbach's move to put Malone in Mary Burke Hall, just to the rear of Foster, had caught the governor by surprise, and that led to the added precaution. Wallace then sat down to a tall glass of iced tea and a medium-rare steak smothered in catsup, with french fries and lettuce. A supporter sent the food from the Stafford. Wallace bantered and quipped with friends, faculty, staff, and aides, including his brothers Jack and Gerald and Seymore Trammell's brother Ap. Hardin returned from his mission to say that General Graham had been consulted and the federals had agreed to the terms. By that time Graham had returned to the Armory under instructions to deploy as many Guardsmen as necessary. Having escorted Freedom Riders out of Montgomery, Graham knew "that you must never send a boy to do a man's job." Translated, it meant force in "such overwhelming numbers" as to intimidate a crowd. By following this principle, he believed it would never be necessary to issue ammunition.[37]

Sizing up the situation at Foster, General Graham sent 100 Guardsmen to the campus. At 3:16 three troop carriers escorted by motorcycle police roared up to the side and rear of Foster. Infantrymen, in green fatigues and carrying M-1 rifles, formed a line up the west side of the auditorium. Another convoy arrived in front of Denny Chimes, across from the president's mansion on University Boulevard. General Graham arrived in a green, unmarked command car. He decided to march a platoon (35 men) to the Auditorium, a Special Forces unit under Colonel Henry Cobb, a classmate of Katzenbach's at Princeton. Graham spotted Lingo and the two saluted and conferred briefly before Graham decided that "the platoon in steel helmets and weapons was an overkill," and left the unit between Farrah and Little halls near University Boulevard.[38] With four sergeants in green berets, he donned his soft cap and moved toward the final confrontation. On the way he huddled with Katzenbach, Norville, and Weaver, who fell in behind the General with the four sergeants bringing up the rear. They strode purposefully toward the wall of state troopers and reporters. The silence was eerie, disturbed only by the soft whirring of cameras and popping flashbulbs. Seeing their approach, Seymore Trammell turned toward the entrance and clapped on his straw hat as a signal for Wallace to take his stand. At 3:30, General Graham, in combat fatigues with the Confederate battle flag of the 31st Dixie Division stitched to his breast pocket, came forward and saluted the governor. Snappily Wallace returned the salute. Graham then said, "It is my sad duty to ask you to step aside, on order of the President of the United States." The words were heartfelt. Earlier General Abrams offered to have a Justice Department staffer write a statement for Graham, but the General said it would not be necessary.[39] The simplicity of his words expressed the sentiments of a majority of the state's moderate whites.

Wallace said, "General, I want to make a statement." Graham replied, "Certainly, sir," then stepped to one side. Speaking from notes scribbled on a spiral calendar pad, Wallace declared, "But for the unwarranted federalization of the Alabama National Guard, I would, at this moment, be your Commander-in-Chief—in fact I am your Commander-in-Chief, and as Governor of this state, I know this is a bitter pill for members of the Alabama National Guard to swallow." Wallace asked that all Alabama citizens remain "calm and restrained." Declaring the National Guard "our brothers," Wallace said, "Alabama is winning this fight against Federal interference because we are awakening the people to the trend toward military dictatorship in this country. I am returning to Montgomery to continue working for constitutional government to benefit all Alabamians—black and white." With that, Wallace and his entourage walked quickly toward waiting patrol cars. The clock showed 3:33 p.m. As the governor's motorcade pulled away, Wallace kept repeating a warm "Come back to see us in Alabama" to a bank of reporters demanding to know whether there would be a press conference. When the motorcade turned onto University Boulevard, students and university staff who had been kept away from the scene showed their approval by applauding the governor. High atop Foster Auditorium, one of the four patrolmen stationed there waved a white flag.

In Bear Bryant's office the trustees congratulated themselves on a peaceful outcome to a ten-year ordeal. A few of the older men shed tears as they saw their beloved institution pass into the new era, but the general feeling was one of relief. Rose and his staff could be satisfied with the elaborate planning that helped make the schoolhouse door the most publicized nonviolent confrontation of the civil rights movement. Even if it was not in any real sense the university's show, it had been staged on the campus and most of the nation came away with the feeling that Alabama's university and its president contrasted favorably with the state's government and its governor.

In Washington Robert Kennedy stood with his staff, arms folded, listening to the drama over radio. He had his coat on now, ready to go to the White House. His expression did not change as the announcer reported that Wallace was leaving the door, but when the commentator relayed Wallace's claim that he was winning the constitutional fight, a grin of disbelief flickered across the Attorney General's face. Kennedy turned to Bob Shuker of Drew Associates and said that the film crew should abandon its plan to shoot at Hickory Hill that evening and instead go to the White House. He provided a driver and car. The speech was on. Meanwhile, Jim Lipscomb took his transistorized camera and rode with Wallace to the Tuscaloosa airport. Wallace continued to look mean-spirited as he stammered that some politicians were going to pay at the polls come next year, that the presidency could not be won without the South.[40]

With Wallace's departure, Jimmy Hood and Vivian Malone entered the schoolhouse door to a spattering of applause. From an upper-floor window a student unfurled an American flag. Registration took less than fifteen minutes but looked almost surreal on the near-empty gym floor. A full cadre of faculty and staff served the two students who had completed most preliminaries earlier. Both had had a chance to change clothes and freshen up between the first and second confrontation. Hood went in first, as always nattily dressed with his snap-brim hat and briefcase. Malone followed, wearing a two-piece pink outfit and, in the style of the early sixties, a short bouffant hair-do with bangs. She looked serious until a reporter drawled, "How do you feel, ma'am?" She answered with a smile. Hood and Malone stayed in the dorms that night, where student leaders had been assigned to make them welcome. Officials worried over a bomb threat against Mary Burke Hall, but a search turned up nothing. That night General Graham slept on a sofa in the lobby. Upstairs, Vivian Malone collected her thoughts alone in her single room. Upstairs, also, slept the general's daughter.[41]

The speech

During Monday's meeting at the White House, Robert Kennedy had urged the President to address the nation. Ted Sorensen thought the speech would be delivered only if there were a setback in Tuscaloosa, then only to explain that the situation was under control and that federal authority would prevail. President Kennedy's parting remark indicated that Sorensen should prepare a draft in case the students had to wait an additional day. The Attorney General, however, continued to believe the President should take the occasion to make his case for civil rights legislation. Sometime before Tuesday afternoon, perhaps Monday evening, he convinced his brother to make such a speech. As soon as Malone and Hood walked inside the door, the Attorney General called to congratulate the President and rang Sorensen to get a draft prepared; if possible, in sufficient time to allow input from Burke Marshall and himself. He suggested that Sorensen bring Louie Martin in on the project.[42] A journalist before joining the Democratic National Committee as liaison to the black community, Martin rivaled Sorensen as a wordsmith. The President himself called the networks and requested fifteen minutes of airtime at 8 o'clock.

Robert Kennedy and Marshall arrived at the White House about an hour before. In the Cabinet room, the President looked remarkably relaxed as Sorensen moved in and out with suggestions and revisions. Sorensen worked deliberately, not frantically. The speech represented the end, not the beginning, of the President's thought on the subject. "It drew," Sorensen remembered, "on at least three years of evolution in his thinking, on at least three months of revolution in the equal rights movement, on at least three weeks of meetings in the White House, on drafts of a new message to Congress, and on his remarks to the mayors on June 9 as well as on the February Civil Rights Message."[43] Shortly after 7 o'clock, Sorensen came in and took notes as the President talked with his brother and Marshall. After Sorensen returned to his typewriter, the President sat alone with his brother at one end of the long table and turned ideas, occasionally dictating notes to his secretary, Evelyn Lincoln. This went on for about twenty minutes. Finally Sorensen returned with the best draft he could make.

With six minutes to spare, a tired, unsmiling President walked into the floodlit Oval Office and seated himself at his desk, still penciling notes as the countdown began. On cue he looked up and said, "Good evening my fellow citizens." He opened with a brief account of the afternoon's events that resulted in "the admission of two clearly qualified young Alabama residents who happened to have been born Negro." He called for every American to "examine his conscience about this and other related incidents." He started flat, almost listless, but then his feeling began to show. He spoke of the nation's commitment "to a world-wide struggle to promote and protect the rights of all who wish to be free," and reminded his listeners that "when Americans are sent to Viet-Nam or West Berlin, we do not ask for whites only." He repeated the things that "ought to be possible" for all citizens—education, public accommodations, the ballot—"in short, every American ought to have the right to be treated as he would wish to be treated, as one would wish his children to be treated. But this is not the case."

By now any awkwardness was gone, fatigue forgotten, as the President moved toward his most memorable lines, the ones contributed by Louis Martin: "The Negro baby born in America today, regardless of the section of the Nation in which he is born, has about one-half as much chance of completing high school as a white baby born in the same place on the same day, one-third as much chance of completing college, one-third as much chance of becoming a professional man, twice as much chance of becoming unemployed, about one-seventh as much chance of earning $10,000 a year, a life expectancy which is seven years shorter, and the prospects of earning only half as much." It was neither a sectional nor a partisan issue. "We are confronted primarily," he emphasized, "with a moral issue. It is as old as the scriptures and is as clear as the American Constitution." Continuing with his litany of broken promises to black Americans and the unacceptable costs of continued racial discrimination, he asked, ". . . who among us would be content to have the color of his skin changed and stand in his place? Who among us would then be content with the counsels of patience and delay?" Having searched the nation's soul, Kennedy outlined his legislative proposals and extemporized a conclusion.

His words did not ring down, as did his inaugural or his Berlin speech or even his commencement address the day before, but, away from the open air, seated in the Oval Office, his speech calmly and resolutely anointed the civil rights struggle with the balm of its own rhetoric. Clearly Martin Luther King's cup overflowed. The SCLC chief had kept up his attacks on the White House even through Tuesday. From Atlanta, King responded instantly. "I have just listened to your speech to the nation," he wrote with more enthusiasm than care. "It was one of the most eloquent[,] profound and unequiv[o-c]al pleas for Justice and Freedom of all men ever made by any President. You spoke passionately to the moral issues involved in the integration struggle."[44] At last the Kennedy administration had skipped into step with a movement that itself teetered on the brink of profound change—change hastened by a flash of gunfire that very night.

Mississippi refrain

Life struck down in its prime is too important not to foreshadow its end. For nine years Medgar Evers, the NAACP's field secretary for the Magnolia State, had endured everything white Mississipians could dish out. He trained his three children to dive for cover "at the sound of danger—a speeding car, or screeching brakes, or gunfire," and for good reason. His ranch-style home on Guynes Street in Jackson had been stoned and fire-bombed; he himself had been cuffed, threatened, clubbed, and shot at. He had not heard the President's speech. He and Gloster Current, down from the national office, attended a meeting at the New Jerusalem Baptist Church. "Everywhere I go," he told Current, "somebody has been following me." As he dropped Current off just after midnight, he said, "I'm tired . . . I want to get home to my family. Then, taking Current's hand, he "held it and held it."[45]

As he drove into the carport, he could see the light of the television set through the picture window, flickering blue and white shapes from a corner of the living room. Myrlie Evers had let the children stay up to watch a late movie. She heard the car and turned on the outside lights. Evers got out carrying a bundle of NAACP T-shirts. He stood for a moment, a World War II veteran too tired to think about the dangerous silhouette he made in the yellowish glow of his own carport. Across the road, hidden in a honeysuckle thicket, a man propped his 30.06 deer rifle on a fence and drew a telescopic bead, his quickening breath visible as it touched the cold steel barrel. He fired one shot, tossed the gun into a patch of weeds, and ran. Unlike Oswald two months before, from a similar distance, the man did not miss. The bullet entered Ever's back, tore out the front of his chest and through the living-room window before spending itself against the refrigerator door. Screaming "Oh my God, my God," Mrs. Evers rushed out, followed closely by her children, who forgot their father's instructions about staying put until danger passed. They cried over and over, "Daddy! Daddy! Daddy, please get up!" Neighbors covered Evers with a blanket and placed him on a mattress in the back of a station wagon. As they drove hopelessly to the hospital, Evers spoke for the first time: "Sit me up." A pause followed, then he gasped, "Turn me loose!" At 1:14 a.m., a doctor pronounced what the tearful neighbors already knew. That night, a nine-year reign of political shootings began. It would not end until May 15, 1972, at a shopping center in Laurel, Maryland. The first four who fell before assassins' bullets inspired the movement that ended legal segregation. Only the last victim, Wallace, survived.

Curtain call for Justice

Hood and Malone woke to the chilling news. Despite the shock, their day passed as normally as could be expected. Overt and self-conscious acts of friendliness helped obscure occasional icy stares or catcalls. They talked to reporters about "bull" sessions and bridge games, as if these accoutrements of college life were a natural part of their environment. Reality, however, constrained them to move like fish in a bowl, alert to shapes and shadows outside their glass container. University officials encouraged them to have one press conference, then limit interviews to reporters and feature writers who were approved to be on campus. The university already forbade students from serving as stringers for the press and warned against discussing racially charged topics. Hood and Malone expected these conditions, and apart from their isolation in single rooms, abided the restrictions well enough.

It was also well enough that they did not know about confusion in their security arrangements. True to his word, Wallace began a phased withdrawal from the campus. Beginning at 3 a.m. on Wednesday the 12th, 300 troopers and special agents pulled out. At 10:26 a.m., Wallace wired the President to say that "at all times since the Federal District Court ordered the admission of Negro students to the University of Alabama, state law enforcement officers under my control have maintained absolute peace and order at the university campus. At 3:33 pm (CST) June 11, 1963 through the use of federal troops you assumed full responsibility for the presence of Negro students and for preserving peace and order on the campus. . . ." To assure "complete co-ordination" between state and federal agents, Wallace announced that an additional 228 state troopers would be off campus by 2 o'clock that same afternoon. The final contingent of 162 would not leave until 2 p.m., Sunday, June 16.[46]

Wallace's pledge of cooperation notwithstanding, the White House learned Wednesday "that the state highway patrol had actually abandoned a large part of its duties around the campus while it was still thought that they were carrying out this security function." Major John T. Sullivan found "13 posts unmanned," "10 check points open or abandoned," "check points at either end of University Avenue manned but persons allowed to pass without being checked for proper ID," and "all check points on north side of perimeter facing Warrior River abandoned."[47] As at Ole Miss, state agents drifted away, but unlike Oxford, Wallace had placed such a clamp on the Tuscaloosa campus that violent groups were not in place to take advantage of the lapse. For the remainder of the year a small contingent of Guardsmen remained to provide security. Nine marshals rotated twelve-hour shifts to protect the students.

Wallace did provide the university and the Justice Department some relief on D-day plus 1. In a telegram to Rose, Wallace said that because President Kennedy had established a "military occupation" of the University, to include all its Centers, "I will not be present on the Huntsville campus tomorrow. However, we will continue relentlessly our fight against forced integration of the University of Alabama."[48] Wallace knew that a repeat performance of the surrender scene would be politically disastrous. At another level, the telegram to Rose was old tricks. Wallace still refused to communicate directly with Washington or Katzenbach. So when the deputy attorney general arrived in Huntsville, he could only suppose what the governor planned. Katzenbach, however, assured Dave McGlathery that he expected a peaceful resolution.

Ever since the decision to make Tuscaloosa "the schoolhouse door," McGlathery had felt isolated. He knew the importance of what he had done and was about to do, but he no longer shared the spotlight. At one time, Charles Morgan brought a proposal from Wallace or the university or both that he register with Malone and Hood at Tuscaloosa, but he declined. The Legal

Defense Fund had offered to include him on the New York trip, but the former Naval air mechanic feared flying and stayed home.[49] The Justice Department installed a hotline from his home to the university in Tuscaloosa, but apart from a call from Jeff Bennett to make sure the connection worked, it was never used. Out of the loop, McGlathery could not escape the consequences of his bold venture. One night-caller told him how ugly he was, especially his thick lips. Somebody from Texas wrote to say that he was on a hit list. As time drew near, he drove home from the Redstone Arsenal each evening feeling a little uneasy, apprehensive about whoever followed or pulled up alongside. When he parked his car and walked to the door, he made himself a moving target, bobbing and weaving as he stepped quickly. Late one night Dave and his wife awoke to a loud boom. The house shook. Thinking the worst had happened, he grabbed his pistol only to find that a drunken truck driver had failed to make a turn, ran up an embankment, and hit the house.[50]

McGlathery saw the Tuscaloosa proceedings on the evening news and admired Kennedy's speech. The story of Medgar Evars greeted him the next morning. His apprehension mounted as a black '57 Chevrolet pulled in behind him on his way to work. After watching the car in his rearview mirror for several blocks, McGlathery headed back to his house, the car still following but more slowly. He called a contact person at Justice who quickly assured him the tailing car was FBI. At the Arsenal, no one said much, not even to joke about the situation. That night he slept fitfully but woke at the customary hour and, as usual, skipped breakfast. Shortly after arriving at the Arsenal, Katzenbach came by. They decided that McGlathery would drive over by himself and go in to register without escort. Katzenbach assured McGlathery that every move would be followed by marshals and that Guardsmen stood ready but would remain out of sight. Katzenbach discounted the prospect of a surprise appearance by Wallace.

On the drive to the one-building extension center, McGlathery felt the comfort of his big, gray 1959 Buick LeSabre, but winced as the thought of its ugly tail fins and wished for a moment that he had not traded in his '61 Ford for better payments. Katzenbach and his aides followed. As McGlathery strode alone toward his schoolhouse door, Katzenbach almost jubilantly described the scene for Robert Kennedy over their radio-phone hookup.[51] Within minutes, eight months of planning, with everything from Voodoo jets to clandestine schemes for using dorm keys, ended in the admission of the last of the three students. McGlathery, Hood, and Malone, by their individual acts of courage, fulfilled the hopes of Lucy and Myers. 'Bama was no longer white.

There was a sidebar. On the morning of Wallace's stand in Tuscaloosa, Robert Muckel, a twenty-nine-year-old white man from Utica, Nebraska, enrolled at Alabama A&M. By his account, Muckel did not know A&M was all-black when he applied for a National Science Foundation summer institute for high-school teachers. When he learned, he decided to come anyway. The president of the college said, "We just assumed anyone coming to A. & M. would be a Negro." A less cautious administrator said, "All I can say is the same thing they said about James Meredith at Mississippi. Mr. Muckel is 'knowingly' our first white student, but we've had many students here over the years who you could not definitely say were white or Negro." A wiser official said that because Muckel enrolled for a summer institute, "so far as we are concerned it might be said that there is no white student registered here."[52]

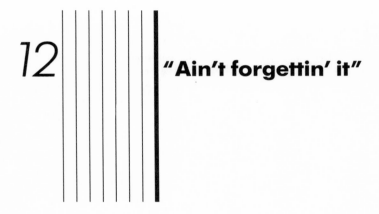

12 "Ain't forgettin' it"

When Wallace and the Kennedys locked horns in Tuscaloosa, few expected a fight to the death, and most expected the one with more muscle to win. Thus Americans saw on TV what they expected, though the conclusions drawn differed between viewers in Massachusetts and those in Alabama. Outside the South, most who watched saw Wallace as Robert Kennedy did; a second-rate figure easily dismissed; but in Ragland and Geneva and Gordo, Alabama, Wallace emerged the victor. He spoke confidently, they observed; he told'em, they reasoned. He stood up. He kept the faith. Shortly after it was over, E.L. "Red" Holland of the *Birmingham News* wrote Ed Guthman to say that Wallace had the clear support of the majority of Alabamians "despite his clear retreat." People seem to have concluded that he "succeeded, despite the Negro students' entry."[1]

So maybe James Baldwin was right. Not that it was, practically speaking, possible, but the Kennedys for all their fine talk needed to crush the Wallaces of the South, banish them from the comity of the Republic, bring them to their knees. It was not to be. Like Lincoln on the eve of his death, President Kennedy appeared content with less. For those who cried peace, the drama ended as it should have. White supremacy, segregation, racism, equality, and the American Creed: all solemnly encased in the symbols of political power; the President renewing the one-hundred-year-old pledge of emancipation, the governor standing for a state's right to subordinate its black citizens. No shots fired, no lives lost in Tuscaloosa, and over the past decade only forty killed in what pundits already styled the South's Second Reconstruction. The scene at Tuscaloosa worked not because nobody won but because of who won and who lost. The national press and most of Alabama's major dailies rushed to note that Wallace surrendered. But that was the way it was supposed to turn out. As a reenactment of Appomattox, the schoolhouse door fulfilled expectations—federal, force-induced surrender followed by a settled conviction that the real cause, white supremacy, was not, indeed, could not, be lost.

The confrontation in Tuscaloosa dramatized the accepted moral and ideological abstractions both of the civil rights movement and of massive resistance and made them political. What the Kennedys and Wallace did in Tuscaloosa dominated public debate for another ten years. The martyrdom of the Kennedys and of Martin Luther King cast the dream of a South in which "all God's children could join hands" into a shimmering moral abstraction. All the while, Wallace euphemized the fears of many white Americans by talking about "bloc voters" and their "pointy-headed" bureaucratic allies in Washington. It was a robust debate, full of promise and threat, but like the first Reconstruction, it drew from a reservoir of moral capital that depleted rapidly. Submerged beneath this debate, but playing upon the fears that gave it life, struggled an underclass of black-power advocates on one side and blue-collar workers of competing ethnic identities and demands on the other.

There were other images in June of 1963, not yet real, but harbingers nonetheless. In the streets of Saigon, on the day Wallace stood in the school house door, a Budhist monk cremated himself, an act of self-immolation that would come to symbolize the end of Pax Americana in that part of the world. In Iran, fanatical followers of the Ayatullah Rouhdolah Khomaini attacked well-dressed men and unveiled women. Incensed zealots yanked one woman from her car, forced her to undress, then beat her to death. In London, Christine Keeler told the world about her tawdry affair with Profumo, while back home, Americans measured their level of moral commitment by debating whether the movie *Cleopatra,* starring Elizabeth Taylor and Richard Burton, or any entertainment was worth a cool $40 million when so much misery abounded.

A pleasing editorial

While symbols and images of the century's decline competed for attention, those whose lives the events in Tuscaloosa most immediately affected carried on. Jimmy Hood woke Wednesday morning and with his usual burst of energy bounded off for class, with marshals hustling to keep up. One student, seeing him take a wrong turn, righted him. He attended his first class, Introduction to Sociology, on the second floor of the new, air-conditioned Music and Speech Building. The class went without incident until a girl suffering from "muscle spasms" fainted and had to be carried out. The young woman's misery ended class early and gave Hood time to drop by the post office in the Union Building. Though early in the term to be checking for mail, a letter from home was possible. He pulled out instead an unmarked, unstamped envelope. Immediately one of the marshals reached for it. After inspecting it, the marshal asked if Hood were sure he wanted to open it. Hood broke the seal and unfolded a handscrawled message that read: "NAACP IS COMMUNIST LED. Wilkins has 7 com. citations. Sec. Spingarn has 50. King has had 60 assoc. with com. front groups. NAACP is for black buzzards. THEY

SHOT A NAACP NIGGR IN JACKSON! DO YOU THINK *YOU* ARE SAFE?"[2] The KKK or their sympathizers still had access to the university post office.

This threat and occasional catcalls did not daunt Hood. He continued to ingratiate himself with classmates who returned his friendly gestures with their own newfound, self-conscious generosity. About ten days into the term, he saw Bob Penny, a graduate student in English who wrote for the student newspaper—most recently a complimentary review of James Baldwin's *The Fire Next Time*. Penny had been a close friend of Mel Meyer. The two of them, along with Hank Black, the new *Crimson-White* editor, had hitchhiked to San Francisco over Christmas break. Hood told Penny about some thoughts he recently put to paper and asked Penny to look them over. Penny told Black. Both agreed that it was a remarkable piece, saying what many white moderates and liberals wanted to hear. When Black talked with Hood about it, Hood seemed enthusiastic to have his ideas appear as a guest editorial. They agreed upon a headline and how to feature the column.[3]

Under the banner "Needed: More Students, Less Pickets," Hood castigated the movement. He talked about the need of the Negro to "meet certain standards and possess certain values" in order to be accepted. He said that he had observed three protest organizations and had reached three conclusions. "First, the leadership . . . is composed of learned men, who have achieved fame outside civil rights. Second, the bulk of the people involved in the organizations are students and uneducated people. Third, these organizations thrive on conflict, and thus, will continue to exist only so long as they are able to instigate conflict." He declared protest "a matter of excitement rather than conviction for most Negroes." As to why the Negro race did not "wake up and go about this thing in a more intelligent way," he blamed the leaders "who benefit from conflict . . . who would lose prestige and money if another way were chosen." In urging education over "sit-ins, lie-ins, swim-ins, etc.," he said, "My position will make me unpopular with the masses of my people, but that has no relevance in what I believe and how I feel about the situation. I honestly believe the big 'unnecessary mess' in which the protest movements have resulted can be solved in the classroom."

Though copyrighted by the *Crimson-White*, the story appeared on same date, June 27, in the *Birmingham News*, albeit with numerous typographical errors. From hindsight Black realized he should have told Hood that the editorial would receive statewide, if not national, attention. Black himself had called Jack Hopper at the *Birmingham News* and told Hopper that he could get the editorial off the *Crimson-White* makeups sent to Birmingham for printing. Black knew the story had news value. He figured that the tip would give both him and the *Crimson-White* a little boost. He did not figure on Hood's reaction; after all Hood had seemed so eager to have the editorial published. On learning that it had appeared in Birmingham, Hood realized instantly that he was in trouble and called Black in considerable agitation. Caught short, Black sputtered an apology and made up a small lie to explain how the editorial might have gotten into the *Birmingham News* on the same day.[4]

Hood, who until then was having "a really great time," should have known better. Somehow he got it into his mind that the editorial, intended for his white classmates, would not go beyond campus. (The editorial would have made state and national wires had Black never told Hopper about it.) The sentiments expressed were respectable among a significant segment of the black middle class but not among those who had supported Hood's cause. It was as if he had never darkened the doors of the SCLC or SNCC or CORE while at Clark College but instead had interned on C.A. Scott's conservative *Atlanta Daily World*. Hood expressed "horror" on learning that George Wallace commented favorably on the editorial; but his "horror" was not half that expressed in Gadsden where marchers whom Hood helped organize the summer before continued demonstrations. On the day Hood's comments appeared, state troopers broke up a sit-in at the courthouse by 350 demonstrators who protested the jailing of 396 from an earlier march. According to some reports, even Hood's family was dismayed. His sister—just fourteen— had been jailed four times already.[5]

Hood had to make amends, so he went home. On the evening of July 16, at a mass rally at the First Baptist Church in the Green Pastures area of East Gadsden, Hood spoke before family and friends. Bernard Lee, one of Martin Luther King's devoted disciples, conducted the rally and introduced Hood. The church was packed, the atmosphere charged. Most knew about Hood's editorial, but in this time of disbelief, they were prepared to believe anything he told them. Three times he denied writing the editorial. On the first, he acknowledged the criticism of the black community, saying members of his own race had called him "everything, excuse the language, from a son of a bitch to a member of the Ku Klux Klan. I did not write the article. I had nothing to do with it. The person who published the article is a cousin of our governor. The person who sold the article to the *Birmingham News* for $500 is a nephew of the governor."[6]

On the second denial, he concocted an elaborate story. "I want you to know that I did not write the article, and I can't prove it. You know why? Because they put in a law at the University that no student no matter who he is can discuss the racial issue." He said that university officials used this ruling to trick him. They set up a press conference for him to explain his editorial, "but they didn't tell me that I couldn't talk about the situation." Five minutes before the press conference, his major professor tearfully warned him about the trick. "I couldn't clear myself. I couldn't say that I didn't write the article because if I did write the article I would have to bring up some facts about the University that people don't really know about." He said that the university criticized the professor who had warned him for giving him an A. "Why? Because they say that he gave me an 'A' because he was afraid of what the President of the United States might say if I got a low grade." Hood said he planned to go to the White House next month to tell the President about the situation in Tuscaloosa. Suborned by the crowd's applause, he compounded perjury with recklessness, at one point declaring, "I hope there is a reporter in here some place and you go back and tell exactly what I've said. . . ." Unfortunately, two state investigators sat in the audience; one with a recorder cradled in his lap. Hood's words wound their way around a reel of tape.

Unaware of the recorder, Hood launched into a tale about Al Lingo coming to his room five minutes before the governor ordered Lingo to pull out. The order to withdraw the troopers, Hood said, came after he (Hood) called Robert Kennedy to report that he was being "treated like a slave" by Lingo's men. Then, by this account, Kennedy called Wallace. This series of events, Hood alleged, brought the Colonel storming into his room, where he proceeded to "curse me for everything he could think of." By now Hood was out of control. While talking about rumors that he would play some sport at the university, he finally realized that he was being recorded. "Well, I can tell you, it will be news tomorrow anyway, I am going to play football at the university." Now completely off-base, he issued his third and final denial. "I did not write the article and I can't prove it until August 31 when I am dismissed from the university as a student." He must have sensed that his words amounted to rope in plenty for a hanging. To laughter and applause he closed, "One thing, they are not going to throw me out for anything other than spitting on the street and I don't spit on the street, I take a paper cup."

Still more denials

Hood was self-destructing. His own nervous energy betrayed him, like an antelope cut from the herd, bounding hopelessly, first in one direction, then another. Wallace moved in for the kill. Already Wallace had vowed publicly to have both students removed once the "feds" were out of the picture. Much to the university's embarrassment, the governor compelled the trustees to appeal Judge Grooms's May 16 ruling admitting Malone and later Hood. It was a frivolous appeal, but Wallace insisted. It demonstrated anew the governor's determination and commitment to his word. University efforts to make Hood feel welcome notwithstanding, these demonstrations by Wallace, and officially by the board of trustees, no doubt played on Hood's already vivid imagination. Moreover, given his naïve belief that the *Crimson-White* editorial would not go beyond campus, subsequent state and national publicity must have seemed like a conspiracy. He could not even count, it now seemed, on sympathetic students such as Hank Black or Bob Penny.

On the night of August 3, Jeff Bennett picked up the phone in his office. The caller warned that the next morning's *Montgomery Advertiser* would carry a story reporting that Hood "may soon be expelled." The *Advertiser*'s crack reporter, Bob Ingram, had received a tip from the Wallace administration that a tape of Hood's speech could be obtained from the Etowah County sheriff's office in Gadsden. Ingram, who previously worked for the Gadsden paper, called the sheriff, who put a copy on the next bus for Montgomery. Bennett's caller read Ingram's article over the phone, a detailed account of which Bennett rapped out on his typewriter.[7] Stunned, Bennett called Dean of Men John Blackburn, and the two agreed to confront Hood the next day.

Hood's performance in Blackburn's office was remarkable. He acknowledged making a speech and even said that he knew of the state investigators' presence. Without blinking he said that every potentially damaging charge in his speech had actually come as questions from the audience. He had simply rebutted these rumors. He said that when asked if Al Lingo had cursed him, he said it was "basically untrue because . . . the security regulations on getting to my room" would have made it impossible. As for his denials about writing the editorial, Hood said that he merely denied giving it to the *Birmingham News*. And so the interview with Bennett and Blackburn continued with Hood saying plausibly that he could not possibly have said the things attributed to him because each of his alleged statements could not have been true. On August 5, Hood signed a statement affirming this account of the speech.[8] His performance persuaded Bennett and Blackburn that Bob Ingram's source had misinterpreted Hood's remarks, probably willfully. Moreover, they now doubted that Wallace had a verbatim recording of the speech.

Before the ink dried on Hood's signature, Bennett reported to the trustees' executive committee that Cecil Jackson, Wallace's legal adviser, claimed to have a recording but as yet "had not been able to produce it for inspection." President Rose decided to call the governor's bluff and asked the executive committee to pass a resolution saying among other things that the committee had "considered news reports concerning a speech made by James Hood" and that "the news reports referred to a tape recording of the Hood speech, but no such recording was available to the Univ[ersity] at the time of the Ex[ecutive] Com[mittee] meeting." The resolution allowed the trustees an out, saying that "information now available did not substantiate the charges reported" but that "the matter has the continuing attention of University officials."[9] The executive committee called for an August 15 meeting of the full board to take up the matter.

That same afternoon, still skeptical, Bennett received a message that Major Cloud of the Department of Public Safety was on his way to Tuscaloosa with the tape. Bennett and Blackburn met in Rose's office. In shock and dismay they listened to the recording and read the transcript.[10] Wallace could now demand that the trustees expel Hood for the same reason they had expelled Lucy. If anything, Hood's charges were more scurrilous than Lucy's. The next day Hood wrote an apology. His words danced in and out of focus like shadows thrown by a campfire. He tried "to clear certain falsity inferred in rumors which had developed without support." When giving the Gadsden speech, he "relied on no formal manuscript, and therefore I became overwhelmed by the circumstances and said things that were not actually what I was intending to say. In essence I was terribly angry with the situation of things, and sought to present the clear facts by using the distorted stories just as they had been told to me. I clearly sought to preference [*sic*] these stories with statements which would have indicated their falsity."[11] His thoughts spun in ever widening gyres of abstraction as he fought through his fabrications to explain what no words could explain.

He delivered his signed apology to Dean Blackburn, and in front of a witness further affirmed that "this statement was written by me of my own volition, and that I have had no assistance in its preparation." He pledged to "adhere to whatever action must be taken in this situation." Blackburn hoped against hope that Hood's apology would suffice, but given the Lucy precedent, there was no way it could. Blackburn scheduled a disciplinary hearing for August 12. In the meantime, Shores negotiated for a less public way of removing Hood. On August 11, Shores sent a letter to the university saying that Hood was withdrawing on the advice of his physician for reasons of health, both mental and physicial. Shores requested postponement of the disciplinary hearing as well as the hearing scheduled before the full board. By telephone poll, the trustees accepted Shores's terms and hoped the governor would go along. He did.[12] Hood's days at the university were over.

As Hood neared the end of his bewildering apology, a truth began to show through. Writing freely now, he said, "In view of the fact, that it is almost impossible for a person to remove himself completely from something with which he is born, and in this case it is my racial background." From this fragmentary thought, he built toward an emotional conclusion. "Such a removal is like taking life itself, I was destined to become one [a Negro], and through the role of one I spoke as one, rather than as the outsider I had set out to be [by attending the University of Alabama]. In fact I know I cannot desert or fail to hear the plea of someone crying, this is my nature, and this is the way I felt that night [in Gadsden]. For I am a Negro and I cannot run away from the fact. I am just as certain to heed the emotional atmosphere by being within as I am to be without."

Hood discovered that he had almost forgotten who he was. "Jimmy," a friend said, "was what you might call a good college kid. His problem? I guess, like all of us, he wanted everyone to like him." In the euphoria of his early acceptance at the university, Hood got to thinking about "being a bridge between two cultures," a bridge that could be improved with better "communication." His problem came in trying to build his bridge from the middle of the stream toward his new white friends, rather than from the shore he so recently left. As he packed his bags for an undisclosed location, he searched once more for words to describe his feelings. "It's hard to find understanding," he told a reporter for *Newsweek*. "I'm in the process of becoming—of *becoming* . . . I'm searching for the methods, for the ways to try to become a part of society that recognizes individuals, that's not Negroes and whites." Watching him pack, the reporter thought, "He hadn't found the way yet; he had fallen between the races, in the middle, alone." Then, almost as an afterthought, Hood said, "I'm just as lost now as I'll ever be in my life."[13]

Two days after Hood's withdrawal, he learned that his father was dying of cancer. The family had kept this information from him. On October 10 he buried the man who had always supported his "cavalier ways."[14] Meanwhile Bennett traveled to Montgomery to give Wallace a report. Bennett took his nine-year-old son along. As Bennett turned to leave the governor's office, Wallace could not resist asking how long it would take "to get the nigger bitch out of the dormitory." The remark was gratuitous and sharp words followed. Stepping out into the corridor, Bennett apologized to his son and told him not to tell his mother about the kind of language he had just heard. Bennett made the tedious journey back to Tuscaloosa both sad and angry; sad for Hood, and angry that his nemesis, Wallace, had scored yet another victory at the expense of the university. When they arrived home, young Jeffrey hit the ground running. "Ma!" he cried, "Daddy and the governor had the worst 'cuss fight' you ever heard, but I think Daddy won."[15]

The mediated president

In early July the *New York Times* sent Gertrude Samuels to do a feature on the university's desegregation for the Sunday magazine. Security precautions continued but were unobtrusive. An occasional army jeep tipped the fact that 100 guardsmen remained on campus. Plainclothes marshals and university police waited outside buildings to escort Malone and, before his withdrawal, Hood. Samuels found the two students laughing and talking in the Supe Store cafeteria between classes. Others carried on at nearby tables in normal tones. "It's fine," Malone said. "We're just two more students here. That's what we hoped we'd be, and that's the way it is. There are no problems at all." She admitted to no social life, but said she did not have time for it anyway, "and then, well, we don't want to appear to be forcing anything and creating any tensions." Hood endorsed what Malone said but "alternated between easy grins and some worried frowns."[16]

While Malone and Hood made the best of their abnormal situation, Frank Rose fretted about his image. He expected Wallace supporters to take swipes as when Senator A.C. Shelton of Calhoun County blasted him for his "$27,500 salary, state-provided home and air-conditioned car,"[17] but he watched closely national press reaction to his leadership. He was sensitive to criticism, even from papers of marginal importance. In early June he wrote the editor of the *Delphos* (Ohio) *Herald* to say, "I would call your attention to the last issue of TIME magazine in which it clearly reveals the tough decisions that I have made and the leadership I have given. There will be a four or five page 'close-up' of my work in LIFE magazine which will be released on Monday, June 10." Later Rose thanked Henry Luce for the manner in which *Life* and *Time* portrayed the situation, saying that they "set the tone for other news media representatives."[18]

Rose had worried about Gertrude Samuels. He told Wyatt Cooper that they "had a great deal of trouble with the woman who came down" from the

New York Times. Alex Pow called the editor of the Sunday magazine, whom he knew, to get her pulled from the story. When the entirely congratulatory feature came out, Rose relaxed and thanked her for "your excellent story and for the fair way in which you treated the University and our two students."[19] From the beginning Rose need not have feared the national press. The overwhelming reaction was favorable. The only flap that occurred during the summer resulted from Rose's own indiscretion, and even then the press reacted favorably.

The incident started when Rowland Evans and Robert Novak learned about some tough talk between Rose and campus security officers. Rose had received word that the Klan planned to burn a cross on campus and, according to Evans and Novak, he summoned the "campus cops" to his office: "Every one of you who cannot follow my instructions to the letter may drop his badge on my desk and resign. My instructions are that if any attempt is made to burn a cross on this campus, I want the man who lights the first match dead at the foot of the cross." The words matched Rose's sentiments, and his way of putting it comported with a side he liked to show. Nonetheless, he spent time in August denying any "shoot to kill" order.[20]

More bombs

Rose's bravura aside, he could not afford to alienate his security force nor wave a red flag in front of the Klan. Security remained the only important thing the university had to do. The rumor mill in Tuscaloosa and around the state ground out endless reports about Hood's and Malone's conduct on campus, most of which charged some abuse of appropriate conduct for Negroes. Dean Healy tried to scotch some of the rumors about Malone: "parents had not been refused requests to visit their daughters in Mary Burke [Hall]"; "not since the first week had mothers been prohibited from going to their daughters' rooms"; "no [other] Negro women in the Residence Hall"; "Miss Malone never entered the corridor bath area"; "at no time has she [Malone] taken the initial step in being seated with other students"; "no white girls have stopped eating in the University dining hall."[21] Stated matter-of-factly, Healy's memorandum spoke worlds about the social trap in which Malone found herself and the spectral fears that drove the wildest fantasies of white fringe groups bent on getting even.

At the same time the university did not need the expense of security. Wallace had said he would not spend state money, which left matters in the hands of the federal government. Most of the cost was for federal forces anyway, marshals and Guardsmen, but the university incurred some expense for its local security detail and for overtime. Rose estimated the additional cost at $100,000 (approximately $400,000 in 1993 dollars). He asked Burke Marshall to intercede with Robert Kennedy to get a grant for one or more of the university's programs that would in turn permit the university "to bear these costs out of their contingency funds." A month later Rose made a more specific proposal for handling reimbursement that the Attorney General endorsed.[22]

If the university felt it could relax, having made it through the summer of 1963, the bombing of Birmingham's Sixteenth Street Baptist Church sobered everyone. On Sunday morning, September 15, "Dynamite" Bob Chambliss settled in front of his TV to learn that the bomb he and accomplices planted the evening before had blown out the northeast corner of the church at 10:22 a.m., killing four young girls. The city and the nation were appalled. Iredell Jenkins, the university's distinguished professor of philosophy, was uncontrollable. "The obvious thing about Birmingham," he told a reporter for *Time,* "is that there's just a lot of goddam white trash that's conglomerated there." George Wallace shared the outrage but drew another conclusion. "There's no integration anywhere in the world," he told the same reporter, "that's working. You can't make it. . . . They threw out prohibition because it didn't work, didn't they?"[23] To Wallace, Chambliss's lunatic deed would be repeated until the lunacy of social engineering stopped.

With this reminder of pent-up rage, the university braced for the fall semester. Students signed pledges forswearing any conduct that might "contribute to disorder."[24] The National Guard stepped up its vigil, but tensions eased after the first weeks passed without incident. Soldiers began to enjoy their soft duty on "Task Force Tusk." Living in dorms, drawing active-duty pay while sometimes holding down civilian jobs as well, made living easy.[25] Besides, there were so many coeds to ogle and comment upon. Then at 3:10 a.m. on the morning of November 16 a bomb went off on Hackberry Lane, about 130 yards from Malone's room. It tore a foot-long hole in the concrete and woke Malone, who went to her window but, seeing nothing, went back to bed. Crank calls poured in, one saying, "Last night the explosion was outside—next it will be inside." The next night, while Malone visited friends at Stillman, another blast went off in the vicinity of the college.[26]

Four days later, after two more blasts, university officials and federal marshals wondered whether Wallace was fulfilling his pledge "to come to Tuscaloosa and remove Malone from the U. of A." The night before, Cecil Jackson called Jeff Bennett and told him "to get Malone out of the dormitory for the safety of the other children." Bennett told Jackson that the university "will not bow to . . . acts of violence." Bennett believed Wallace either ordered the bombings or at least was using them as a pretext for getting rid of Malone. He thought Wallace could stop the terror and went to Montgomery to tell him so. Wallace stared right past Bennett's bluster and asked what Bennett "would do if 4 or 5 carloads of Alabama Highway Patrolmen attempted to remove Malone from Mary Burke Hall." Later, Bennett and Rose asked the deputy U.S. marshal what he would do if confronted by Wallace's troopers. The marshal replied lamely that his men "would not use their weapons" but "would do all in their power to resist them."[27]

That same night Rose called John Caddell to complain about Wallace's threats. Caddell called Wallace the next morning "and told him they [state troopers] will not take Malone out of the school and they don't want his legal advisers to be harassing Rose and Bennett with telephone calls—if Wallace has anything to say—he should contact him or someone on the Board of Trustees." Wallace was in a feisty, confident mood. He had just returned from a speaking tour of hostile New England campuses. At Harvard, Dartmouth, Smith, and Brown, he provoked the students by his very presence and delighted in their tumultuous reactions. On return he made additional sport of twitting Rose and Bennett. Eventually five National Guardsmen were arrested for the bombings, which led General Graham to speculate that the bombers were engaged in "a juvenile plot to have the detachment continued on duty." After all, it had been a cushy assignment. The bombings coincided with their scheduled demobilization.[28]

Whether a prank or more seriously motivated, the play between Wallace and university officials ended abruptly. The day after Caddell told Wallace to back off, Oswald took aim from the Texas School Book Depository and plunged the nation into mourning.

Small steps

For Vivian Malone the shooting of the President underscored more dramatically than all the other terrorist acts since June 11 the precarious position in which she found herself. Anyone could kill her at any time. Quiet perserverance, however, was her long suit, and she continued a steady course. She had half a dozen or so good friends on campus and often went to Stillman for parties and a more relaxed social life. She got to know one Stillman student, Michael Jones, whom the university employed as a driver for her, well enough to marry him after graduation. She maintained outstanding grades and, despite switching majors from accounting to personnel management, graduated on schedule in the spring of 1965. For its part, the university continued to monitor her security closely and to do all in its power to make her stay on campus as pleasant as circumstances permitted.

The university took small, safe steps to surround Malone with more black students. On the anniversary of the stand in the schoolhouse door, two more women joined her. By the spring of 1965 ten black students strolled the campus. The university's cautious progress owed as much to its own attitudes as to surveillance by Wallace. When the "colored professional staff" sought to desegregate the cafeteria in the Basic Science Building of the Medical College in Birmingham, officials held them at arm's length. When they complained about foot-dragging, Bennett sent a note to Rose saying that the university would proceed with its plans to desegregate "but with notice to these people [Negro employees] that if they issue any more quasi-deadlines we would do nothing."[29]

Another incident, involving John W. Nixon, a prominent black dentist from Birmingham and president of the recently revived state NAACP, underscored the limits of tolerance. Nixon wrote an article saying the time had come for blacks to apply to Alabama's dental school. Appearing in a black professional journal, the article extolled the liberality of Dr. Joseph Volker, the school's dean, and indicated the school was at last prepared to accept blacks. Unfortunately, Nixon said that until recently the school had not been worth integrating. Nixon's temerity brought a stern rebuke from Charles A. "Scotty" McCallum, an associate dean in the medical college, and occasioned a letter of apology from Nixon.[30] In time Nixon would be recognized as a staunch supporter and ally of the university, but in 1964 he could be scolded for effrontery.

The ideal always had been for whites to control desegregation, much as William Faulkner advised in the wake of the Lucy episode. During that crisis, an Alabama student had written the novelist, asking what to do. Faulkner took a page from *Intruder in the Dust* and urged students to get out in front of segregation's collapse. "I vote," he advised, "that we ourselves choose to abolish it, if for no other reason than by voluntarily giving the Negro the chance for whatever equality he is capable of, we will stay on top; he will owe us gratitude; where if his equality is forced on us by law, compulsion from the outside, he will be on top from being the victor, the winner against opposition." Faulkner conceded that his view was "the simple expediency of the matter, apart from the morality of it." His letter arrived after the Lucy crisis and remained unpublished until two days before the stand in the schoolhouse door, when the *Crimson-White* made it front-page reading.[31]

Belated efforts by the university to control the pace and procedure of its desegregation, though understandable, appeared churlish. Old patterns did not die easily. Almost two years after the schoolhouse door, the university told Marvin Carroll that it could not forgive or forget his youthful indiscretion, and thus he could not take night courses at Huntsville.[32] Hood also attempted to re-enroll, only to be told in March of 1965 that he would have to go through the postponed disciplinary hearing.[33] These actions proved not so much a mean spirit as a continuing powerlessness to do the right thing. Almost certainly the press would have zeroed in on Carroll's and Hood's attempts to enroll with the equal certainty of heat from politicians. Besides, the university could note with interest that two black students were doing excellent graduate work at the Birmingham center and that three businessmen were attending a special non-credit course in investments—"one of whom will be recognized, Arthur Shores."[34]

Having done one's damndest

The usual festivities surrounded Vivian Malone's graduation. Her family obliged reporters with statements about their pride and happiness; but for the most part, no one singled them out—a condition that suited their wishes. The occasion had its ironies. Among those receiving honorary degrees were

Marion Beirne Spragins, a Huntsville banker who helped thwart desegregation on that city's campus in 1958, and the indefatigable segregationist Gessner McCorvey, one of Malone's fellow Mobilians. As president pro tem of the board, McCorvey played an active role in the final drama at Foster Auditorium. By then Caddell and Blount were the most influential board members, but McCorvey, because of his impeccable segregationist credentials, became a chief actor. When the board needed to explain why it had to comply with the court order, McCorvey issued the statement. When the board needed to intercede with Wallace, again McCorvey was the man.

The grand old gentleman took it upon himself to expand upon everyone's virtues and to make everyone feel good about what had to be done. On the day after Wallace's stand he sat down to write just about everyone involved. McCorvey addressed one letter to Rose, Bennett, Pow, and Bealle commending them for their wonderful work and also singling out Blount and Caddell among the trustees, "while, of course, all of 'us Trustees' were doing our best."[35] In a letter to Wallace, he praised the governor for his magnificent performance and only regretted that the board could not have done more to help. "However," he added, "I think that we handled the matter just as well as we possibly could and certainly arranged things so that you could 'have your say.' "[36] It remained for Blount to sum up McCorvey's role. "I share your opinion and feeling," Blount wrote, "for the leadership that Dr. Rose and his team provided, but the essential ingredient has been supplied by you and that is to do the body blocking or the clearing of the path in the field of public opinion in our State."[37]

Blount may have overstated, but McCorvey's Dixiecrat connections helped the university. Moreover, unlike his tried and true friend Hill Ferguson, McCorvey voted consistently to comply with the law. In the end he accepted the verdict on segregation. He had not lost his turn-of-the-century antipathy for the Negro race. In a letter to Frank Rose, he enclosed a picture of Cleve McDowell, the black who desegregated the University of Mississippi's law school, and said, ". . . I think you will agree with me that this negro should be expelled on account of his looks, if for nothing else. The idea of 'things' like that associating with our boys and girls really makes me 'boil over.' "[38] Still, McCorvey overcame personal revulsion to help his university bring about a peaceful resolution. In his congratulatory letter to the Rose team, McCorvey lamented that he had not been "born fifty years earlier and thus missed many things that have worried me greatly. However, somebody has *got* to 'carry on,' and we have *got* to leave behind us the best possible country that we can leave for our children and grandchildren." Then, as if catching himself from yielding too much to necessity, he added the hope that Negro students will "have sense enough to conduct themselves in a way that will not be offensive to our people."[39]

McCorvey did one other thing to help: he shored up spirits during the final court appeal. Everyone knew it was hopeless, but Wallace insisted. Andrew Thomas, always sticking to facts, wrote a letter in which the facts virtually begged the university to stop its senseless litigation. A three-judge panel—Walter Gewin of Tuscaloosa, John Minor Wisdom of New Orleans, and Griffin Bell of Atlanta—was to hear the appeal from Judge Groom's May 16 desegregation order. They had just ruled against the Birmingham school board, and Thomas reasoned, a fortiori, that their decision "indicates clearly the action which will be taken when our appeal is presented. . . ."[40] The appeal was ludicrous, alleging that Grooms had refused unreasonably to grant the university a delay until "the state of unrest in race relations in the State of Alabama has materially improved."[41] Of course, Grooms had relied on Wallace's statements that law and order would be preserved and had noted that setting aside a court order because of potential violence would vitiate any court order a community did not like. Moreover, the schoolhouse door drama was already past, and there had been no violence.

McCorvey tried unsuccessfully to get Wallace to agree to a dismissal of the appeal. Caddell tried Cecil Jackson, who said "that neither he nor the Governor had any illusions about our chances of winning the appeal," but the governor did not want it dismissed. Jackson said "that the brief and argument could be very short and superficial." "[A]nything the Court of Appeals might say," reasoned Jackson, "would be vastly better for the Governor's position than publicity to the effect that we had given up the fight."[42] Under pressure, the board relented and prepared for a March 3, 1964, hearing. The court rendered its decision on March 13. In a letter to McCorvey, Andrew Thomas said the "judgment is brief and reads as follows: 'Per Curiam: The judgment of the district court is AFFIRMED.' " Thomas added, "We are gratified at the brevity of the judgment rendered by the Court of Appeals and particularly that no criticism was voiced of the action on the part of you and the other members of the Board of Trustees for taking this appeal."[43]

Thomas requested that no effort be made to pursue the leftover cases of Sandy English and Marvin Carroll, both of whom had withdrawn their applications. McCorvey agreed to " 'let sleeping dogs lie.' " Then in the understatement of the whole ordeal, McCorvey said, ". . . we have really done our damndest."[44]

One life at a time

No matter how the larger drama played for historians, the events of 1943 to 1963 affected individual lives in different and deeply personal ways. The major players—lawyers, university officials, and politicians—suffered few adverse consequences. Most prospered. The students who braved the odds built their lives from their own inner resources; if all did not prosper materially, they uniformly constructed useful lives. Nathaniel Colley, the young army captain from Snow Hill, Alabama, whose letter of application caused the first serious stir, went on to Yale Law School and practiced law in Sacramento, California.

Colley, who died May 20, 1992, stayed in touch with Alabama as a trustee of his alma mater, Tuskegee Institute.

Pollie Anne Myers left on a cold February night in 1956 for Detroit. Two master's degrees from Wayne State University earned her employment in Detroit's public school system. When I first contacted her in 1985, she was on leave teaching in Nigeria. She and her children get back to Birmingham occasionally, more often when her father was living, and she still thinks about retiring to Alabama's Magic City—the weather being more hospitable.

Lucy's marriage to the Reverend Hugh Lawrence Foster in April 1956 moved her to Texas. Because of her high profile and because public school desegregation failed to make substantial progress until much later, she found it difficult to find employment as a teacher. By the time race became less a factor, she had fallen into the age trap. As a result, most of her desire to teach has been satisfied as a supply teacher. The Fosters now live in the Bessemer area of Birmingham, where he served as pastor of the New Zion Baptist Church. They have four children, two of whom have attended the University of Alabama. In May 1988, the board of trustees removed her expulsion, and in the summer of 1989 she completed her first course in the College of Education. Now in her early sixties, she teaches full time in the Birmingham school system and occasionally talks to interested groups about her experiences.

Ruby Steadman Peters, the woman who at age forty sought admission to the Birmingham branch campus only to have her application discouraged by some of Hill Ferguson's professional and business associates, completed her undergraduate degree at Alabama State University in 1960. She received a master's degree in special education from the University of Alabama in 1970 and headed a school for the learning disabled in Birmingham. She died of cancer in 1981. Billy Joe Nabors graduated from Howard University Law School in Washington, D.C. He did not answer a letter to his address in the Marshall Islands, where alumni records from Talladega College showed he was serving as a U.S. attorney. Joseph Louis Epps, whose mother's mental condition was used by the university to force him to withdraw his application, graduated from Morehouse College in Atlanta. He attended medical school for a year in Washington and later studied theology in Rochester, New York. At last account, he was living in Birmingham.

Leonard Wilson, whose expulsion from the university for leading the student demonstrations in 1956 was viewed by many as an expedient to justify the expulsion of Lucy, lives near Carbon Hill, Alabama, a small town near what Alabamians refer to as the Free State of Winston; so named because Winston County voted to secede from the Confederacy. Wilson has a successful business career and keeps his hands in politics. He stayed on as executive secretary of the Alabama Citizens' Councils until that group dissolved in 1969. Once a Democrat, he is now a Republican. When I visited Wilson's home in 1975, he invited me into his library where on one wall he kept works ranging from Eldridge Cleaver's *Soul on Ice* to Martin Luther King's *Why We Can't Wait*.

On the other wall, works advocating white supremacy lined the shelves. He winked and said, "Even here they don't mix." On July 4, each year, he hosts a watermelon cutting where friends and associates from massive resistance days gather. His son and my oldest son graduated from high school together, though I did not make the connection until I spotted Wilson at the graduation exercises.

In 1965 James Hood tried one more time to enroll at the university, but the Gadsden speech thwarted his efforts. He graduated from Wayne State University in 1970 and earned master's degrees in criminology from Michigan State University and in sociology from Cambridge University in England. He was involved in Coleman Young's rise to power in Detroit and eventually became a deputy police commissioner. He moved to Madison, Wisconsin, in 1980 and now chairs a department of criminal justice and security at the Madison Area Vocational College. He continues to hold many of the controversial views on the civil rights movement expressed in the *Crimson-White* editorial of June 1963. He believes Wallace made a constitutional stand for states' rights at the schoolhouse door and gives the governor full credit for living up to his pledge. He does not view Wallace's action as racially motivated.

Despite her exemplary record, Vivian Malone did not receive a single job offer from a business in Alabama. For the next four years she worked in Washington, first with the Justice Department and later with the Veteran's Administration. In 1969 she transferred to Atlanta, where her husband, Mack Jones, attended medical school at Emory, and in 1971 she joined the Environmental Protection Agency. She later became the first woman to head the Voter Education Project, Inc., before returning to the EPA in 1986 as its regional equal opportunity officer. She is also a successful entrepreneur, having established Real Estate Professionals, Inc., a commercial and residential real estate company in downtown Atlanta. She is a partner in the Chicken George fast-food franchise and serves as President of Metro Medical, Inc., a medical equipment supplier in Atlanta. Her husband is an obstetrician-gynecologist, and her two children attend Howard University. Though she has returned to her alma mater several times to speak, very little ties her to the university she helped change in such a profound way.

Dave McGlathery and Marvin Carroll stayed on with NASA. McGlathery failed his first course at Huntsville, but subsequently completed over forty hours of coursework in engineering and applied mathematics. He credits his study at UAH with career enhancement and especially with two publications in professional journals. He remains active as a pastor. Carroll recently retired from NASA and is in private consulting.

The lawyers have all done well. Arthur Shores's persistence and bulldog tenacity won victory after victory as the courts applied more systematically the principles of equality to his cases. He enjoys remembering how the white lawyers squirmed and contorted to make their convoluted arguments for segregation stick. His bravery cost him two dynamite explosions at his home. Shores also collected his fees through those years. Being associated with the

Legal Defense Fund meant he could charge the going rate with reasonable certainty of payment. His practice continued to expand. He maintained the respect of the white community and became Birmingham's first black city councilman. In 1975, the University of Alabama awarded him an honorary doctorate.

Charles Morgan collected his fees also. He gained fame as a young and increasingly apostate Birmingham lawyer who gave a courageous speech in the wake of the Sixteenth Street Church bombing, a speech in which he laid blame for the tragic deaths on all of Birmingham's white citizens, not simply the followers of Bull Connor. This final act of racial betrayal resulted in Morgan's departure from Alabama for Atlanta. He joined the ACLU and soon became its chief. But apostasy and representation of the downtrodden seldom meant destitution for lawyers who identified with civil rights issues. In representing McGlathery and Carroll, which by Morgan's own estimates required little work, Morgan billed the Rights of Conscience Committee of the American Friends Service Committee $10,000—about $40,000 in present dollars. After leaving the ACLU, Morgan established his own law firm in Washington where he finds clients on the other side of affirmative action questions—recently, Sears.[45]

Of course, lawyers are not in the charity business, and when a few brave souls accepted civil rights cases in the fifties, they took on high-risk ventures. Charles Morgan forfeited a promising political career and left the community in which he grew up. He and others like him could not have known for sure that civil rights was a cresting legal tide and that many of them would ride it to fortune. Similarly, the lawyers at the Legal Defense Fund in New York profited. Constance Motley and Robert Carter moved on to the federal bench. Thurgood Marshall became the nation's first black Supreme Court justice.

University officials also did well. Carmichael suffered the outrages of the Lucy debacle, an experience that left permanent scars, but he enjoyed his Ford Foundation study on higher education in the English Commonwealth nations and semi-retirement in his and Mae's beloved Asheville. Occasionally he corresponded with Jeff Bennett to keep up with things in Alabama. He returned to Tuscaloosa in 1964 for a symposium titled "The Deep South in Transformation." His paper, "Training for Responsible Citizenship," suffered from the same idealistic abstractions that marked his administration.[46] Carmichael died in Asheville on September 26, 1966. He was seventy-four. Mae had preceded him in death.

Jeff Bennett chafed under Rose's leadership, and at the first opportunity headed for Washington. He had impressed the Kennedy administration, and Sargent Shriver asked him to consider an assistant directorship in the Peace Corp. He wound up taking a job as policy coordinator for the Health Services and Mental Health Administration in Washington and later became vice chancellor and president of the University of the South in Sewanee, Tennessee. In the late seventies he returned to Tuscaloosa as professor of law and assistant to the chancellor of the newly created university system. In 1983 he retired to Orange Beach on the Gulf Coast of Alabama. Alex Pow also left, becoming president of Western Carolina University before being felled by a stroke. He returned to Tuscaloosa where he continues to reside.

Frank Rose stayed on at the Capstone for six more years. The university continued to grow in the fertile soil of sixties prosperity. By 1966 the three campuses had 17,000 students, 400 of them black. Rose still had clout and, as in the schoolhouse door confrontation, occasionally used it for constructive ends. With Bennett's adroit help in Montgomery, he blocked a speakers' ban bill in the state legislature, but lost credit for that by turning around the next year and personally preventing William Sloan Coffin from coming on campus. Rose's overbearing style and tendency to misstate facts began to work against him. With top staffers deserting, his last major fund-raising campaign came unglued. The trustees accepted his resignation with regret and relief: regret that a man who had served them so well had to go and relief that he went so easily. His letter of resignation duly noted the improvement in facilities and faculty, but skeptics pointed to equal if not greater gains at comparable institutions. Rose became president of the General Computing Corporation in Washington, D.C., and later established a private affirmative action consulting firm. He died in Lexington, Kentucky, on February 1, 1991.

Two days after Denise McNair, Carole Robertson, Cynthia Wesley, and Addie Mae Collins died in the ruins of the Sixteenth Street Baptist Church, Hill Ferguson sent Frank Rose "a copy of the Introduction and Index of my latest compilation," which he titled, "From Miss Lucy, '56 to Miss Vivian, '63." The compilation drifted from talk of the university's "Tragic Era," meaning 1956 to 1963, to discussions of mongrelization and miscegenation laws. Rose thanked the old man for "furnishing me this information" and filed it under "Crank Letters."[47] Ferguson was long beyond his usefulness, but for his years of service, the university in 1972 named its new student union building the Ferguson Center. Appropriately, it overlooks McCorvey Drive.

Of those who went through the university's ordeal of desegregation, none profited more than George Wallace. His symbolic defiance at Foster Auditorium led straight to Saunders Hall at Harvard where he earned the derision of America's pampered elite and the admiration of America's "average man." The confrontation with Kennedy gave him three runs at the presidency. Wallace broke the old Roosevelt Democratic coalition by winning the endorsement of working-class whites and their allies among the shopkeepers, merchants, and less upwardly mobile professionals. Eventually Ronald Reagan would win with these former Democrats, but it was Wallace who broke them loose.

After the assassination attempt which put him in a wheelchair for life, Wallace dropped from the national scene but continued to hold the affections of most Alabamians. Ever the consummate politician, he maneuvered in these later years to capture the very vote he had once castigated, the "bloc vote," his euphemism for blacks. They responded: in part because Alabama's Republicans gave them an unattractive alternative and in part because Wallace showed genuine signs of conversion on the race question. Cynics said he was just trying to get his karma right before meeting his maker. Others believed it was the true Wallace, the compassionate Wallace of the fifties at last showing through. Whatever, he openly acknowledged his fight to maintain segregation as a mistake and apologized. By a huge majority, blacks gave their votes for governor in 1982 to their old enemy and new friend. Wallace had won his last election. His paralyzed body, wracked with pain, would not allow another campaign.

In Hale County an old black man leaned forward against his cane, the better to hear the reporter.[48] The year was 1985. No, he didn't think he would vote for Wallace again. "I did the last time," he said, "but I think I'm gonna let him go this time."

"Why?" asked the reporter.

"Well, because he did some dirty things," the man said.

"What dirty things?"

"Well, he stood in that schoolhouse door to keep blacks out," the man said.

"But that was more than twenty years ago."

"I know," he said. "But I ain't forgettin' it."

Epilogue

All stories should end one page before the epilogue, but this story begs a footnote. Just as the forces of segregation exacted a future in American politics from their surrender at the schoolhouse door, so did the national conscience prevail, and nowhere more importantly to this author than at the schoolhouse door itself.

Throughout its ordeal with desegregation, the University of Alabama never lost its fingerhold on the future. To be sure, it bargained and compromised with the forces of resistance, but in the end, it purchased a better posterity. It did so fitfully, awkwardly, furtively, playing both ends against the middle, not always realizing that *it* was the middle. For most of the university's leadership, the principal motive was not to stop the inevitable but to buy time for an eventual and peaceful resolution. Treading water in a stormy sea, they reasoned, was less risky than swimming for an uncertain shore. Though the university did not escape embarrassment, it survived with its conscience intact and stirred. Within two years of Wallace's stand, Robert Kennedy was back on campus for Emphasis Week and James Brown held a stage denied to Ray Charles. As change accelerated, the recently unthinkable became self-consciously acceptable.

Today, the university's 10 percent African-American enrollment (15 percent in-state) makes it a regional leader and fifteenth among the nation's 155 doctoral degree granting institutions. The campuses it spawned in Birmingham and Hunstville show equally impressive results, the University of Alabama System collectively ranking eleventh among doctoral institutions. The university's problems with diversity continue, but they resemble national problems rather than the older parochial issues that plague some institutions in the state and region.

On December 27, 1991, Judge Harold S. Murphy, presiding for the Northern District of Alabama, reached a decision in the decade-long "vestiges of segregation" case brought against the state of Alabama by the Justice Department and the state's historically black public universities. Though finding generally for the plaintiffs, Judge Murphy singled out the Tuscaloosa campus for commendation. "The ignominious image," he wrote, "of Governor Wallace barring black enrollment at the University of Alabama is emblazoned on the American consciousness as a memorial to all that is wrong and pernicious with racial segregation. That image, however, is beginning to fade, not because of the passage of time, but because of the university's affirmative efforts to deal positively with its segregative past. The university has made giant strides towards eliminating the policies defended by Governor Wallace's 'stand in the school-house door' and is today, in many respects, on the fore of university race relations nationwide."

The waters remain troubled. The consensus that formed around Martin Luther King's arc of justice has long since given way to a cacophony of voices, some the anguished protest of dried-up dreams, others, more extreme, rivaling the hatred of white supremacy. Mythologies of race and ethnicity crowd the transcendant ideology of equality. But as an ideal, King's vision of inclusive justice finds refuge in universities, for they exist not to establish orthodoxies but to challenge them. They debunk, confute, festoon themselves in irony, even to the point of an occasional silliness. They strain toward the assumption that knowledge is common property and that it is produced best in a community open to all. This simple notion, that knowledge derived from a just society is superior knowledge, has steadied universities against those currents that would bear us to the worst of our past.

On May 9, 1992, forty years after she first applied to the University of Alabama, Autherine Lucy Foster received her master of arts degree in education. Her daughter, Grazia, who obtained her bachelor's degree in corporate finance, joined her in commencement. The audience, which included many who attended college in those tumultuous days, gave Mrs. Foster the warmest applause of the general exercises and a standing ovation during divisional ceremonies. The university named an endowed scholarship in her honor and unveiled a portrait of her in the Ferguson Center, overlooking the most trafficked spot on campus. The inscription reads: "Her initiative and courage won the right for students of all races to attend the University." It was a day to celebrate.

Bibliography

Primary Sources

Manuscript Collections

Alabama Department of Archives and History
    Civil Rights and Segregation Files
    Governor's Papers, 1940–1964: Frank Dixon; James E. Folsom; John Patterson; Gordon Persons; Chauncey Sparks; George C. Wallace
    George Wallace Vertical Clipping Files
Atlanta University Center:
    Southern Regional Council Papers
Birmingham Public Library
    Bishop C. C. J. Carpenter Papers
    Hill Ferguson Papers
    Emory O. Jackson Papers
    Southern Regional Council Papers
Boston University Library
    Allan Knight Chalmers Papers
Burr and Forman Law Firm
    University of Alabama Desegregation Files
John Fitzgerald Kennedy Library (JFKL)
    John Fitzgerald Kennedy, President's Office File
    Robert Francis Kennedy Papers
    Burke Marshall Papers
    Lee C. White Papers
Library of Congress
    National Association for the Advancement of Colored People Papers
    Nation Urban League Papers
Miles College Archives, Birmingham
NAACP Legal Defense Fund File, New York
Talladega College Historical Collections
    Arthur Gray Papers

Bibliography

        Arthur D. Shores Papers
        Trustees File
University of Alabama Library
        James Jefferson Bennett Papers
        Lee Bidgood Papers
        Louis Corson Papers
        Oliver Cromwell Carmichael Papers
        John M. Gallalee Papers
        Autherine Lucy File
        James H. Newman Papers
        Frank A. Rose Papers
        News Bureau Clipping Files
University of Alabama Systems Office
        Official Minutes
        Trustees Files
University of South Alabama
        John L. LeFlore Papers
Private Collections
        Emmett Gribbin Papers
        Duncan Hunter Papers
        George LeMaistre Papers
        Harry Shaffer Papers

Interviews

Clarence W. Allgood, 1985, Birmingham
Waverly Barbe, by S. Cloud, 3/20/75, Tuscaloosa
Emily Barrett, by C. H. Ziff, 7/31/75, Tuscaloosa
Rufus Bealle, by R. P. Bolt, Jr., 5/1/75, Tuscaloosa
James Jefferson Bennett, 7/14/83, 5/16/88, and 6/23/89, Orange Beach, Ala.
Scott Henry Black, 10/6/89, Birmingham
Earl Blaik, JFKL
Buford Boone, by C. H. Ziff, 4/28/75, Tuscaloosa
Edward O. Brown, by P. L. Williamson, 4/10/75, Tuscaloosa
John A. Caddell, 6/29/89, Decatur
Marvin Phillips Carroll, 6/29/89, Huntsville
John Cashin, 8/16/88, Huntsville
Nelson Cole, by K. Parcus, 4/7/75, Birmingham
Nathaniel S. Colley, 7/5/89 (phone), Sacramento, Calif.
Mason Davis, 11/12/85, Birmingham
Joseph F. Dolan, by C. T. Morrissey, JFKL
Sarah Healy Fenton, by L. Hill, 6/10/88
Autherine Lucy Foster, 3/19/75, 4/2/75, and 4/30/75, Birmingham
Henry V. Graham, 9/1/89, Birmingham
Margaret Green, 3/31/89 (phone), Tuscaloosa
Emmett Gribbin, by G. B. Holt, 1975, Tuscaloosa

Harlan Hobart Grooms, 1985, Birmingham
Annabel D. Hagood, by P. Linder, 2/21/75, Tuscaloosa
Taylor Hardin, 9/26/89 (phone), Montgomery
Alfred C. Harrison, 9/21/89 (phone), Opelika, Ala.
Leigh Harrison, by P. Linder, 4/22/75, Tuscaloosa
H. Donald Hays, by P. Bolt, 4/2/75, Tuscaloosa
Nathaniel Howard, by A. Desantis, 1987
James A. Hood, 9/12/89, Tuscaloosa
Duncan Hunter, 1/12/89, Tuscaloosa
Ruby Hurley, by W. M. Minor, 3/8/76, Atlanta
Iredell Jenkins, by C. E. Richardson III, 3/19/75, Tuscaloosa
Vivian Malone Jones, 7/11/89, Atlanta
William Jones, 6/29/89, Huntsville
Nicholas deB. Katzenbach, 9/13/89 and 10/7/91 (phone), Morristown, N.J.
Robert F. Kennedy, by A. Lewis, JFKL
Jean Lyda, by S. Cloud, 2/20/75, Tuscaloosa
Henry McCain, 12/5/85, Birmingham
Dave M. McGlathery, 1/27/89, Huntsville
Burke Marshall, by C. Elebash and J. Terry, University of Alabama
Burke Marshall, by L. J. Hackman, JFKL
Hubert Mate, by C. E. Richardson III, 4/28/75, Tuscaloosa
David Mathews, by G. B. Holt, 1975, Tuscaloosa
John Bruce Medaris, 6/7/89, (phone), Highlands, N.C.
Charles Morgan, Jr., 8/4/89, Washington, D.C.
James L. Nisbet, 1/31/75, Tuscaloosa
Frank Nix, 4/19/90, Birmingham
John S. Pancake, by M. Rogers, 2/20/75, Tuscaloosa
Eris Paul, 1/22/89, Elba
Pollie Anne Myers Pinkins, 5/26/85, Birmingham, 6/4/85, Detroit
Alex Pow, 5/12/88, Tuscaloosa
John Frazer Ramsey, by K. Parcus, 2/20/75, Tuscaloosa
M. L. Roberts, 2/14/75, Tuscaloosa
Frank A. Rose, with C. Curry, 6/10/88, Tuscaloosa
John Rutland, 11/18/85, Birmingham
Arthur D. Shores, 11/18/85, Birmingham
Donald Strong, by P. L. Williamson, 2/13/75, Tuscaloosa
Joseph A. Volker, 11/21/85, Birmingham
Jean Warren, 7/1/85, Tuscaloosa
Leonard Ray Wilson, 3/28/75, Carbon Hill

Magazine and Newspaper Articles

Paul Anthony and Fred Routh, "Southern Resistance Forces," *Phylon* 18 (Spring, 1957), 50–58.
Hodding Carter, "Racial Crisis in the Deep South," *Saturday Evening Post,* Dec. 17, 1955, pp. 26–27, 75–76.

Bibliography

Joseph Kraft, "Riot Squad for the New Frontier," *Harpers* (Aug. 1963), 69–75.
Wayne Phillips, "Tuscaloosa: A Tense Drama Unfolds," *New York Times Magazine,*
Feb. 26, 1956, pp. 9, 47–50.
Gertrude Samuels, "Alabama U.: A Story of 2 Among 4,000," *New York Times Maga-*
zine, July 28, 1963, pp. 11, 49–50, 54.
Nora Sayer, "Barred at the Schoolhouse Door," *The Progressive* (July 1984), 15–19.
Bernard Taper, "A Reporter at Large: A Meeting in Atlanta," *The New Yorker,* Mar. 17,
1956, pp. 78–121.

Newspapers

Alabama Citizen
Alabama Journal
Atlanta Constitution
Birmingham News
Birmingham Post-Herald
Birmingham World
Montgomery Advertiser
New York Herald Tribune
New York Post
New York Times
Tuscaloosa News
University of Alabama *Crimson White*
University of Alabama *The Mahout*
University of Alabama *The Rammer Jammer*

Pamphlets, Memoirs, Speeches, and Miscellaneous Writings

Alabama College *Bulletin,* "The Fiftieth Anniversary of the Founding of the College"
(Oct. 1946).
Barnett, Lincoln. *The Universe and Dr. Einstein.* New York: Harper & Brothers,
1948.
Carmichael, Oliver Cromwell. "Education and Southern Progress," *Bulletin* 29, No. 2
(Oct. 1936). Alabama College.
———, "Racial Tensions: A Study in Human Reactions," An Address at the Univer-
sity of New Hampshire, May 14, 1959, pub. by University of New Hampshire
(July 1959).
Coan, Roger [Charles Mandeville]. *University.* New York: Exposition Press, 1973.
"Governor's Committee on Higher Education for Negroes in Alabama," (April 20,
1949), Alabama Department of Archives and History.
Edgar W. Knight, "A Study of Higher Education for Negroes in Alabama," Montgom-
ery (June 1940), Alabama Department of Archives and History.
"Papers on Marengo County," n.a., Birmingham Public Library.
Rowe, Gary Thomas. *My Undercover Years with the Ku Klux Klan.* New York: Bantam
Books, 1976.

Secondary Sources

Books

Ashmore, Harry S. Foreword by Owen J. Roberts. *An Epitaph for Dixie.* New York: W. W. Norton, 1957.

———. *The Negro and the Schools.* Chapel Hill: Univ. of North Carolina Press, 1954.

Barnard, Hollinger F., ed. *Outside the Magic Circle: The Autobiography of Virginia Durr.* Foreword by Studs Terkel. University: Univ. of Alabama Press, 1985.

Barnard, William D. *Dixiecrats and Democrats: Alabama Politics 1942–1952.* University: Univ. of Alabama Press, 1974.

Bartley, Numan V. *The Rise of Massive Resistance: Race and Politics in the South during the 1950s.* Baton Rouge: Louisiana State Univ. Press, 1969.

Bass, Jack. *Unlikely Heroes: The Dramatic Story of the Southern Judges of the Fifth Circuit Who Translated the Supreme Court's Brown Decision into a Revolution for Equality.* New York: Simon and Schuster, 1981.

Bass, Jack, and Walter De Vries. *The Transformation of Southern Politics: Social Change and Political Consequence Since 1945.* New York: New American Library, 1976.

Belknap, Michael R. *Federal Law and Southern Order: Racial Violence and Constitutional Conflict in the Post-Brown South.* Athens: Univ. of Georgia Press, 1987.

Bond, Horace Mann. *Social and Economic Influences on the Public Education of Negroes in Alabama, 1865–1930.* Washington, D.C.: Associated Publishers, 1939.

Branch, Taylor. *Parting the Waters: America in the King Years, 1954–63.* New York: Simon and Schuster, 1988.

Brauer, Carl M. *John F. Kennedy and the Second Reconstruction.* Contemporary American History Series, William E. Leuchtenberg, ed. New York: Columbia Univ. Press, 1977.

Bryant, Paul W., and John Underwood. *Bear: The Hard Life and Good Times of Alabama's Coach Bryant.* Boston: Little, Brown, 1974.

Bulloch, Henry Allen. *A History of Negro Education in the South: From 1619 to the Present.* Cambridge: Harvard Univ. Press, 1967.

Burk, Robert Frederick. *The Eisenhower Administration and Black Civil Rights.* Knoxville: Univ. of Tennessee Press, 1984.

Cagin, Seth, and Philip Dray. *We Are Not Afraid: The Story of Goodman, Schwerner, and Chaney and the Civil Rights Campaign for Mississippi.* New York: Macmillan, 1988.

Carmichael, John Leslie, et al. *The Saga of an American Family: For Almost Two-Hundred Years (1785–1982).* Birmingham, Ala.: Typewritten and photocopy, Birmingham Public Library, 1982.

Carmichael, Oliver Cromwell. *Universities: Commonwealth and American: A Comparative Study.* Freeport, N.Y.: Books for Libraries Press, 1959.

———. *Graduate Education: A Critique and a Program.* New York: Harper & Brothers, 1961.

Carmichael, Oliver Cromwell, Jr. *New York Establishes a State University: A Case Study in the Processes of Policy Formation.* Nashville: Vanderbilt Univ. Press, 1955.

Carmichael, Omer, and Weldon James. *The Louisville Story.* New York: Simon and Schuster, 1957.

Bibliography

Carroll, Peter N. *Famous in America: The Passion to Succeed*. New York: E. P. Dutton, 1985.

Carter, Dan T. *Scottsboro: A Tragedy of the American South*. New York: Oxford Univ. Press, 1969.

Cason, Clarence. *90° in the Shade*. Intro. by Wayne Flynt. University: Univ. of Alabama Press, 1983.

Clarke, Jacqueline Johnson. *These Rights They Seek*. Washington, D.C.: Public Affairs Press, 1962.

Cobb, James C. *The Selling of the South: The Southern Crusade for Industrial Development, 1936–1980*. Baton Rouge and London: Louisiana State Univ. Press, 1982.

Conkin, Paul K. *Gone with the Ivy: A Biography of Vanderbilt University*. Knoxville: Univ. of Tennessee Press, 1985.

Couch, W. T., ed. *Culture in the South*. Chapel Hill: Univ. of North Carolina Press, 1934.

Dyer, Thomas G. *The University of Georgia: A Bicentennial History, 1785–1985*. Athens: Univ. of Georgia Press, 1985.

Eisenhower, Dwight D. *The White House Years: Mandate for Change, 1953–1956*. New York: Doubleday, 1963.

Elliott, Sr., Carl, and Michael D'Orso. *The Cost of Courage: The Journey of an American Congressman*. New York: Doubleday, 1992.

Frady, Marshall. *Wallace*. New York: World Publishing, 1968.

———. *Southerners: A Journalist's Odyssey*. New York: New American Library, 1980.

Franklin, Jimmie Lewis. *Back to Birmingham: Richard Arrington, Jr., and His Times*. Tuscaloosa: Univ. of Alabama Press, 1989.

Garrow, David J. *Bearing the Cross: Martin Luther King, Jr., and the Southern Christian Leadership Conference*. New York: William Morrow, 1986.

———, ed. *The Montgomery Bus Boycott and the Women Who Started It: The Memoir of Jo Ann Gibson Robinson*. Knoxville: Univ. of Tennessee Press, 1987.

———. *Protest at Selma: Martin Luther King, Jr., and the Voting Rights Act of 1965*. New Haven: Yale Univ. Press, 1978.

Gitlin, Todd. *The Sixties: Years of Hope, Days of Rage*. New York: Bantam Books, 1987.

Goldfield, David R. *Black, White, and Southern: Race Relations and Southern Culture 1940 to the Present*. Baton Rouge: Louisiana State Univ. Press, 1990.

Grafton, Carl, and Anne Permaloff. *Big Mules and Branchheads: James E. Folsom and Political Power in Alabama*. Athens: Univ. of Georgia Press, 1985.

Graham, Hugh Davis. *The Civil Rights Era: Origins and Developments of National Policy*. New York: Oxford Univ. Press, 1990.

Graves, John Temple. *The Fighting South*. Intro. by Fred Hobson. University: Univ. of Alabama Press, 1985.

Greenhaw, Wayne. *Watch Out for George Wallace*. Englewood Cliffs, N.J.: Prentice-Hall, 1976.

Griffith, L. *Alabama College, 1896–1969*. 1969.

Guthman, Edwin. *We Band of Brothers: A Memoir of Robert F. Kennedy*. New York: Harper and Row, 1971.

Guthman, Edwin O., and Jeffrey Shulman, ed., *Robert Kennedy in His Own Words: The Unpublished Recollections of the Kennedy Years,* with Foreword by Arthur M. Schlesinger, Jr. New York: Bantam Books, 1988.

Hames, Carl Martin. *Hill Ferguson: His Life and Works.* University: Univ. of Alabama Press, 1978.

Hamilton, Virginia Van der Veer. *Lister Hill: Statesman from the South.* Chapel Hill: Univ. of North Carolina Press, 1987.

Haws, Robert, ed. *The Age of Segregation: Race Relations in the South 1890–1945.* Jackson: Univ. Press of Mississippi, 1978.

Highsaw, Robert B., ed. *The Deep South in Transition: A Symposium.* University: Univ. of Alabama Press, 1964.

Hollis, Daniel Webster, III. *An Alabama Newspaper Tradition: Grover C. Hall and the Hall Family.* University: Univ. of Alabama Press, 1983.

Jacoway, Elizabeth, and David R. Colburn, eds. *Southern Businessmen and Desegregation.* Baton Rouge and London: Louisiana State Univ. Press, 1982.

Jones, Bill. *The Wallace Story.* Northport, Ala.: American Southern Publishing, 1966.

Jones, Maxine D., and Joe M. Richardson. *Talladega College: The First Century.* Tuscaloosa: Univ. of Alabama Press, 1990.

Key, Jr., V. O. *Southern Politics in State and Nation.* New York: Alfred A. Knopf, 1949.

King, Martin Luther, Jr. *Why We Can't Wait.* New York: Harper and Row, 1963.

Kirby, James. *Fumble: Bear Bryant, Wally Butts, and the Great College Football Scandal.* New York: Harcourt Brace Jovanovich, 1986.

Kluger, Richard. *Simple Justice: The History of Brown v. Board of Education and Black America's Struggle for Equality.* New York: Alfred A. Knopf, 1976.

Knopke, Harry J., Robert J. Norrell, and Ronald W. Rogers, eds. *Opening Doors: Perspectives on Race Relations in Contemporary America.* Tuscaloosa: Univ. of Alabama Press, 1991.

Lewis, Anthony, and *The New York Times. Portrait of a Decade: The Second American Revolution.* New York: Random House, 1964.

McDougall, Walter A. . . . *the Heavens and the Earth: A Political History of the Space Age.* New York: Basic Books, 1985.

McLaurin, Melton, and Michael Thomason. *Mobile: The Life and Times of a Great Southern City.* Woodland Hills, Calif.: Windsor Publications, 1981.

McMillen, Neil R. *The Citizens' Council: Organized Resistance to the Second Reconstruction, 1954–64.* Urbana: Univ. of Illinois Press, 1971.

Marshall, Burke. Foreword by Robert F. Kennedy. *Federalism and Civil Rights.* New York: Columbia Univ. Press, 1964.

Medaris, John B. *Countdown for Decision.* New York: 1960.

Mellown, Robert Oliver. *The University of Alabama: A Guide to the Campus.* Tuscaloosa: Univ. of Alabama Press, 1988.

Morgan, Charles, Jr. *A Time to Speak.* New York: Harper and Row, 1964.

———. *One Man, One Voice.* New York: Holt, Rinehart and Winston, 1979.

Muse, Benjamin. *The American Negro Revolution: From Nonviolence to Black Power, 1963–1967.* Bloomington: Indiana Univ. Press, 1968.

———. *Ten Years of Prelude: The Story of Integration since the Supreme Court's 1954 Decision.* New York: Viking, 1964.

———. *Virginia's Massive Resistance.* Bloomington: Indiana Univ. Press, 1961.

Navasky, Victor S. *Kennedy Justice.* New York: Atheneum, 1971.

Norrell, Robert J. *Reaping the Whirlwind: The Civil Rights Movement in Tuskegee.* New York: Alfred A. Knopf, 1985.

Bibliography

O'Reilly, Kenneth. *Racial Matters: The FBI's Secret File on Black America, 1960–1972.* New York: Free Press, 1989.

Raines, Howell. *My Soul Is Rested: Movement Days in the Deep South Remembered.* New York: G. P. Putnam's Sons, 1977.

Salisbury, Harrison E. *A Time of Change: A Reporter's Tale of Our Time.* New York: Harper & Row, 1988.

Salmond, John. *A Southern Rebel: The Life and Times of Aubrey Willis Williams, 1890–1965.* The Fred W. Morrison Series in Southern Studies. Chapel Hill: Univ. of North Carolina Press, 1983.

Sansing, David G. *Making Haste Slowly: The Troubled History of Higher Education in Mississippi.* Jackson and London: Univ. Press of Mississippi, 1990.

Schlesinger, Arthur M., Jr. *Robert Kennedy and His Times.* Boston: Houghton Mifflin, 1978.

———. *A Thousand Days: John F. Kennedy in the White House.* Boston: Houghton Mifflin, 1965.

Sherrill, Robert. *Gothic Politics in the Deep South.* New York: Grossman, 1968.

Sims, George E. *The Little Man's Big Friend: James E. Folsom in Alabama Politics, 1946–1958.* University: Univ. of Alabama Press, 1985.

Singal, Daniel Joseph. *The War Within: From Victorian to Modernist thought in the South, 1919–1945.* Fred W. Morrison Series in Southern Studies. Chapel Hill: Univ. of North Carolina Press, 1982.

Sorensen, Theodore C. *Kennedy.* New York: Harper & Row, 1965.

Sosna, Morton. *In Search of the Silent South: Southern Liberals and the Race Issue.* Contemporary American History Series. William E. Leuchtenberg, ed. New York: Columbia Univ. Press, 1977.

Southern, David W. *Gunnar Myrdal and Black-White Relations: The Use and Abuse of an American Dilemma, 1944–1969.* Baton Rouge: Louisiana State Univ. Press, 1987.

Talese, Gay. *The Kingdom and the Power.* New York: World Book Publishing, 1969.

Taylor, Sandra Baxley. *Me'n George: A Story of George Wallace and His Number One Crony Oscar Harper.* Mobile: Greenberry Publishing, 1988.

Tindall, George B. *The Emergence of the New South, 1913–1945.* A History of the South, Vol. X, Wendell Holmes Stephenson and E. Merton Coulter, eds. Baton Rouge: Louisiana State Univ. Press, 1967.

Tushnet, Mark V. *The NAACP's Legal Strategy against Segregated Education, 1925–1950.* Chapel Hill: Univ. of North Carolina Press, 1987.

Watson, Mary Ann. *The Expanding Vista: American Television in the Kennedy Years.* New York: Oxford Univ. Press, 1990.

Weatherby, W. J. *James Baldwin: Artist on Fire.* New York: Donald I. Fine, 1989.

Whitfield, Stephen J. *A Death in the Delta: The Story of Emmett Till.* New York: Free Press, 1988.

Williamson, Joel. *The Crucible of Race: Black-White Relations in the American South Since Emancipation.* New York: Oxford Univ. Press, 1984.

Woodward, C. Vann. *The Strange Career of Jim Crow.* New York: Oxford Univ. Press, 1966.

Wright, Lawrence. *In the New World: Growing Up with America, 1960–1984.* New York: Alfred A. Knopf, 1988.

Yarbrough, Tinsley E. *Judge Frank Johnson and Human Rights in Alabama.* University: Univ. of Alabama Press, 1981.

Dissertations and Theses

Autrey, Dorothy A. "The National Association for the Advancement of Colored People in Alabama, 1913–1952." Ph.D. diss., University of Notre Dame, 1985.

Clarke, Jacqueline Johnson. "Goals and Techniques in Three Negro Civil-Rights Organizations in Alabama." Ph.D. diss., Ohio State University, 1960.

Cochran, Lynda Dempsey. "Arthur Davis Shores: Advocate for Freedom." Master's thesis, Georgia Southern College, 1977.

Corley, Robert Gaines. "The Quest for Racial Harmony: Race Relations in Birmingham, Alabama, 1947–1963." Ph.D. diss., University of Virginia, 1979.

Elliff, John Thomas. "The United States Department of Justice and Individual Rights, 1936–1962." Ph.D. diss., Harvard University, 1967.

Gilliam, Thomas J. "The Second Folsom Administration: The Destruction of Alabama Liberalism, 1954–1958." Ph.D. diss., Auburn University, 1975.

Harris, J. Tyra. "Alabama Reaction to the *Brown* Decision, 1954–1956: A Case Study in Early Massive Resistance." Ph.D. diss., Middle Tennessee State University, 1978.

Jones, George Dan, Jr. "The Last Door Swings Open: A Historical Chronology of the Desegregation of the University of Alabama." Ed.D diss., University of Alabama, 1982.

Mitchell, Ann. "Keep 'Bama White." MA thesis, Georgia Southern College, 1971.

Schlundt, Ronald Alan. "Civil Rights Policies in the Eisenhower Years." Ph.D. diss., Rice University, 1973.

Thompson, Ouida Jan Gregory. "A History of the Alabama Council on Human Relations, 1920–1968." Ph.D. diss., Auburn University, 1983.

Articles

Fairclough, Adam. "The Preachers and the People: The Origins and Early Years of the Southern Christian Leadership Conference, 1955–1959." *The Journal of Southern History* 52 (Aug. 1986), 403–40.

Mayer, Michael S. "With Much Deliberation and Some Speed: Eisenhower and the Brown Decision." *The Journal of Southern History* 52 (Feb. 1986), 43–76.

Norrell, Robert J. "Caste in Steel: Jim Crow Careers in Birmingham, Alabama." *Journal of American History* 58 (1986), 669–94.

Thornton, J. Mills, III. "Challenge and Response in the Montgomery Bus Boycott of 1955–1956." *The Alabama Review* (July 1980), 163–235.

Watson, Mary Ann. "Adventures in Reporting: John Kennedy and the Cinema Verite Television Documentaries of Drew Associates." *Film and History* (forthcoming).

Notes

Chapter 1. Beginnings

1. Reminiscences which follow from Pollie Myers Pinkins, interview by author 26 May and 4 June 1985.

2. *Birmingham World,* 4 Jan. 1952.

3. *Ibid.,* 1 Jan. 1952.

4. Reminiscences which follow from Autherine Lucy Foster, interview by author, 2 April 1975.

5. See George Brown Tindall, *The Emergence of the New South: 1913–1914,* in Wendell Homes Stephenson and E. Merton Coulter, eds., *A History of the South,* vol. 10 (Baton Rouge: Louisiana State Univ. Press, 1967), 560–62.

6. See Daniel Joseph Singal, *The War Within: From Victorian to Modernist Thought in the South, 1919–1945* (Chapel Hill: Univ. of North Carolina Press, 1982).

7. Harry H. Smith to Raymond R. Paty, ca. June 1945, John M. Gallalee Papers, University of Alabama Library (hereafter cited as Gallalee Papers, UAL).

8. Lee Bidgood to Paty, 14 June 1945, *ibid.*

9. Raymond R. Paty, "Notes on the Report Titled 'A Study of Higher Education for Negroes in Alabama,' " *ibid.* For original report see Edgar W. Knight, "A Study of Higher Education for Negroes in Alabama" (Montgomery, June 1940), in Alabama Department of Archives and History, Montgomery, Alabama (hereafter cited as ADAH).

10. E.G. McGehee, Jr., to Paty, 19 Sept. 1944, Gallalee Papers, UAL.

11. Correspondence between Ralph E. Adams and E. Shell, Jan. and Feb. 1945; Adams to Leroy H. Jackson, 7 July 1945. See also Adams to Cora Price Grier, 16 May 1945, *ibid.*

12. Brewer Dixon to Ralph E. Adams, March 30, 1945; A.B. Moore to Ralph Adams and to H.W. McElreath, 11 Feb. 1944, *ibid.*

13. Nathaniel S. Colley to the Registrar, 11 March 1946, *ibid.* Colley, phone interview by author, 5 July 1989.

14. Adams to Colley, 21 March 1946, Gallalee Papers, UAL.

15. William H. Hepburn to Marion Smith, 23 March 1946, and Smith to Hepburn, 25 March 1946, *ibid.*

16. John E. Adams to Ralph E. Adams, 4 April 1946, and Ralph Adams to John Adams, 18 April 1946, *ibid.*

17. John Adams to Ralph Adams, 20 April 1946, *ibid.*

18. Ralph Adams to Logan Martin, 26 May 1947, and Martin to Adams, 24 June 1946, *ibid.*

19. John B. Scott to Rufus Bealle, 14 July 1953, *ibid.*

20. Suzanne Rau Wolfe, *The University of Alabama: A Pictorial History* (Tuscaloosa: Univ. of Alabama Press, 1983), 184–85.

21. Borden Burr to Noble Hendrix, 27 May 1952, Gallalee Papers, UAL, and Wolfe, *University,* 184–85.

22. John M. Gallalee to Gaines S. Dobbins, 10 Jan. 1951, Gallalee Papers, UAL.

23. *New York Times,* 22 Aug. 1948. Original "Student Opinion Survey" in Gallalee Papers, UAL.

24. For Morrison Williams's role see J. Tyra Harris, "Alabama Reaction to the *Brown* Decision, 1954–1956: A Case Study in Early Massive Resistance" (Ph.D. diss., Middle Tennessee State University, 1978), 251–52. Harvey's comments appeared in the *Birmingham Post-Herald,* 5 Oct. 1950, and those of the Auburn editor in the *Birmingham Post-Herald,* 29 Oct. 1950. Noble B. Hendrix to Gallalee, 31 Oct. 1950, Gallalee Papers, UAL.

25. *Birmingham World,* 15 Feb. 1952, and *Birmingham Post-Herald,* 16 April 1951.

26. E.O. Jackson to Ruby Lee Jackson Gaines, 21 Aug. 1965, Emory Overton Jackson Papers, Birmingham Public Library (hereafter cited as Jackson Papers, BPL).

27. FBI correspondence with Jackson from FBI files loaned to author by Patrick S. Washburn. See Washburn, *The Federal Government's Investigation of the Black Press During World War II* (New York: Oxford Univ. Press, 1986).

28. Citation, Honorary Degree, Morehouse College, 1 June 1965, Jackson Papers, BPL.

29. *Birmingham World,* 5 Sept. 1952.

30. Jackson to R.L. Alford, 11 April 1952, Birmingham File, NAACP Papers, Library of Congress, and *Birmingham World,* 21 March 1952.

31. *Birmingham World,* 18 Jan. and 10 June 1952.

32. *Birmingham World,* 15 Feb. 1952.

33. Ruby Hurley, interview by Anne Mitchell, 4 June 1970, in Mitchell, "Keep 'Bama White" (M.A. thesis, Georgia Southern College, June 1971).

34. Arthur D. Shores, interview by Mitchell, 28 May 1970, in Mitchell, "Keep 'Bama White."

35. Arthur D. Shores, *ibid.*

36. Clarence W. Allgood, interview by author, 28 May 1986.

37. William F. Adams, interview by Mitchell, 15 June 1970, in Mitchell, "Keep 'Bama White."

38. Transcript of Record, United States Court of Appeals, Fifth Circuit, William F. Adams, Appellant, *vs.* Autherine J. Lucy and Polly [*sic*] Anne Myers, Appelles, p. 121.

39. Gallalee to Shores, 22 Oct. 1952, Arthur Davis Shores Papers, Talladega College Historical Collections, Talladega, Alabama.

40. *Birmingham World,* 23 Sept. 1952.

41. James L. Nisbet, interview by author, 31 Jan. 1975, and clipping files, Gallalee Papers, UAL.

Chapter 2. The Players

1. *Crimson-White,* 5 May 1953.

2. Jean Warren, interview by author, 1 July 1985.

3. *Alabama Journal,* 1 June 1953, and *St. Louis Globe-Democrat,* 3 June 1953.

4. *Newsweek,* 18 May 1953, and *New York Herald-Tribune,* 13 Dec. 1953.

5. John Leslie Carmichael, *The Saga of an American Family,* private printing, Birmingham Public Library, 1977. Much of the biographical information on Carmichael comes from this manuscript prepared by his brother, John Leslie.

6. Hollinger F. Barnard, ed., with foreword by Studs Terkel, *Outside the Magic Circle: The Autobiography of Virginia Durr* (University: Univ. of Alabama Press, 1985), 285.

7. Suzanne Rau Wolfe, *The University of Alabama: A Pictorial History* (University: Univ. of Alabama Press, 1983), 194.

8. McCorvey to Carmichael, 19 May 1954, Oliver Cromwell Carmichael Papers, University of Alabama Library (hereafter cited as Carmichael Papers, UAL).

9. Eris Paul, interview by author, 15 Jan. 1989. Memorandum to Carmichael, subject: Eris Paul, July 31, 1954, author unknown, Carmichael Papers, UAL.

10. Carl Martin Hames, *Hill Ferguson: His Life and Works* (University: Univ. of Alabama Press, 1978). Much of the biographical information that follows comes from this source.

11. W.J. Cash, *The Mind of the South* (New York: Alfred A. Knopf, 1941), 334.

12. Ferguson to Carmichael, 28 July 1953, Carmichael Papers, UAL.

13. John Scanlon to Carmichael, 28 July 1953, and Carmichael to Alvin C. Eurich, 9 Sept. 1953, *ibid.*

14. Frontis H. Moore to Carmichael, 22 Oct. 1953, *ibid.* For correspondence on Ferguson's appointment to the board, see J.K. Dixon to Thomas C. McClellan, 11 Nov., and McClellan to Dixon, 14 Nov. 1919, Thomas C. McClellan Papers, Alabama Department of Archives and History, Montgomery.

15. For discussion of Governor Person's letter and the response, see J. Tyra Harris, "Alabama Reaction to the *Brown* Decision, 1954–56: A Case Study in Early Massive Resistance" (Ph.D. diss., Middle Tennessee State University, 1978), 89–103. Persons to Carmichael, 18 Dec. 1953, and Carmichael to Persons, 4 Jan. 1954, Carmichael Papers, UAL.

16. Harry S. Ashmore, with foreword by Owen J. Roberts, *The Negro and the Schools* (Chapel Hill: Univ. of North Carolina Press, 1954), 68–69. Carmichael to Alvin Eurich, 3 Feb. 1954, and Carmichael to Charlotte Kohler, 14 Sept. 1954, Carmichael Papers, UAL.

17. Carmichael to Wilton B. Persons, 1 March 1954, *ibid.*

18. "Memorandum for the President," 1 March 1954, *ibid.*

19. Carmichael to Wilton B. Persons, 8 Sept. 1954, and 21 Jan. 1955; Persons to Carmichael, 14 Feb. 1955, *ibid.* See Michael S. Mayer, "With Much Deliberation and Some Speed: Eisenhower and the *Brown* Decision," *Journal of Southern History* 52 (Feb. 1986), 43–76.

20. *Montgomery Advertiser,* 2 June 1954.

21. See John Temple Graves, with intro. by Freb Hobson, *The Fighting South* (University: Univ. of Alabama Press, 1985).

22. Carmichael to J.T. Graves, 23 July 1954, Carmichael Papers, UAL. Graves to Carmichael, 24 and 25 July 1954, *ibid.*

23. Barnard, *Outside the Magic Circle,* 127.

24. Harry S. Ashmore, *An Epitaph for Dixie* (New York: W.W. Norton, 1958), 22.

25. Virginia Durr to Carmichael, 14 Sept. 1954. Carmichael Papers, UAL.

26. Carmichael to Alvin Eurich, 2 Sept. 1954, *ibid.*

27. Carmichael to Paul J. Braisted, 13 Jan. 1955 and 3 July 1956; and Braisted to Carmichael, 14 Jan. 1956, *ibid.*

28. See Robert Gaines Corley, "The Quest for Racial Harmony: Race Relations in Birmingham, Alabama, 1947–1963" (Ph.D. diss., University of Virginia, 1979), 100–102. See also file #4/14/2/65, Arthur Gray Papers, Talladega College Historical Collections (hereafter cited as TCHC).

29. Carmichael to Joe Adams, 5 April 1956, and Adams to Carmichael, 21 March 1956, Carmichael Papers, UAL.

30. Jean Warren, interview by author, 1 July 1985, and Carmichael files, February–April 1956, *ibid.*

31. Willis F. Kern to Carmichael, 29 March and 5 May 1955; Jefferson Bennett to Kern, 1 April 1955; Bennett to Carmichael, 21 May 1955; and Carmichael to Kern, 12 May 1955; *ibid.*

32. *Alabama Magazine,* 10 June 1955.

33. Bennett to Carmichael, 3 June 1955, Carmichael Papers, UAL.

34. Carmichael to William H. Hoover, 24 June 1955, *ibid.*

35. Carmichael to Ferguson, 27 June 1955, *ibid.*

36. Carmichael to Charlton Ogburn, 19 Sept. 1955, *ibid.*

37. John W. Davis to Carmichael, 17 June 1955, and Carmichael to Davis, 25 June 1955, *ibid.*

38. Emory O. Jackson, "Report of the Chairman of the Alabama Scholarship and Educational Fund," 7 June 1953, Arthur Davis Shores Papers, TCHC.

39. W.C. Patton to Gloster Current, 11 Sept. 1952, NAACP Papers, Library of Congress (hereafter cited as LC).

40. Gloster B. Current to W.C. Patton, 14 Oct. 1952, *ibid.*

41. *Birmingham World,* 7 Oct. 1952.

42. Jackson to Campaign Committee, Alabama Scholarship and Educational Fund, 9 Jan. 1953; W.C. Patton to Gloster B. Current, 11 Nov. 1953; NAACP's Alabama Scholarship and Educational Fund, Financial Statement, prepared by E.O. Jackson, 7 July 1953; and W.C. Patton to Roy Wilkins, 18 July 1953; NAACP Papers, LC.

43. Report to Executive Board, Birmingham Branch, NAACP, prepared by Ruby Hurley, 27 April 1953; and Gloster B. Current to White, 16 July 1953, *ibid.*

44. Shores to Marshall, 15 Dec. 1952 and 25 Feb. 1953, Shores Papers, TCHC.

45. Shores to Marshall, ca. May/June 1953, *ibid.*

46. In Transcript of Record, United States Court of Appeals, Fifth Circuit, William F. Adams, Appellant, v. Autherine J. Lucy and Polly [*sic*] Anne Myers, Appelles, P. 146.

47. Shores to Marshall, 8 and 9 June 1953, Shores Papers, TCHC.

48. Shores to Marshall, 13 June 1953, and to Robert Carter, 27 June 1953, *ibid.*

49. Andrew J. Thomas to R.E. Steiner, Jr., 20 July 1953, Carmichael Papers, UAL.

50. Shores to Robert Carter, 9 July 1953, Shores Papers, TCHC.

51. *Birmingham Post-Herald,* 16 July 1953.

52. Harlan Hobart Grooms, "Segregation, Desegregation, Integration, Resegregation," a manuscript in the possession of Grooms, written ca. 1980, pp. 1–2.

53. Jackson to R.W. Hayden, 18 June 1955, NAACP Papers, LC.

54. Ferguson to Steiner, Carmichael, and Gordon Palmer, 10 June 1955, Steiner to Ferguson, 11 June 1955, Carmichael Papers, UAL.

55. Andrew Thomas to Steiner, 22 May 1954, *ibid.*

56. Arthur Shores, interview by Ann Mitchell, 28 May 1970, in Ann Mitchell, "Keep 'Bama White" (MA thesis, Georgia Southern College, 1971), 19.

57. *Birmingham World,* 1 July 1955.

58. *Alabama: The News Magazine of the Deep South,* 8 July 1955.

59. Agnes N. Studemeier to O.H. [*sic*] Carmichael, 31 May 1955, and Carmichael to Studemeier, 16 June 1955, Carmichael Papers, UAL.

60. *Birmingham News,* 2 July 1955.

61. Jackson to Dr. S.A. Rodgers, Jr., 4 July 1955, NAACP Papers, LC.

62. Jackson to Roy Wilkins, 4 July 1955; Wilkins to Jackson, 14 July 1955; and Herbert L. Wright to Jackson, 15 July 1955; *ibid.*

63. Jackson to Herbert L. Wright, 16 July 1955, *ibid.*

64. Pollie Myers Hudson to whom it may concern, 21 July 1955, *ibid.*

65. Herbert L. Wright to Pollie Myers Hudson, 28 July 1955, *ibid.*

66. Justine Smadbeck to Herbert L. Wright, 31 Aug. 1955, and Wright to Roy Wilkins, 6 Sept. 1955, *ibid.*

67. Jackson to Mrs. G.R. Lee, ca. Oct. 1955, *ibid.*

68. Pollie Myers Pinkins, interview by author, 4 June 1985, and Jackson to Herbert L. Wright, 14 Jan. 1956, NAACP Papers, LC.

69. Henry Myers, interivew by author, ca. Jan. 1985.

70. Ferguson to Carmichael, 12 Aug. 1955, Carmichael Papers, UAL.

71. Bob Hughes to Harold Fleming, 20 June 1960, Southern Regional Council Papers, Birmingham Public Library.

72. Minutes, Executive Committee Meeting, University of Alabama Board of Trustees, 16 July 1955, Trustees Files, University of Alabama System Office, Tuscaloosa (hereafter cited as UA System).

73. Andrew Thomas to Robert Steiner, 28 July 1955, Carmichael Papers, UAL.

74. Ferguson to Trustees, 29 Aug. 1955, and McCorvey to Ferguson, 1 Sept. 1955, *ibid.*

75. Leigh Harrison to Carmichael, 16 Sept. 1955, *ibid.*

76. Ferguson to Carmichael, 11 Feb. 1955, *ibid.*

77. McCorvey to Ferguson, 1 Sept. 1955, and J. Rufus Bealle to Carmichael, 11 Oct. 1955, *ibid.*

78. Grooms, "Segregation," 8.

79. Steiner to Ferguson, 9 Sept. 1955, Carmichael Papers, UAL.

80. Ferguson to Board of Trustees, 14 Sept. 1955, Autherine Lucy File, Folsom Papers, Alabama Department of Archives and History, Montgomery.

81. Grooms, "Segregation," 9.

82. Thomas to Steiner, 10 Oct. 1955, Carmichael Papers, UAL.

83. Steiner to Thomas, 1 Aug. 1955, *ibid.*

84. Minutes, Adjourned Homecoming Meeting, Board of Trustees, 12 Nov. 1955, Trustees Files, UA Systems, Tuscaloosa.

85. Gordon Palmer to Steiner and Ferguson, 26 July 1955, Carmichael Papers, UAL.

86. Mintues, Board Meeting, 12 Nov. 1955, Trustees Files, UA System, Tuscaloosa.

87. Notes in Carmichael's hand on the 12 Nov. 1955 Board of Trustees Meeting,

Carmichael Papers, UAL; Rufus Bealle to Carmichael, 28 Nov. 1955; Carmichael to Bealle, 5 Dec. 1955; and Ferguson to Bealle, 27 Dec. 1955; Minutes and Records, Board of Trustees Meeting, 12 Nov. 1955, Trustees Files, UA System, Tuscaloosa.

Chapter 3. An Uneasy Calm

1. See Daivd J. Garrow, *Bearing the Cross: Martin Luther King, Jr., and the Southern Christian Leadership Conference* (New York: William Morrow, 1986), 56–58.

2. Garrow, *Bearing*, 31.

3. Autherine Lucy Foster, interview by author, 2 April 1975.

4. See *Birmingham World*, 27 Jan. 1956. Quotation from *Birmimgham News*, 26 Jan. 1956.

5. Hill Ferguson to Board of Trustees, 9 Jan. 1956, Oliver Cromwell Carmichael Papers, University of Alabama Library (hereafter cited as Carmichael Papers, UAL).

6. Typed notes on action by Board of Trustees, 21 Jan. 1956, *ibid.*

7. John L. Blackburn to student, 8 Nov. 1956, *ibid.*

8. See Thomas J. Gilliam, "The Second Folsom Administration: The Destruction of Alabama Liberalism, 1954–1958" (Ph.D. diss., Auburn University, 1975).

9. *Birmingham News*, 30 Jan. 1956.

10. Herbert L. Wright to Emory O. Jackson, 30 Jan. 1956, NAACP Papers, Library of Congress (hereafter cited as LC).

11. *Birmingham News*, 1 Feb. 1956.

12. *Birmingham Post-Herald*, 1 Feb. 1956. See also Autherine Lucy Foster, interview by author, 2 April 1975.

13. *Tuscaloosa News*, 2 Feb. 1956.

14. William F. Adams, interview in Ann Mitchell, "Keep 'Bama White" (M.A. thesis, Georgia Southern College, 1970), 36.

15. Much of the information in this paragraph and the three that follow comes from "My Observations on the Autherine Lucy Controversy," a manuscript prepared at the request of Dean Sarah L. Healy by M.L. Roberts, Jr., 20 Feb. 1956, copy in the author's possession.

16. *Birmingham News*, 1 March 1956.

17. *Ibid.*

18. Autherine Lucy Foster, interivew by author, 2 April 1975.

19. Emory O. Jackson, interview in Mitchell, "Keep 'Bama White," 37.

20. Andrews J. Thomas and Frontis H. Moore to R.E. Steiner, 2 Feb. 1956, Carmichael Papers, UAL.

21. *Ibid.* See also Garrow, *Bearing*, 25–26.

22. Carmichael to Alvin Eurich, 2 Feb. 1956, Carmichael Papers, UAL.

23. *Birmingham Post-Herald*, 27 Jan. 1956.

24. In addition to newspapers, much of the information about the first day's activities, thoughts, and reflections come from interviews with Autherine Lucy Foster by author, 2 April 1975; Emily Barrett by C.C. Ziff, 31 July 1975; and Jean Lyda by S. Cloud, 20 Feb. 1975.

25. Quotation and information about Wilson from Leonard R. Wilson, interview by author, 28 March 1975, unless otherwise noted.

26. "That Defiant Sophomore," *U.S. News and World Report,* 26 March 1956, p. 25.

27. Southern Regional Council, "A Survey of Resistance Groups in Alabama," July 1956, copy in Birmingham Public Library (hereafter cited as BPL).

28. See Mitchell, "Keep 'Bama White," 50.

29. Southern Regional Council, "A Survey of Resistance Groups in Alabama," July 1956, BPL.

30. In addition to newspapers, much of the information on Saturday night demonstrations and Sunday's preparations from "Notes on Demonstrations: Saturday Night Demonstrations, February 4, 1956," and "The Autherine Lucy Episode: A Case Study," in J. Jefferson Bennett Papers, UAL. Also "Autherine Lucy," a manuscript prepared by Bennett (ca. 1982) in author's possession.

31. John L. Blackburn, interview by C.E. Richardson III, April 1976.

Chapter 4. The Mob Is King

1. Emory O. Jackson, interview in Ann Mitchell, "Keep 'Bama White" (M.A. thesis, Georgia Southern College, 1971), 58–59.

2. Unless noted, the events from 6–10 February 1956 are from the following newspapers: *Tuscaloosa News, Birmingham News, Birmingham Post-Herald, Birmingham World, Montgomery Advertiser,* and the *New York Times*.

3. Jefferson Bennett, "Autherine Lucy," lecture presented at the University of Alabama, 1982, p. 12.

4. Jean Lyda, interview by Susan Cloud, 20 Feb. 1975.

5. Bennett, "Autherine Lucy," 15.

6. Lyda interview.

7. *New York Post,* 4 March 1956; *Birmingham Post-Herald,* 7 Feb. 1956.

8. Bob Hughes to Harold Fleming, 18 Aug. 1959, Southern Regional Council Papers, xerox copies, Birmingham Public Library (hereafter cited as SRC Papers, BPL); *Tuscaloosa News,* 30 Jan. 1956. On 27 Oct. 1956, Hartley was sentenced to five years probation for violation of U.S.C. 18-1708, Embezzled Mail Matter.

9. *New York Times,* 10 and 26 Feb. 1956; *Tuscaloosa News,* 15 Feb. 1956.

10. M.L. Roberts, Jr., to Sarah L. Healy, 20 Feb. 1956, statement prepared at request of Healy "concerning my observations of and participation in the events connected with the enrolling of Autherine Lucy." Copy in author's possession.

11. *Ibid.*

12. Autherine Lucy Foster, interview by author, 13 and 19 March and 2 April 1975. Subsequent references to Lucy's first-hand experience are from this interview.

13. Robert Emmet Gribbin, interview by Gennetta B. Holt, March 1975.

14. Statement of Tuscaloosa Police Seargent Robert C. Sawyer, ca. 10 Feb. 1956, Burr-Forman Law Offices, Birmingham (hereafter cited as BF).

15. Emily Barrett, interview by C. Ziff, 31 July 1975.

16. Nathaniel Howard, Jr., interview by Alan Desantis, July 1987.

17. Telegrams to Folsom from Ruby Hurley and Roy Wilkins, 6 Feb. 1956, and from G.A. Rogers, Jr., state president of the NAACP, 8 Feb. 1956, Governor's Papers, Autherine Lucy File, Alabama Department of Archives and History, Montgomery (hereafter cited as ADAH). Part of the controversy over non-use of the Guard rests on erroneous information in the *New York Times* that Carmichael had asked the governor for troops and been refused, 11 Feb. 1956.

18. Bennett, "Autherine Lucy," 19.

19. *Tuscaloosa News,* 8 Feb. 1956.

20. Murray Kempton, *New York Post,* 8 Feb. 1956.

21. *Ibid.*

22. Nelson Cole, *Crimson-White,* 7 Feb. 1956.

23. Albert Horn interview, 17 Sept. 1970, in Mitchell, "Keep 'Bama White," 99–100.

24. Annabel Hagood, interview by Patricia Linder, 21 Feb. 1975. Holt also received congratulatory telegrams from Ralph Bunche and Willie Mays. See *Birmingham News,* 14 Feb. 1956.

25. Donald Strong, interview by Phil Williamson, 13 Feb. 1975.

26. Andrew Thomas to Robert E. Steiner, 4 Feb. 1956; Gessner McCorvey to Board of Trustees, 7 Feb. 1956; and Steiner to Board of Trustees, 7 Feb. 1956; all in Carmichael Papers, UAL.

27. Handwritten notes, Board of Trustees Meeting, 8 Feb. 1956, Trustees Files, University of Alabama System Office, Tuscaloosa (hereafter cited as UAS).

28. See Mitchell, "Keep 'Bama White," 118.

29. Pollie Myers Pinkins, interview by author, 26 May 1985.

30. Pollie Myers Hudson, handwritten notes, Arthur Davis Shores Papers, Talladega College Historical Collections. On October 25, 1956, months after Pollie had gone to Detroit, her mother, Alice, was fined $100 and costs and given a suspended 180-day jail sentence for selling gin in soft-drink bottles at the Waterfront Cafe, a place where Pollie also worked for a time. The judge found her father, Henry, not guilty of the same charge. *Birmingham News,* 26 Oct. 1956.

31. Autherine Lucy Foster, interview by author, 13 and 19 March and 2 April 1975.

32. Pollie Myers Pinkins, interview by author, 4 June 1985.

33. The conspiracy charge may be found in the state's major dailies and the NAACP Legal Defense Fund Files, New York, folders 1328–1344 *passim,* microfilm copy in author's possession.

34. *Autherine J. Lucy v. William F. Adams,* 18 Jan. 1957, xerox copy in BF, and also in NAACP Legal Defense Fund Files, New York, folders 1328–1344, microfilm copy. Testimony of Davenport Smith, news director, WBRC radio, is on pp. 24–44 of the transcript.

35. Arthur Davis Shores, interview by author, 18 Nov. 1985. Bennett, "Autherine Lucy," 22. Bennett writes, "Constance Motley . . . wrote that affidavit and staged those public presentations. She was an extremely aggressive female lawyer, but she lost her head that time." Interviews with Ruby Hurley, 4 June 1970, and Emory O. Jackson, 24 June 1970, in Mitchell, "Keep 'Bama White," 157, and Ruby Hurley, interview by Wanda Madison Minor, 8 March 1976.

36. *Tuscaloosa News,* 7 Feb. 1956.

37. *Ibid.,* 11 Feb. 1956.

38. *New York Herald-Tribune,* 9 Feb. 1956, and *Tuscaloosa News,* 10 Feb. 1956. Carmichael released the following statement through the University News Bureau: "I cannot understand how anyone could infer from my remarks that the return of Autherine Lucy to the campus is prerequisite to my staying." Carmichael Papers, UAL.

39. John A. Caddell to Carmichael, 11 Feb. 1956, Carmichael Papers, UAL.

40. *New York Times,* 21 Feb. 1956.

41. *South: The Magazine of Dixie,* 13 Feb. 1956.

Chapter 5. A Gift of Peace

1. Gay Talese, *The Kingdom and the Power* (New York: World Publishing, 1969), 234.

2. *New York Times Magazine,* 26 Feb. 1956.

3. *South: The News Magazine of Dixie,* 20 Feb. 1956.

4. *New York Times,* 19 Feb. 1956.

5. *Tuscaloosa News,* 18 Feb. 1956. Descriptions of the meeting that follow come from this edition.

6. *New York Times,* 26 Feb. 1956.

7. *Montgomery Advertiser,* 16 July 1956.

8. See George E. Sims, *The Little Man's Big Friend: James E. Folsom in Alabama Politics, 1946–1958* (Tuscaloosa: Univ. of Alabama Press, 1985), and Thomas J. Gilliam, "The Second Folsom Administration: The Destruction of Alabama Liberalism, 1954–1958" (Ph.D. diss., Auburn University, 1975).

9. *Birmingham Post-Herald,* 11 Feb. 1956. On February 7, Roy Wilkins tried to get the Justice Department involved but received a noncommittal assurance that events were being followed "closely." Arthur B. Caldwell to Wilkins, 23 Feb. 1956, NAACP Papers, Library of Congress (hereafter cited as LC). When in 1963 Burke Marshall consulted Justice files to find out how the Eisenhower administration handled the Lucy situation, he found no file. Jack Bass, *Unlikely Heroes: The Dramatic Story of the Southern Judges of the Fifth Circuit Who Translated the Supreme Court's Brown Decision into a Revolution for Equality* (New York: Simon and Schuster, 1981), 181.

10. *Tuscaloosa News,* 15 Feb. 1956. For King's assessment of Nixon see David J. Garrow, *Bearing the Cross: Martin Luther King, Jr. and the Southern Christian Leadership Conference* (New York: William Morrow, 1986), 119.

11. Stallworth to Carmichael, n.d., and Durr to Carmichael, 16 Feb. 1956, Oliver Cromwell Carmichael Papers, University of Alabama Library (hereafter cited as Carmichael Papers, UAL).

12. *Birmingham News,* 20 March 1956.

13. *Tuscaloosa News,* 15 and 16 Feb. 1956.

14. *Birmingham News,* 2, 4, 15, and 19 March 1956.

15. *Tuscaloosa News,* 16 and 17 Feb. 1956.

16. Ollie Atkins to Carmichael, 18 Feb., and Larry Racies to Carmichael, 17 Feb. 1956, James H. Newman Papers, UAL. Atkins to Carmichael, 16 July, and Carmichael to Atkins, 24 July 1956. Carmichael handwritten notes and typescript, "To Members of the Press." Lawrence Spivak to Carmichael, 28 Feb., and Carmichael to Spivak, 3 March 1956, Carmichael Papers, UAL. *Tuscaloosa News,* 9 Feb. 1956.

17. Steiner to Thomas, 10 Feb. 1956, *ibid. Birmingham Post-Herald,* 21 Feb. 1956.

18. Caddell to Thomas and Moore, 10 Feb. 1956; Thoms to Steiner, 13 Feb. 1956; and Thomas to McCorvey, 22 Feb. 1956; Burr-Forman Law Offices, Birmingham (hereafter cited as BF). McCorvey to Thomas and Moore, 17 Feb. 1956, Carmichael Papers, UAL. Minutes and Notes, Executive Committee, 16 and 19 Feb. 1956, Trustees File, University of Alabama System, Tuscaloosa (hereafter cited as UAS).

19. Jefferson Bennett, "Notes on Campus Security," J. Jefferson Bennett Papers, UAL.

20. *Tuscaloosa News,* 17 Feb. 1956, and *New York Times,* 18 Feb. 1956.

21. Thomas to Steiner, 22 Aug. 1956, and note in Thomas's hand, [ca. 28 Feb. 1956], BF.

22. Notes, 28 Feb. 1956, Board of Trustees Meeting, Trustees File, UAS.

23. Rufus Bealle to Trustees, 6 March; Caddell to Bealle, 7 March; and Dixon to Beale, 7 March 1956; Trustees File, UAS. *Birmingham News,* 28 Feb. 1956.

24. Ruby Hurley, interview by Wanda M. Minor, 8 March 1956. Unless otherwise

noted, information about the trial comes from a variety of newspapers, mostly local, and the trial transcript.

25. Emily Barrett, interview by Cecelia Hale Ziff, 31 July 1975.

26. *Tuscaloosa News,* 28 Feb. 1956.

27. Emmett Gribbin to Dear Folks, 29 Feb. 1956, in Gribbin's personal papers.

28. *New York Times,* 2 March 1956, and *Time Magazine,* 12 March 1956.

29. *Tuscaloosa News,* 2 March 1956.

30. After delays, Grooms set the hearing on expulsion for January 18, 1957. From NBC file footage titled "Lucy's Lawyers News Conference," No. A-710, and from the testimony of Davenport Smith, news director for Birmingham's WBRC, the university established only "that the news conference was held in the office of the N.A.A.C.P. and lasted almost forty minutes and that in additon to Arthur Shores and Constance Motley reading from the motion or petition which had been filed in court that day, they made numerous comments thereon in the form principally of answering questions propounded to them by the various news gatherers. . . ." Memorandum, Andrew J. Thomas, 8 Jan. 1957. The NBC footage showed that Motley read verbatim from the conspiracy motion filed earlier that day in court. Davenport Smith followed her verbatim statement with questions to determine whether she meant to say "assimilate" an air of riot or "simulate" (she meant "assimilate"). Despite the restraint showed by NAACP lawyers in talking about the conspiracy charge and notwithstanding their claim that motions filed in court were privileged, on January 24, 1957, Judge Grooms construed narrowly (Atlanta News Publishing Co. vs. Medlock, 123 Ga. 714, 51 S.E. 756) the NAACP's right to reiterate privileged filings and held that the university, in expelling Lucy, acted within its lawful police powers to discipline students. The NAACP filed for a rehearing on April 10, 1958. When they failed to show, Grooms dismissed the case with prejudice. Letters, filings, and notes in BF.

31. *Birmingham Post-Herald,* 2 March 1956; *Baltimore Afro-American,* 10 March 1956.

32. *New York Times,* 2 March 1956.

33. Constance Motley, interview by Nora Sayers, 14 Sept. 1983, p. 23, copy in author's possession.

34. *New York Times,* 11 March 1956. *Tuscaloosa News,* 11 March 1956.

35. *New York Times,* 11 March 1956.

36. Bennett, "Autherine Lucy," 1983.

37. Steiner to Thomas, 10 Feb. 1956, Carmichael Papers, UAL.

38. *Birmingham News,* 3 March 1956.

39. *Ibid.,* 1, 3, 13 March and 12 April 1956.

40. "Message to Dr. Carmichael from Robert Earl Chambliss, 2505 32nd Avenue, N., Birmingham, Alabama," ca. 6 March 1956, Carmichael Papers, UAL.

41. *Birmingham Post-Herald,* 7 March 1956. Murray Kempton, "Hymns of Hate," *New York Post,* 7 March 1956.

42. *Ibid.*

43. Dean Louis Corson to Leonard R. Wilson, 15 March 1956, and "Statement of John T. Sullivan Relating to Jack Winfield," Carmichael Papers, UAL; *Tuscaloosa News,* 12 March 1956; *Birmingham News,* 12 March 1956.

44. *Crimson-White,* 6 March 1956.

45. Dallas County Alumni Chapter to Wilson, ca. 12 March 1956, Carmichael Papers, UAL. *Birmingham News,* 13 March 1956; *Tuscaloosa News,* 15 March 1956.

46. Leonard R. Wilson, interview by author, spring 1976.

47. Lillian Smith to Carmichael, 12 March 1956, Carmichael Papers, UAL.

48. *New York Times,* 26 Feb. 1956.

49. Autherine Lucy Foster, interview by author, 2 April 1975.

50. *Birmingham World,* 6 March 1956; *New York Times,* 8 March 1956.

51. *Birmingham News,* 11 March and 23 April 1956; *Birmingham World,* 23 March 1956. Memo to Marshall from Gloster B. Current and from Current to Wilkins, 26 April 1956, and A.L. Foster to Herbert Wright, 19 Sept. 1956, NAACP Papers, LC.

52. *Birmingham World,* 18 Aug. 1956.

53. Phillips to Bennett, 16 April 1956, Bennett Papers, UAL.

54. Bennett to Phillips, 1 May 1956, *ibid.*

55. Kempton, "Hymns," *New York Post,* 7 March 1956.

56. *Birmingham News,* 4, 11, and 18 April 1956; *New York Times,* 13 March 1956.

57. Bennett to Carmichael, 18 Sept. 1956, Bennett Papers, UAL.

58. John A. Morsell to Kelly Burnett, 12 March 1956, NAACP Papers, LC.

59. Marshall to A.L. Foster, 15 Feb. 1957, Arthur Davis Shores Papers, Talladega College Historical Collections, Talladega, Alabama.

60. *Birmingham World,* n.d.

Chapter 6. And Four to Go

1. Bernard Taper, "A Reporter at Large: A Meeting in Atlanta," *The New Yorker,* 17 March 1956, p. 106.

2. Henry McCain, interview by author, Dec. 1985.

3. On a motion by William Key, seconded by Eris Paul, the board authorized the admission of Peters should she be qualified. On March 9, 1956, Ferguson continued his oppostion to Peters in a telegram to Carmichael: "With all the tension and unrest that now prevails, I think it would be suicide to provoke another incident. Hence I urge no action on Peters case." Oliver Cromwell Carmichael Papers, University of Alabama Library (hereafter cited as Carmichael Papers, UAL).

4. Richard T. Eastwood hand-delivered letter to Carmichael, 7 March 1956, with memorandum of record on conversation with Peters attached, *ibid.*

5. J.R. Morton to Carmichael, 10 March 1956, *ibid.*

6. Ferguson to Carmichael, 26 July 1955, and Carmichael to Ferguson, 27 July 1955, *ibid.*

7. Ferguson letters to Judge Clarence W. Allgood and to James A. Green, 19 March 1956, *ibid.*

8. McCain interview.

9. Dan T. Carter, *Scottsboro: A Tragedy of the American South* (New York: Oxford Univ. Press, 1971), 335–90 *passim.*

10. Allan Knight Chalmers to Arthur Gray, 9 and 21 March 1956, Trustees File, Talledega College Historical Collections (hereafter cited as TCHC).

11. John Rutland, interview by author, 18 Nov. 1985.

12. Whitsett to Chalmers, 6 March 1956, Allan Knight Chalmers Papers, Boston University (hereafter cited as Chalmers Papers, BU). A.K. Chalmers, "Hope in Mississippi," *Interracial Review: A Journal for Christian Democracy* 29 (Jan. 1956), 5–6.

13. Chalmers to Billy Nabors, 22 Aug. 1956, Trustees File, TCHC. *Birmingham World,* 18 May 1956. See also Chalmers to Carmichael, 26 Sept. and 8 Oct. 1956, Carmichael Papers, UAL.

14. Arthur D. Gray to Philip M. Widenhouse, 17 April 1956, Arthur D. Gray Papers, TCHC.

15. Brewer Dixon to Carmichael, 11 June 1956, Carmichael Papers, UAL.

16. Memorandum of record by A.K. Chalmers, "Situation on Unviversity of Alabama—August, 1956." Chalmers to Carmichael, 4 Aug. 1956, Trustees File, TCHC. Rutland interview.

17. Mason Davis, interview by author, 12 Nov. 1985.

18. Carmichael to Chalmers, 7 Aug. 1956, Carmichael Papers, UAL. Mentioned in Chalmers's "Situation" memorandum, Aug. 1956. Brewer Dixon to William F. Adams, 17 Aug. and to Hill Ferguson, 5 Sept. 1956, Carmichael Papers, UAL.

19. Chalmers to Nabors with blind copy to Arthur Gray, 22 Aug. 1956, Trustees File, TCHC.

20. Chalmers, "Situation," and Davis interview.

21. Chalmers to Gray, 29 Aug. 1956, Trustees File, TCHC.

22. Rutland interview.

23. Emmet Gribbin to John Burgess, ca. Feb. 1956, Gribbin private papers.

24. Burgess to Gribbin, ca. March 1956, *ibid.*

25. Joseph Louis Epps's mother to Carmichael, 31 July; Carmichael to Ms. Epps, 2 Aug.; Epps to Carmichael, 4 Sept.; and Dean Adams to Epps, 14 Sept. 1956; Carmichael Papers, UAL.

26. Dixon to Ferguson and Ferguson to Carmichael, 5 Sept. 1956, Carmichael to Ferguson, 6 Sept. 1956, *ibid.*

27. Notes recorded by Jefferson Bennett, 6–12 Sept. 1956, James Jefferson Bennett Papers, UAL. See also notes from Bennett for Governor Folsom as recorded by Ralph Hammond, University of Alabama File, 1 Oct. 1955–30 Sept. 1956, Alabama Department of Archives and History, Montgomery.

28. Ruby Epps to Folsom, 5 Aug. 1956, Governor's Papers, University of Alabama File, 1 Oct. 1955–30 Sept. 1956, *ibid.*

29. F. Edward Lund to Jefferson Bennett, 10 Sept. 1956, *ibid.*

30. Jefferson Bennett to Edward Lund, 14 Sept. 1956 and 9 Oct. 1959, *ibid.*

31. John Leslie Carmichael, *Saga of an American Family* (Privately printed, 1982, Birmingham Public Library), 284–89.

32. Steiner to Carmichael, 23 Oct.; Carmichael to Steiner, 26 Oct. 1956; and Ferguson to Andrew Thomas, 18 Oct. 1956; Carmichael Papers, UAL.

33. Carmichael to Alfreda Mosscrop and to Harvie Branscomb, both 5 March 1956, *ibid.*

34. Carmichael to Isadore I. Kastin, 16 March 1956, and to C. Edward Peterson, 25 Feb. 1956, *ibid.*

35. Jean Warren, interview by author, 1 July 1985.

36. Handwritten notes by Carmichael on agenda for 8 Feb. 1956, University Council Meeting, and York Willbern to Carmichael, 10 Feb. 1956, Carmichael Papers, UAL.

37. Department of Psychology to Theodore M. Newcomb, president of the American Psychological Association, 8 May 1956. M. Leigh Harrison to Carmichael, 12 July 1955 and also 4 Jan. 1956, *ibid.*

38. Albert P. Benderson to Arthur Shores, 6 March 1956, Arthur D. Shores Papers, TCHC. Typescript prepared for *Ebony* magazine by Harry G. Shaffer, in author's possession, p. 16. Portions of the manuscript were published in *Ebony,* Feb. 1957.

39. John W. Pendleton to Carmichael, 15 Feb. 1956, Carmichael Papers, UAL.

40. Linus Pauling to Robert B. Scott, 11 Feb. 1956, *ibid.*

41. Carmichael to Linus Pauling, 16 Feb., and Pauling to Carmichael, 6 March 1956, *ibid.*

42. Ferguson to Carmichael, 27 April 1956, *ibid*.

43. Correspondence and draft copy of a statement on academic freedom by the presidents of Alabama colleges, never submitted, 1 July 1956, *ibid*.

44. Steiner to Carmichael, 2 and 6 June, Carmichael to Steiner, 5 and 9 June 1956, *ibid*.

45. Thomas S. Lawson to Carmichael, 5 June 1956, *ibid*.

46. Carmichael to Steiner, 9 June 1956, *ibid*.

47. Dixon to Carmichael, 8 June 1956, *ibid*.

48. Lawson to Carmichael, 11 June 1956, *ibid*.

49. Carmichael to F.B. Cawley, 31 March 1956, *ibid*.

50. *Time Magazine,* 19 Nov. 1956.

51. *Tuscaloosa News,* 24 Feb. 1956, and Wallace D. Malone to Carmichael, 20 March 1956, Carmichael Papers, UAL.

52. Carmichael to Wallace Malone, 22 March 1956, *ibid*.

53. Paul K. Conkin, *Gone with the Ivy: A Biography of Vanderbilt University* (Knoxville: Univ. of Tennessee Press, 1985), 404–5 and 428.

Chapter 7. Interlude

1. Walter W. Anderson, Jr., to George Mitchell, 8 Feb. 1956, Southern Regional Council Papers, Birmingham Public Library (hereafter cited as SRC, BPL).

2. Quarterly Report, Alabama Council on Human Relations, 1st Quarter, 1957, SRC, BPL.

3. *Montgomery Advertiser,* 26 Aug. 1957.

4. Bennett to Willbern, 17 Sept. 1957, Jefferson Bennett Papers, University of Alabama Library (hereafter cited as UAL).

5. Bennett confidential memo to Newman, 16 April 1957, Bennett Papers, UAL.

6. John Caddell, interview by author, 29 June 1989, and Jefferson Bennett, interview by author, 23 June 1989.

7. *Birmingham News,* 6 Oct. 1957.

8. Rose to Caddell, 6 Dec. 1957, Frank A. Rose Papers, UAL. For board discussion of Rose's hiring and contract details see notes on board meetings of 5 Sept. 1957 and 14 Feb. and 22 Sept. 1959.

9. Caddell to Rose, 9 Dec. 1957, Rose Papers, UAL, and Caddell interview, 29 June 1989.

10. For discussion of $5 million campaign see notes on trustees meeting, 30 May 1958, UAS.

11. Frank Rose, interview by author and Carolyn Curry, 10 June 1988. Rose first encouraged this view in press releases during the June 1963 stand in the schoolhouse door.

12. *Birmingham News,* 2 Dec. 1957.

13. Iredell Jenkins, "Segregation and the Professor," *The Yale Review* (Winter 1957), 311–20. Quoted material from 319–20.

14. Bennett to Rose, 3 Oct. 1957, Bennett Papers, UAL.

15. Patrick W. Richardson to Rose, 1 Oct. 1958, Rose Papers, UAL.

16. *New York Times,* 30 Dec. 1958. See also Jack Bass, *Unlikely Heroes,* 225.

17. Handwritten notes on Executive Committee of Board of Trustees, 29 May 1959, UAS. Department of Public Safety Annual Reports, Fiscal Year 1960–61 and 1962,

Governors' Files, ADAH. Records in Bennett's file, ca. May 1960, indicate that at least fourteen black applicants had applied for admission March 22–24, 1960. Handwritten notes state: "Highway Patrol (Investigator Allen says [applicants numbered] 2 & 5 (Floyd Willis Coleman and Marzette Watts) are on disciplinary probation at Ala. State." "Chief Ruppenthal is learning home addresses and home states." Bennett Papers, UAL.

Chapter 8. Class of '65

1. Bennett to Rose, 2 and 9 Feb. 1961, J. Jefferson Bennett Papers, University of Alabama Library (hereafter cited as Bennett Papers, UAL).

2. Rose to John Patterson, 17 Feb. 1961, Frank A. Rose Papers, University of Alabama Library (hereafter cited as Rose Papers, UAL).

3. *Crimson-White*, 26 April and 10 May 1961. Benjamin Muse, Confidential Memorandum, "Visit to North Alabama, Birmingham, and Tuscaloosa, June 18–28, 1961," 13 July 1961, copy in Alabama Council on Human Relations Papers, Birmingham Public Library (hereafter cited as ACHR, BPL).

4. *Crimson-White*, 14 June 1961.

5. See Robert J. Norrell, *A Promising Field: Engineering at Alabama, 1837–1987* (Tuscaloosa: Univ. of Alabama Press, 1990); Walter A. McDougall, *The Heavens and the Earth: A Political History of the Space Age* (New York: Basic Books, 1985); and John Bruce Medaris, *Countdown for Decision* (New York, 1960).

6. Mandeville File (4 folders), Box #4, Bennett Papers, UAL.

7. Roger Coan [Charles Mandeville], *University* (New York: Exposition Press, 1973); interviews by author with Margaret Green, 31 March 1989, and Alex Pow, 12 May 1988.

8. Robert H. Garner, interview by author, 27 Feb. 1989.

9. Alex Pow to Rose, 21 April, and Rose to Pow, 8 May 1961, Rose Papers, UAL.

10. Taylor Branch, *Parting the Waters, America in the King Years, 1954–1963* (New York: Simon and Schuster, 1988), 474–76. For the Kennedy administration perspective see Burke Marshall, with foreword by Robert F. Kennedy, *Federalism and Civil Rights* (New York: Columbia Univ. Press, 1964), 63–74.

11. Copy of Robert H. Garner to John D. McGuire [*sic*], 26 May 1961, Bennett Papers, UAL.

12. Norrell, *A Promising Field*, 152; and Rose to Paul Siegel, 6 Feb. 1961, Rose Papers, UAL.

13. Muse, "Visit . . . ," 13 July 1961, copy in ACHR Papers, BPL.

14. Muse, "Dangerous Situation in Birmingham: General Observations," 11 Jan. 1962, copy in ACHR Papers, BPL.

15. See Charles Morgan, *A Time to Speak* (New York: Harper & Row, 1964), 123–27.

16. See Muse, "Dangerous Situation," and Branch, *Parting the Waters*, 644–46.

17. For an account of King and Kennard see David G. Sansing, *Making Haste Slowly: The Troubled History of Higher Education in Mississippi* (Jackson: Univ. Press of Mississippi, 1990), 144–54; for the Ole Miss crisis see Branch, *Parting the Waters*, 633–72, and Sansing, *Making Haste*, 156–95.

18. Anecdote by Ina and Tom Leonard.

19. Paul Rilling to Leslie W. Dunbar, 5 March 1962, Southern Regional Council Papers, BPL (hereafter cited as SRC Papers).

20. Winton M. Blount to Rose, 11 Oct., and Rose to Blount, 31 Oct. 1962, Rose Papers, UAL.

21. *New York Times,* 26 Nov. 1962. Mrs. George LeMaistre to Marge Manderson, 14 Nov., and Paul Anthony to Nat Welch, 27 Dec. 1962, reel #140, SRC Papers, Atlanta University Center.

22. Jim Buford to LeMaistre, 24 May 1963, George LeMaistre Papers, copy in possession of author.

23. ACHR Quarterly Report, Sept., Oct., and Nov. 1962, SRC Papers, BPL, and Governors Files, Segregation, 1962–63, Alabama Department of Archives and History, Montgomery (hereafter cited as ADAH).

24. Draft memo to Rose from Bennett, in Bennett's hand, n.d., Bennett Papers, UAL.

25. The State of the University (draft), 30 Oct. 1962, Bennett Papers, UAL. President's Report to the Trustees, 9 Nov. 1962.

26. ACHR Quarterly Report, Mar.–May 1962, SRC Papers, BPL.

27. Bennett to Rose, 30 Aug. 1962, Rose Papers, UAL. Handwritten notes on Board of Trustees Executive Committee Meeting, 16 Aug. 1962, University of Alabama System Files (hereafter cited as UAS). *Crimson-White,* 20 Sept. 1962.

28. *Crimson-White,* 20 Sept. 1962.

29. *Ibid.,* 27 Sept. 1962.

30. Rose to Thomas S. Lawson, 3 Oct. 1962, Rose Papers, UAL.

31. *Crimson-White,* 11 Oct. 1962; Lawson to Rose, 5 Oct.; Laseter to Rose, 4 Oct.; Brewer Dixon to Walter Givhan, 5 Nov.; Rose to Ehney Camp, 5 Oct. 1962; all in *ibid.*

32. The detective's reports in Box #4, Bennett Papers, UAL.

33. Notes in Bennett's hand, 19 Nov. 1962, *ibid.*

34. Bennett to Wallace, 30 May 1962, *ibid.*

35. Handwritten notes on Board of Trustees Homecoming Meeting, 9–10 Nov. 1962, UAS.

36. Patterson to Rose, 6 Feb., and Rose to Patterson, 21 Feb. 1962, Rose Papers, UAL. The information on W.C. George comes from an Associated Press release by Donald Etinger, dateline Montgomery, and appears in a clipping from the *Atlanta Constitution,* ca. April 1962, in James A. Hood scrapbook.

37. Rose's statement about 213 applicants comes from the handwritten notes on the 9–10 Nov. 1962 Board Meeting: "Caddell—Credit to Gov & Dr Rose for maint seg. But this Res[olution] gives us strength in Fed. Court / Rose—not take prerogative of Bd 213 not admitted." The number 213 is about right. In spring 1962, sixty-two Alabama A&M students made application, most to the university's nursing school in Birmingham, box #4, Bennett Papers, UAL. Notes taken at 23 Feb. 1963 board meeting reveal that "236 applications" were received but not processed, UAS.

38. *Crimson-White,* 6 Dec. 1962. J.F. Volker to Bennett, 12 Dec., and Bennett to Volker, 18 Dec. 1962, Bennett Papers, UAL.

39. Burke Marshall, memo to file, 30 Nov. 1962, box #17, Burke Marshall Papers, John F. Kennedy Library (hereafter cited as JFKL).

40. Brooks Hays, memo for the President, 30 Nov. 1962, box #18, Marshall Papers, JFKL.

41. Bennett to Marten ten Hoor, 11 Oct. 1962, Bennett Papers, UAL.

42. McCorvey to Rose, 5 Dec. 1962, Rose Papers, UAL.

43. *Alabama Journal,* 28 Dec. 1962. McCorvey to J. Miller Bonner, 7 Jan. 1963, Rose Papers, UAL.

Notes

Chapter 9. Final Cast

1. George R. Doster, Jr., Brigadier General ANG, to George C. Wallace, 23 Jan. 1963, and Babs H. Deal to Commanding Officer, ANG, 19 Dec. 1962, Governor's Files, Alabama Department of History and Archives, Montgomery (hereafter cited as ADAH). See also *South: The Magazine of Dixie,* "Bobby's Voodoo," 4 Feb. 1963.

2. Edward D. McLaughlin, Jr., to Burke Marshall, 1 May 1963. See also E.L. Holland, Jr., to Marshall, 30 April 1962. Burke Marshall Papers, John F. Kennedy Presidential Library, Boston (hereafter cited as Marshall Papers, JFKL).

3. William G. Jones, interview by author, 29 June 1989.

4. Paul Anthony to Leslie W. Dunbar, 21 Jan. 1963, Southern Regional Council Papers, Birmingham Public Library (hereafter cited as SRC, BPL).

5. *South: The Magazine of Dixie,* 18 Feb. 1963.

6. Handwritten notes, Board of Trustees meeting, 23 Feb. 1963, University of Alabama System Office, Tuscaloosa (hereafter cited as UAS).

7. Williams to McCorvey, 5 Feb. 1963, J. Jefferson Bennett Papers, University of Alabama Library (hereafter cited as UAL).

8. Rose to Maj. W.R. Jones, 16 Jan. 1963, Frank A. Rose Papers, UAL.

9. Rose to Wallace, 1 May 1963, *ibid.;* Frank A. Rose, interview by author and Carolyn Curry, 10 June 1988; John A. Caddell to Rose, 18 June 1963, Rose Papers, UAL; and Robert F. Kennedy, interview by Anthony Lewis, 520–21, Oral History Collection, JFKL.

10. Bennett to Marshall, 4 March 1963, and Marshall memo to file, 5 March 1963, Marshall Papers, JFKL. "Confidential Notes on Events and Meetings Involving Carroll & McGlathery," ca. March 1963, SRC, BPL.

11. Dave M. McGlathery, interview by author, 27 Jan. 1989.

12. Marvin P. Carroll, interview by author, 29 June 1989, and John Cashin, interview by author, Aug. 1988.

13. Investigative reports, Burr-Forman files, Birmingham (hereafter cited as BF).

14. Raymond Ettinger to Monroe, 5 Feb. and 6 March 1963, and "Confidential Notes on Events and Meetings involving Carroll and McGlathery," SRC, BPL.

15. Ettinger to Monroe, 5 Feb. 1963, SRC, BPL.

16. "Confidential Notes . . . ," SRC, BPL.

17. Carroll and McGlathery, Affidavit to Support Motion for Preliminary Injunction, 8 May 1963, USDC, Northern District of Alabama.

18. LeFlore to Thurgood Marshall, 25 Aug. 1952; Juanita [Lucy] to John [LeFlore], 1 July 1953; and LeFlore to Lucy, 15 Feb. 1956; LeFlore Papers, University of South Alabama Archives. See also Melton McLaurin and Michael Thomason, *Mobile: The Life and Times of a Great Southern City* (Woodland Hills, Calif.: Windsor Publications, 1981).

19. Vivian Malone Jones, interview by author, 11 July 1989. Other biographical information from news and magazine portraits of Malone.

20. Matthew F. McNulty, Jr., to Bennett, 23 Aug. 1961, being a memorandum of record on the application of Carlena Chapman, Bennett Papers, UAL.

21. Rose interview, 10 June 1988. King and the SCLC had little to do with desegregation in higher education. The university had a surfeit of qualified applicants already, Vivian Malone among others.

22. Jones interview, 11 July 1989.

23. *New York Times,* 12 June 1963, and *Newsweek,* 26 Aug. 1963.

24. Hood interview, 12 Sept. 1989, and undated clippings, ca. April 1962, Hood scrapbook.

25. Rose to W.R. Dunn, 16 Jan. 1963, Rose Papers, UAL.

26. Cashin interview, Aug. 1988, Tom Dent, interview by author, Oct. 1988.

27. From handwritten notes on board meeting, 18 March 1963, UAS.

28. See James Kirby, *Fumble: Bear Bryant, Wally Butts, and the Great College Football Scandal* (New York: Harcourt Brace Jovanovich, 1986).

29. Board Meeting, Nov. 1960, UAS. *Tuscaloosa News,* 3 Nov. 1960, p. 1.

30. Bill Jones, *The Wallace Story* (Northport, Ala.: American Southern Publishing, 1966), 50–54. Bill Jones interview, 29 June 1989. Wallace denies talking with the Citizens' Council of Jackson, Mississippi, about strategy. He says the only thing he got from them was a New York contact through which he received $20,000, a sum he in turn distributed among the private segregation academies "in order to get votes." Wallace interview, 23 Oct. 1991.

31. Marshall, Memorandum to the File, re: University of Alabama, 19 March 1963, Marshall Papers, JFKL.

32. Macon L. Weaver to Marshall, 19 Nov. and 18 Dec. 1962; V. L. Jansen, Jr., to Marshall, 17 Dec. 1962; and David Vann to Robert F. Kennedy, 19 Jan. 1963; Marshall Papers, JFKL.

33. *South: The Magazine of Dixie,* 29 April 1963.

34. Jones, *Wallace,* 81.

35. *Newsweek,* 6 May 1963, p. 29.

36. Robert F. Kennedy, interview by Anthony Lewis, JFKL.

37. "Transcript of Conversation between Attorney General Robert F. Kennedy and Governor Wallace," Montgomery, 25 April 1963, Robert F. Kennedy Papers, JFKL.

38. Edwin O. Guthman and Jeffrey Shulman, eds., *Robert Kennedy in His Own Words: The Unpublished Recollections of the Kennedy Years* (New York: Bantam Books, 1988), 186.

39. Frank M. Johnson, Jr., to RFK, 29 April 1963; Charles Morgan, Jr., to Marshall, 30 May 1963; transcript of the Wallace April 25 TV appearance also in Marshall Papers, RFK to John C. Godbold and to Truman Hobbs, 9 May 1963, RFK Papers; all in JFKL. Quotation from USC Law School speech from Marcia G. Synott, "Federalism Vindicated: University Desegregation in South Carolina and Alabama, 1962–1963," American Historical Association Annual Meeting, 27–30 Dec. 1987, p. 20. Wallace leaked the transcript to *South: The Magazine of Dixie,* 13 May 1963, p. 6. The magazine's report was fair.

40. Edgar A. Brown, Pres. Pro Tem, S.C. State Senate, to Pat Young, Democratic National Committee, 15 April 1963; John K. Cauthen to LeMaistre, Pritchett, and Warner, 5 April 1963; Le Maistre to Dr. Robert C. Edwards, 9 April 1963; all in Marshall Papers, JFKL. See also file in Box #4, Bennett Papers, UAL, and George A. LeMaistre personal papers.

41. Marshall to Attorney General, 9 April 1963, Marshall Papers, JFKL.

42. Copies of the examination and essay are in Box #4, Bennett Papers, UAL. See also Lincoln Barnett, *The Universe and Dr. Einstein* (New York: Harper & Brothers, 1948).

43. Notes on Board Meeting, 30 April 1963, UAS.

Chapter 10. Three to Make Ready

1. *Birmingham News,* 9 March 1963. Lawrence Wright, *In the New World: Growing Up with America, 1960–1984* (New York: Afred A. Knopf, 1987), 36–37; Taylor Branch, *Parting the Waters: America in the King Years, 1954–63* (New York: Simon and Schuster, 1988), 656 and 667; Neil R. McMillen, *The Citizens' Council: Organized Resistance to the Second Reconstruction, 1954–64* (Urbana: Univ. of Illinois Press, 1971), 196, 275, and 344–45.

2. Alabama Council on Human Relations Quarterly Report, March, April, May 1963, p. 2, Southern Regional Council Papers, Birmingham Public Library (hereafter cited as SRC, BPL).

3. Martin Luther King, Jr., *Why We Can't Wait* (New York: Harper and Row, 1964), 64–65. In addition to King, *Why We Can't Wait,* the Birmingham story is drawn principally from Branch, *Parting the Waters;* David J. Garrow, *Bearing the Cross: Martin Luther King, Jr. and the Southern Christian Leadership Conference* (New York: William Morrow, 1986); Robert Gaines Corley, "The Quest for Racial Harmony: Race Relations in Birmingham, Alabama, 1947–1963" (Ph.D. diss., University of Virginia, 1979); and unpublished manuscript by author, "Letter from the Birmingham Jail: A Twenty-fifth Year Retrospective."

4. Howell Raines, *My Soul Is Rested: Movement Days in the Deep South Remembered* (New York: G.P. Putnam's Sons, 1977), 176.

5. Buford Boone to George Wallace, 13 March 1963, Frank A. Rose Papers, University of Alabama Library (hereafter cited as UAL).

6. Wright, *In the New World,* 38.

7. Mate deposition, Burr-Forman Files, Birmingham (hereafter cited as BF).

8. *Birmingham News,* 19 May 1963; *Time,* 24 May 1963, pp. 22–23.

9. Memorandum for Lee C. White, 17 May 1963, Box #19, Lee C. White Papers, John F. Kennedy Presidential Library (hereafter cited as JFKL).

10. Bill Jones, *The Wallace Story* (Northport, Ala.: American Southern Publishing, 1966), 83.

11. Handwritten notes, Board of Trustees Meeting, 19 May 1963, University of Alabama System Office, Tuscaloosa (hereafter cited as UAS).

12. *Birmingham News,* 21 May 1963.

13. *Ibid.,* 21 and 22 May 1963.

14. King, Jr., *Why We Can't Wait,* 107.

15. Memo to File, based on *Birmingham Post-Herald* report, 22 May 1963, Burke Marshall Papers, JFKL.

16. John A. Caddell, interview by author, 29 June 1989.

17. Correspondence file on Winton Blount and Business Strategy, Marshall Papers, JFKL. For summary of results see William H. Orrick, Jr., memo for the A.G., 6 June 1963, Box #17, Marshall Papers, JFKL.

18. Memorandum of Conversations, Treasury Department, 29 May 63, pp. 4–5, Marshall Papers, JFKL.

19. Edwin O. Guthman and Jeffrey Shulman, eds., *Robert F. Kennedy in His Own Words: The Unpublished Recollections of the Kennedy Years* (New York: Bantam Books, 1988), 189–90.

20. Robert Sherrill, *Gothic Politics in the Deep South: Stars of the New Confederacy* (New York: Grossman, 1968), 262, and Peter N. Carroll, *Famous in America: The Passion to Succeed* (New York: E.P. Dutton, 1985), 32.

21. *Birmingham News,* 22 and 23 May 1963. E.L. "Red" Holland of the *Birmingham News* stayed in touch with the Kennedy administration through Edwin O. Guthman, RFK's press aide.

22. Jefferson Bennett, interview by author, 16 May 1988. Desegregation of the University of Alabama, Racial Matters, 31 May 1963, Marshall Papers, JFKL.

23. Jones, *The Wallace Story,* 85–87, and Carroll, *Famous in America,* 45–46.

24. Wayne Greenhaw, *Watch Out for George Wallace* (Englewood Cliffs, N.J.: Prentice-Hall, 1976), 1–3 and 114, and Daniel Webster Hollis III, *An Alabama Newspaper Tradition: Grover C. Hall and the Hall Family* (Tuscaloosa: Univ. of Alabama Press, 1983).

25. Greenhaw, *Watch Out,* 1–2.

26. *Birmingham News,* 3 June 1963.

27. Jones, *The Wallace Story,* 85–87, and Carroll, *Famous in America,* 45–46.

28. Gay Talese, *The Kingdom and the Power* (New York: World Publishing, 1969), 349–50.

29. Greenhaw, *Watch Out,* 3.

30. James E. (Ed) Horton, Jr., 1st Dist. Senator, Lauderdale and Limestone Cos., farmer and past president Alabama Cattlemen's Association. Horton was also in trouble for filibustering a road bill that allowed Wallace to appoint attorneys in each county for condemnation proceedings. A rider also permitted Wallace to appoint attorneys for segregation cases. *Birmingham News,* 25, 29, and 31 May 1963.

31. *Ibid.,* 4 June 1963.

32. *Montgomery Advertiser,* 6 June 1963.

33. See Mary Ann Watson, *The Expanding Vista: American Television in the Kennedy Years* (New York: Oxford Univ. Press, 1990).

34. *Ibid.,* 145.

35. *Ibid.*

36. Jones, *The Wallace Story,* 85–86, and *Birmingham News,* 30 May 1963.

37. *Birminghan News,* 3 June 1963.

38. Jefferson Bennett interview, 16 May 1988.

39. *Birmingham News,* 3 and 4 June 1963.

40. *Ibid.,* 5 June 1963.

41. *Ibid.,* 6 June 1963, and *Alabama Journal,* 6 June 1963.

42. *Birmingham News,* 5 June 1963.

43. *Ibid.,* 30 May, and 6 and 8 June 1963.

44. Jones, *The Wallace Story,* 91–94.

45. D. Robert Owen to Marshall, re: Summary, University of Alabama, 6 June 1963, Box 17, Marshall Papers, JFKL.

46. Gary Thomas Rowe, *My Undercover Years with the Ku Klux Klan* (New York: Bantam Books, 1976), 81–91.

47. *Ibid.,* 86, and Raines, *My Soul Is Rested,* 328.

48. *Birmingham News,* 9 June 1963.

49. George C. Wallace telegram to the President, 9 June 1963, box 73, Theodore Sorenson Papers, JFKL.

50. "Desegregation of the University of Alabama: Racial Matters," FBI Report, Special Agents Coleman D. Gary III and Pierce A. Pratt, 31 May 1963, Boxes 17 and 18, Burke Marshall Papers, JFKL.

51. "Record of Conference Held on Friday, May 24, 1963, at the Hotel Stafford, Tuscaloosa, Alabama," Bennett Papers, UAL.

52. "University of Alabama News Media Policies and Procedures for 1963 Summer Session Admission of Negro Students at Tuscaloosa and Huntsville," Confidential, 1 June 1963, Bennett Papers, UAL, and *Publishers' Auxiliary,* 8 June 1963.

53. JFK telegram to Wallace, 10 June 1963, Sorenson Papers, JFKL.

54. D. Robert Owen to Marshall, re: Summary, University of Alabama, 6 June 1963, Marshall Papers, JFKL.

55. Nicholas deB. Katzenbach to Attorney General, 31 May 1963, Box 10, RFK Papers, JFKL.

56. *Birmingham News,* 9 June 1963.

57. Jefferson Bennett interview, 16 May 1988.

58. Frank Rose told Jeff Prugh of the *Los Angeles Times* that the students were enrolled in Judge Lynne's chambers, *Los Angeles Times,* 11 June 1978. Prugh read more than justified into statements about preparations in Lynne's chambers.

59. Tom Dent, interview by author, Oct. and Nov. 1988. Vivian Malone Jones, interview by author, 11 July 1989.

60. Guthman and Shulman, eds., *Robert Kennedy,* 188–89.

61. Raines, *My Soul Is Rested,* 330.

62. Note in BF.

63. Notes in *ibid.*

64. Record of a phone conversation between Marvin Phillips Carroll and Hubert Mate, 6 Jan. 1965, Bennett Papers, UAL.

Chapter 11. A Moral Issue

1. Brewer Dixon to Rufus Bealle, 21 May 1963, file #185, University of Alabama System Office, Tuscaloosa (hereafter cited as UAS).

2. Alex Pow, interview by author, 12 May 1988, and Jefferson Bennett, interview by author, 16 May 1988.

3. Jean Warren, interview by author, 1 July 1985.

4. Pow interview, and J. Jefferson Bennett, interview by Meredith Walker, 10 June 1988.

5. Bill Jones, *The Wallace Story,* (Northport, Ala.: American Southern Publishing, 1966), 96. Bill Jones, interview by author, 29 June 1989.

6. *Alabama Journal,* 6 June 1963, and *Birmingham News,* 6 June 1963.

7. "Desegregation of the University of Alabama," Racial Matters, FBI Report, 31 May 1963, Box 18, Burke Marshall papers, John F. Kennedy Presidential Library, Boston (hereafter cited as JFKL).

8. Minutes of an Adjourned Meeting of the Annual Meeting of the Board of Trustees of the University of Alabama, 10 June 1963, UAS.

9. Bennett, interview by author, 16 May 1988; John A. Caddell, interview by author, 29 June 1989; and an informal conversation with Winton "Red" Blount, ca. April 1985.

10. Howell Raines, *My Soul Is Rested: Movement Days in the Deep South Remembered* (New York: G.P. Putnam's Sons, 1977), 339.

11. "Tuscaloosa—Tuesday, June 11, 1963," Box 10, Robert F. Kennedy Papers, JFKL.

12. The conversation between RFK and Katzenbach, plus cuttings that follow, are taken from Drew Associates, "Crisis."

13. Raines, *My Soul Is Rested,* 331.

14. Mary Ann Watson, *The Expanding Vista: American Television in the Kennedy Years* (New York: Oxford Univ. Press, 1990), 148. The *Crisis* documentary takes liberties with chronology. The exchange with Kerry Kennedy, for instance, appears in Tuesday footage when surrounding evidence and Katzenbach's recollection indicates that it occurred on Monday.

15. Edwin O. Guthman and Jeffrey Shulman, eds., *Robert Kennedy in His Own Words: The Unpublished Recollections of the Kennedy Years* (New York: Bantam Books, 1988), 223–26. There are many accounts of the meeting because Baldwin and others present took special delight in publicizing how they bearded the Attorney General on civil rights. Among other accounts see David J. Garrow, *Bearing the Cross: Martin Luther King, Jr., and the Southern Christian Leadership Conference* (New York: William Morrow, 1986), 268–69.

16. Unsigned memorandum to John F. Kennedy, 20 May 1963, re Legislative Possibilities, Box 53, JFK Presidential Office File, JFKL.

17. *Birmingham News,* 28 May 1963.

18. *Baltimore Sun,* 10 June 1963; *Newsweek,* 24 June 1963.

19. Arthur M. Schlesinger, Jr., *A Thousand Days: John F. Kennedy in the White House* (New York: Houghton Mifflin, 1965), 901–2.

20. A legend has developed that the speech was ordered and drafted at the last hour, Tuesday, June 11. See Theodore C. Sorensen, *Kennedy* (New York: Harper and Row, 1965), 493–96. However, in *Crisis: Behind a Presidential Commitment,* the documentary by Drew Associates, the President, during the Monday afternoon meeting of the 10th, directs Sorensen to prepare a draft, and Robert Kennedy clearly refers to a draft that already existed. RFK may have had in mind material developed by Louis E. Martin, a black journalist and member of the Democratic National Committee who advised the President on civil rights matters, especially the recruitment of blacks.

21. Malone, interview by author, 11 July 1989. The question of sleep was addressed in Bennett interview, 16 May 1988, and Katzenbach, interview in Raines, *My Soul Is Rested,* 341.

22. Raines, *My Soul Is Rested,* 342.

23. Robert Sherrill, *Gothic Politics in the Deep South: Stars of the New Confederacy* (New York: Grossman, 1968), 261.

24. Jones, *The Wallace Story,* 103.

25. The route and schedule for Tuesday is drawn from a variety of often contradictory sources. Part of the confusion in accounts resulted from continual improvisation by Katzenbach's team. A document filed by planners in Tuscaloosa on Monday, but arriving in Washington on Tuesday, illustrates significant variance between what they thought would happen as of Monday morning and what actually transpired. "Tuscaloosa—Tuesday, June 11, 1963," Box 10, RFK Papers, JFKL. No doubt some of this planning information reached the press and colored their accounts. What follows comes close to resolving competing claims. The church where they stopped is to Hood's best recollection the Enterprise Baptist Church, about fifteen miles from the campus. James A. Hood, interview by author, 12 Sept. 1989.

26. Watson, *Expanding Vistas,* 149.

27. The description of the encounter is drawn from a variety of sources, especially the Drew Associates *Crisis* documentary, Jones, *The Wallace Story,* and newspapers and news magazines.

28. Alex Pow, interview by author, 12 May 1988.

29. Pow mentions the request and refusal of two-way general-area radio communication, but did not know the reason for Justice's decision to deny the request, *ibid.*

30. Raines, *My Soul Is Rested,* 340–41. Interestingly, Katzenbach remembers Rose as "a great hero," and Jeff Bennett as "more difficult." Katzenbach interview, 7 Oct. 1991.

31. Drew Associates, *Crisis.*

32. "Telephone Conversation between General Harrison and General McGowan, 1205 Hours, 11 June 1963," in Governors' Files, George C. Wallace, Alabama Department of Archives and History, Montgomery (hereafter cited as ADAH).

33. Henry V. Graham, typescript reminiscence, p. 24, in author's possession.

34. Henry V. Graham, interview by author, 1 Sept. 1989.

35. Graham reminiscence, p. 25, and Raines, *My Soul Is Rested,* 341. Taylor Hardin does not remember delivering the message to Graham. Hardin, phone interview by author, 20 Sept. 1989. General Alfred C. Harrison also was with the governor and thinks it consistent that Hardin would have been the one to communicate the message to Graham. Harrison, phone interview by author, 21 Sept. 1989. Whoever delivered the message, it is clear to Graham and to Katzenbach that Wallace communicated his intentions and wishes to Graham upon his arrival in Tuscaloosa.

36. *Tuscaloosa News,* 12 June 1989.

37. Graham reminiscence, pp. 24–29.

38. *Ibid.*

39. *Ibid.*

40. Watson, *Expanding Vistas,* 149, and Drew Associates, *Crisis.*

41. Graham reminiscence, pp. 27–29, and Graham interview, 1 Sept. 1989.

42. Drew Associates, *Crisis.*

43. Sorensen, *Kennedy,* 495.

44. Quoted in Taylor Branch, *Parting the Waters: America in the King Years, 1954–1963* (New York: Simon and Schuster, 1988), 824.

45. The account of Evers's death is drawn from a variety of sources. Quoted material from *Newsweek,* 24 June 1963. Byron dela Beckwith was charged with the shooting but acquitted by an all-white jury.

46. Wallace telegram to JFK, 12 June 1963, Governors' Files, George C. Wallace, ADAH.

47. JEN to Ed Guthman, 15 June 1963, Box 10, RFK Papers, JFKL.

48. Wallace telegram to Rose, 12 June 1963, Governors' Files, George C. Wallace, ADAH.

49. Charles Morgan, interview by author, 4 Aug. 1989.

50. And following description of McGlathery's experience in McGlathery, interview by author, 27 Jan. 1989.

51. Drew Associates, *Crisis.*

52. *New York Times,* 12 June 1963.

Chapter 12. "Ain't forgettin' it"

1. E.L. Holland to Ed [Guthman], ca. June 1963, Box 17, Burke Marshall Papers, John F. Kennedy Presidential Library, Boston (hereafter cited as JFKL).

2. James A. Hood's scrapbook in Hood's possession.

3. "Hood's Statement to Blackburn and Bennett," 4 Aug. 1963, Bennett Papers, University of Alabama Library (hereafter cited as UAL).

4. Scott Henry Black, interview by author, 5 Oct. 1989.

5. *Montgomery Advertiser,* 4 Aug. 1963, and *Newsweek,* 26 Aug. 1963.

6. All quotations from Hood's speech are from a transcript in Bennett Papers, UAL. Excerpts from a copy of same appear in *Montgomery Advertiser,* 4 Aug. 1963.

7. "Rec'd on 8/3/63 by Phone," caption in Bennett's hand and report typed on legal paper, Bennett Papers, UAL. Bob Ingram, phone interview by author, 18 Oct. 1989.

8. "Hood's Statement to Blackburn and Bennett," Bennett Papers, UAL.

9. From handwritten notes on Executive Committee meeting, Board of Trustees, 5 Aug. 1963, and from the resolution as distributed under the same date, Trustee Files, University of Alabama System Office, Tuscaloosa (hereafter cited as UAS).

10. Bennett, interview by author, 16 May 1988.

11. James A. Hood, "Apology," 6 Aug. 1963, Bennett Papers, UAL.

12. Correspondence and details of Hood's withdrawal in file folder marked "Comments and Speeches (James A. Hood)," Bennett Papers, UAL.

13. *Newsweek,* 26 Aug. 1963.

14. Hood interview, 12 Sept. 1989.

15. Bennett interview, 23 June 1989.

16. *New York Times Magazine,* 28 July 1963.

17. *South: Magazine of Dixie,* 13 May 1963, p. 6.

18. Rose to Murray Cohen, 27 June 1963; Rose to Henry M. Johnson, Jr., 27 June 1963; Rose to Henry R. Luce, 21 June 1963; Rose Papers, UAL.

19. Rose to Cooper, 12 July, and Rose to Samuels, 31 July 1963, *ibid.* Alex Pow interview, 12 May 1988.

20. Evans and Novak, *New York Herald-Tribune,* 2 Aug. 1963. Bennett to Nelson O. Fuller, 26 Aug. 1963, Rose Papers, UAL.

21. Sarah Healy to Alex Pow, 3 July 1963, Rose Papers, UAL.

22. Marshall to R.F. Kennedy, 26 June 1963, Box 10, RFK Papers, JFKL; Marshall to R.F. Kennedy, 26 July 1963, Box 17, Marshall Papers, JFKL.

23. *Time,* 27 Sept. 1963, pp. 18 and 21.

24. Student Pledge, 26 Sept. 1963; and J.L. Blackburn to John B. Oakes, 16 Oct. 1963, Rose Papers, JFKL.

25. *Newsweek,* 30 Dec. 1963, p. 17.

26. J.W. Cameron to Tuscaloosa File, 16 Nov. 1963, Box 17, Marshall Papers, JFKL.

27. *Ibid.,* 21 Nov. 1963.

28. *Ibid.,* and *Newsweek,* 30 Dec. 1963, 17.

29. The Negro Employees memo to Joseph F. Volker, 9 Oct. 1963, and Bennett to Rose, 17 Oct. 1963, Bennett Papers, UAL.

30. "NDA [National Dental Association] Zone 4, Vice-President Quarterly Memo," n.d.; Charles A. McCallum to John W. Nixon, 6 May 1964; and Nixon to McCallum, 15 June 1964; Rose Papers, UAL.

31. Faulkner to David Kirk, 8 March 1956, in *Crimson-White,* 9 June 1963. Kirk was a student of O.B. Emerson, a popular teacher of English literature. Emerson made a copy of the letter available to Scott Henry "Hank" Black, editor of the *C-W.* Black interview, 5 Oct. 1989.

32. Record of phone conversation between Marvin Phillips Carroll and Hubert E. Mate, 6 Jan. 1965, Bennett Papers, UAL.

33. John L. Blackburn to Hood, 3 March 1965, *ibid.*

34. "Status Report on Black Students" to Bennett, 21 May 1964, *ibid.*

35. McCorvey to Rose, Bennett, Pow, and Bealle, 12 June 1963, *ibid.*

36. McCorvey to Wallace, 12 June 1963, Rose Papers, UAL.

37. Winton Blount to McCorvey, 19 June 1963, *ibid.*

38. McCorvey to Rose, 4 Oct. 1963, *ibid.*

39. McCorvey to Rose, Bennett, Pow, and Bealle, 12 June 1963, Bennett Papers, UAL.

40. Thomas to Caddell, 9 Sept. 1963, *ibid.*

41. Gessner T. McCorvey et al, *v.* Autherine J. Lucy et al., Appeal from the United States District Court for the Northern District of Alabama, 5th Circuit U.S. Court of Appeals, No. 20, 898.

42. Caddell to Thomas, 13 Sept. 1963, Bennett Papers, UAL.

43. Thomas to McCorvey, 16 March 1964, *ibid.*

44. McCorvey to Thomas, 30 March 1964, *ibid.*

45. Morgan to Eleanor Eaton, 24 June 1963, Southern Regional Council Papers, Birmingham Public Library. Morgan, interview by author, 4 Aug. 1989.

46. In Robert B. Highsaw, ed., *The Deep South in Transformation: A Symposium* (University: Univ. of Alabama Press, 1964), 47–56.

47. Ferguson to Rose, 17 Sept. 1963, Rose Papers, UAL.

48. *Birmingham News,* 13 Oct. 1985, p. C-1.

Index

Index

Index

Index

Index